Analytic Models for

MANAGERIAL
and ENGINEERING
ECONOMICS

Reinhold Industrial Engineering and Management Sciences Textbook Series

CONSULTING EDITOR: Professor Robert N. Lehrer
School of Industrial Engineering
Georgia Institute of Technology

IN PRESS

Management of Improvement: Concepts, Organization, and Strategy, *by Robert N. Lehrer*

Project Management with CPM and PERT, *by Joseph J. Moder and Cecil R. Phillips*

IN PREPARATION

Introduction to Digital Computer Applications, *by Andrew G. Favret*

Systems Engineering: Analysis and Design of Production Systems, *by Salah E. Elmaghraby*

Consulting Editor's Statement

It is a particular pleasure to introduce Dr. Schweyer's book—not only because it is an excellent and creative presentation, but also because it is the first book to be issued in the new REINHOLD INDUSTRIAL ENGINEERING AND MANAGEMENT SCIENCES TEXTBOOK SERIES.

Dr. Schweyer's presentation fills a long felt need for a basic introductory treatment of economic concepts, analytical methods and models, and their use as aids in decision making in business and industry. His book has been tailored to the needs of both technical and nontechnical students. His writing possesses clarity and preciseness, and features generous use of actual examples to reinforce concepts, principles, and procedures. I am confident that it will be well received by both students and professors—and by many managers and engineers with an interest in better decision making.

The REINHOLD INDUSTRIAL ENGINEERING AND MANAGEMENT SCIENCES TEXTBOOK SERIES is devoted to the development of much needed text material combining science, engineering, and management in a unique and as yet untried way. The key characteristics of the series is an *effective* use of concepts, knowledge, and methods stemming from modern science and technology, so as to achieve workable solutions to problems within the real world setting from which they arise. Some titles will emphasize recent advances in theory, while others will provide updated information based upon past practice in the light of current tested experience. They will originate from many fields, and will cut across such varied topics and academic disciplines as engineering science, mathematics, the behavioral sciences, and many areas of functional business administration and management.

Other books currently in preparation cover project management based upon critical path methods and PERT, the management of improvement activities, the methodology and tools for designing integrated production systems, an analysis of attempts to rationalize and "research" the R & D process, basic concepts of management, and an introduction to digital computer applications. The series is of considerable breadth as this list indicates. Additional titles will be forthcoming.

Atlanta, Georgia
January, 1964

ROBERT N. LEHRER

Analytic Models for

MANAGERIAL
and ENGINEERING
ECONOMICS

Herbert E. Schweyer

University of Florida
Gainesville, Florida

New York
REINHOLD PUBLISHING CORPORATION
Chapman & Hall, Ltd., London

PREFACE

This text is written especially for business and technical students with an interest in the application of economic theory to real world situations. The objective is to demonstrate by elementary mathematics how a simple mathematical model can be utilized effectively in making practical business and production decisions.

Persons responsible for curricula leading to a major in business or those interested in engineering economy will find this text useful in presenting a variety of subject matter otherwise found only in special courses in each area. Because of the broad coverage, only the elementary principles are presented, and the student is directed to more advanced texts for a comprehensive treatment where desired.

The ever-increasing application of mathematics to business and production problems, and the development of high speed computers, have necessitated a better understanding of mathematical models and of their effective application. This text does not consider the details of computer programming and the actual utilization of computers for calculation. It concentrates upon the formulation of economic relationships encountered in real business and production problems in order to give the reader an appreciation of what can be done in developing and using models. These models may be employed by management to support its decisions and policies. With this support any final decision may then be made with more confidence because a quantitative evaluation is available to confirm intuition.

A model as used in this text is a mathematical expression, a formalized tabulation, or a drawing that state the interaction of the variables that apply to a specific problem. The function of the model is (1) to disclose as precisely as possible the significance of the several variables in an economic context, and (2) to provide a mathematical expression that can be manipulated, by computers or otherwise, to supply management with desired information in making decisions.

The broad general subject of production economy and economics of the firm that is covered here comprises: (1) organization from the accounting records of the firm the data that are appropriate for reconciliation with economic theory, (2) utilization of this background material in the development of mathematical models for an economic analysis of a proposed problem, and (3) application of these models to decision making by management based on some criterion or selected conditions to be met.

The presentation is an elementary one without rigorous mathematical proof, since the emphasis is on useful models. The proofs may be found in more advanced texts. Accordingly, only a background of first year mathematics including the rudiments of calculus is required for most sections. Certain more advanced sections in the latter part of the book perhaps will be understood more readily if the reader has had additional mathematics, but in these cases the necessary mathematics is included in the development of the model, or the required result is stated without proof.

It should be emphasized that this is a simplified presentation and gives neither a complete discussion of all concepts presented nor a comprehensive coverage of all kinds of models utilized in production analysis. Considerable space has been employed for illustrative problems which should help the reader to grasp more quickly the converting of accounting data to a form where management may more readily formulate policies and make decisions. However, these procedures are not substitutes for experience and mature judgment. Instead, they must be considered as powerful tools that are a means for verifying intuitive plans, thus providing more confidence in final decisions.

The book is written so that many chapters may be considered separately and in any selected order with omission of certain chapters where desired. As a college text it is adaptable for any juniors or seniors with an interest in the real world economic analysis of operations, for evaluating revenues, costs, and profits to establish the best operation of a business firm.

Where this text is used in the classroom, the instructor is invited to contact the publisher for suggested course outlines and other material to assist him in presenting the material to students.

Acknowledgment is made for the assistance of Dr. F. P. May, Dr. F. Poska, Dr. D. B. Smith, and Mr. Mario Ariet in preparing portions of the manuscript and to McGraw-Hill Book Co., Inc., for permission to use certain limited amount of material from a previous book by the present author.

The author would appreciate having any errors of commission or omission brought to his attention.

Gainesville, Florida   H. E. SCHWEYER
January, 1964

CONTENTS

Annual Costs. Alternatives with Different Lives. The Continuous Service Model. Example for Perpetual Service. Replacements of the Same Kind. Replacements of a New Kind. Simple Replacement Example. Extended Replacement Example. Variations in Replacement Analysis. Comment on Taxes. The Rent or Buy Example. Irreducible Factors in Economic Analyses. Real World Practice. Summary. Problems

PRODUCTION ECONOMY MODELS

There are many factors to consider in applying pure economic theory to the problem of accounting for practical dollars in the real world. These variable factors are often considered qualitatively by management in the intuitive determination of policies. However, management need not limit itself to intuitive formulations. Recent advances in production analysis have resulted in the development of mathematical and analytical models designed to provide management personnel with more confidence in making such decisions. It is the purpose of this book to give a broad presentation of typical basic models and to demonstrate how they apply to real world problems.

MATHEMATICAL MODELS

Economic theory can be reconciled with accounting data by using mathematical relations to provide approximate quantitative numbers for the production personnel who are setting up models for decision-making purposes. There are numerous types of models, some of which utilize principles differing entirely in concept from those employed in other models. Many of these models have been in use for years. However, additional types of models involving statistical and stochastic (probability) considerations have been added recently to management's evaluation techniques under the elegant term of "operations research." From a business and economy viewpoint these new additions are simply the application of complex mathematics to production problems. Developments in mathematics and computing machines permits analysis of production problems heretofore considered to be so complex as to be not worth the effort involved in their solution.

The elements in the development of a model are illustrated in Fig. 1-1. Briefly, a mathematical model is an algebraic equation (or equations) or a specified procedure for the evaluation of some criterion (such as expense, profit, quantity, time of operation, etc.) that depends on other operating conditions that are subject to control by management. By inspection or further analysis of the model, one can find what conditions are required for an "optimum or best policy," that is, a policy producing the best approximation of the desired result for the criterion. Thus, these mathematical treatments are a powerful aid in the establishment of a policy or a decision regarding production and utilization. For example, management may wish to establish a number of production centers spread over a certain geographical area. From a consideration of population, distances from a warehouse, capital requirements, capital available, etc., management might seek to determine the *optimum number* of centers to use as a criterion of maximum annual dollar profit. The analysis might show five plants to be optimum, and if this were in accord with other experience and judgment, making a *decision* to establish five production centers would be logical.

In general, many simplifications of the real world problem are made to avoid complicated equations in setting up the model. Thus, the model is at best an approximation. This does not detract from its utility, however, if the simplifications do not invalidate the solution. In fact, one very important function of the model is to put all factors in perspective with respect to their importance. Development of a model shows that some economic variables at first considered to be important are often of minor consequence and *vice versa*.

For the situations constituting the major portion of this text, the models are quite simple and represent the application of elementary mathematics. The emphasis is on the reasoning behind the model rather than on a completed "how to do it" chart or equation. If the reader understands the underlying principles of mathematical models in general, he can develop new models for new situations. These new models may then be employed with confidence and with a full knowledge of their limitations, since the user will have an understanding of the basis upon which they rest.

DEVELOPING A MODEL

Almost all models for decision making follow the same general form of development shown in Fig. 1-1. A production economy problem arises for which a decision is necessary. The problem then is defined within limitations (step 1). There is a certain amount of information or data that can be made available relating to the problem and controlling a decision. The available data and basic economic principles are assembled (step 2). These are employed in some form of mathematical expression (step 3) under the specified set of conditions prescribed in step 1. This expression is then

Step 1. Define the problem and specify its limitations

Step 2. Assemble the economic and accounting data that apply

Step 3. Establish the criterion equation (The equation that fits the conditions of the problem and the answer sought)

Step 4. Evaluate the criterion equation for the best value or values of the variables that are controllable by management

Step 5. From results in step 4 establish a policy or make a decision

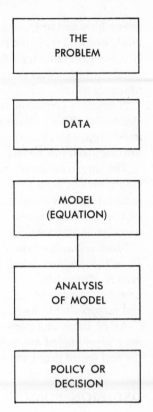

Fig. 1-1. Production economy model development and use.

subjected to mathematical analysis (step 4), after which management makes a decision (step 5).

Although all steps in this procedure are important, steps 3 and 4 are usually the most critical. In many cases the problems can be classified as a type for which models or equations have been proposed previously. For such cases, the analysis is straightforward. However, in other cases a specific relationship must be developed. For this purpose, a study of the models applying in other cases is most helpful and will aid in setting up the new model.

In step 1 where the problem is defined, limitations and assumptions are made to simplify the mathematical analysis. If all minor details are included, the mathematical analysis may get too complicated and become too troublesome or costly to solve. Accordingly those factors which, with reasonable certainty, are known to have only a minor influence are generally omitted from the mathematical treatment in order to keep it simple. If, after a preliminary solution is obtained, the effect of such minor factors on the solution can be tested, and where some doubt exists as to the significance of their effect, adjustments can then be made.

Numerous economic relations may be assembled in step 2 in setting up the model. Those relations covering the value of money, interest, and depreciation, as well as the influence of the economic state, prices, costs, etc., are discussed in Chapters 2 through 5.

The interrelation of market demands and price relationships from the standpoint of how these affect earnings, profits, recovery of capital, and accounting for capital is covered in Chapters 6 through 10. In this portion of the book the more simple types of models usually presented in real world economics studies are discussed.

More advanced type of models are considered in Chapters 11 through 16, which include profitability evaluations allowing for the time value of money, the economics in cyclic processes, and the influence of uncertainty on the model, and in the final chapter there is a discussion of financial factors that influence decisions.

All of these chapters provide a variety of illustrative problems which may be employed as guides for the reader in using the material in the text for real world problems.

DECISION POLICY

Upon development of the model and its analysis in step 4, a final step remains, to make a decision or establish a policy based on the results available. However, this management response is not made solely upon the basis of the results of factors included within the model, but upon all factors involved, some of which may be external to those in the mathematical model. (For example, a study may indicate that it is uneconomical to try to recover the final traces of an oil in a waste stream discharging to a river. Yet, the state law may be such that recovery is mandatory. Here some factor other than economy dictates the decision or policy.)

However, where the model results are reasonable and a logical confirmation of experienced judgment, then the model has provided a strong aid to validate a production decision, and the decision can be made with the assurance that it is the correct one. The importance of considering all factors that might significantly influence a decision cannot be overemphasized. For instance, the numerical results given by a mathematical model may not always be the best answer because of the vagaries of nature or the perfidies of human beings, both of which are external to the model. Thus there is always justification for considering common sense and experience in rendering judgment in any given situation.

INTRODUCTORY EXAMPLE

In order to indicate the real world application of mathematical models for decision making, a simple example will be given. This example demonstrates the objectives in this book in the most elementary manner.

The management of a company, called the AB Company, has adopted a policy for investing in new proposed ventures as follows: When the ratio of investment to the sum of (depreciation plus profit) is less than 2, then AB Company will initiate the venture. A simple mathematical model may be proposed as follows:

$$n_c = \frac{I}{Y + D} \qquad (1\text{-}1)$$

where n_c = the criterion upon which the decision is based
$\quad\quad I$ = investment dollars
$\quad\quad Y$ = annual profit dollars
and $\quad D$ = annual depreciation dollars (assumed constant for this example)

Thus for all levels of investment and profit and for all depreciation procedures employed, Eq. (1-1) may be used to obtain numerical answers. The equation may be programmed for computer evaluation over a wide range of investments and operating conditions so as to obtain numerical answers in an automatic manner. Based on managerial policy for making a decision, all conditions giving values of n_c greater than 2 may then be discarded. From those giving a value of less than 2, management can proceed to a decision which may involve consideration of the magnitude of the value for n_c and of other factors, perhaps including those external to the model. These other factors for example may be economic, legal, psychological, or scientific.

Equation (1-1) is considered to be one measure of evaluating profitability in the form of payout time. This will be amplified in subsequent chapters.

In order to provide the data for the model in Eq. (1-1), one may need to resort to other models to obtain the data to insert in the equation. The preliminary models in this book are tools utilized by management. They may vary from the balance sheet (which is a simple arithmetic model) to a highly sophisticated probability relation. It is the purpose of this text to demonstrate the broad range of types of models at an elementary level. From these developments and basic tools, the reader may then advance to the more complex forms with a confidence in understanding the concepts that are basic to sound decision making.

One of the major difficulties in setting up a model is finding suitable data that are sufficiently valid for use in the equations that relate the variables under study. Often such data are nonexistent and must be estimated or at best are so limited that considerable extrapolation or interpolation is required. However, even under these conditions, employing such information is better than doing nothing, and in many cases a completely satisfactory model for decision making can be obtained with full knowledge of the limitations of the model because of the nature of the data being used.

Throughout this text illustrations are employed to demonstrate the application of a concept to a real world problem. In some instances the data utilized are selected to illustrate a point. However, the reader must consider all pertinent information in his own applications of these illustrative models. Where a particular item or variable is a major component of the model, its real world values obviously should be of a high degree of validity to ensure that the model will give a true answer. On the other hand, if certain items are of minor influence, considerable latitude in their accuracy is permissible. It is the responsibility of the one proposing the model to describe the assumptions inherent in the analysis and to define the limits of applicability.

INTEREST
CALCULATIONS

2

In economic theory a major factor is the rent that should be paid for the use of money. This is interest and in the real world the magnitude of this factor must be considered in making economic analyses. It is the purpose of this chapter to define and illustrate the quantitative relations among interest and capital in applied economics. In itself interest is a mathematical model, but far more important is its inclusion in other more complicated models. Interest has a variety of uses and therefore the concepts of this chapter are quite significant.

The need for investment capital in organizing and operating a firm is self evident, but the interest on that investment capital is just as vital. Management has the dual responsibility of protecting initial investment capital and also of paying *interest* for its use. If capital is borrowed from a commercial bank, interest charges appear as a direct charge against the cost of operations. On the other hand, if capital is obtained by selling stock, management is obligated to pay the stockholder interest in the form of *dividends* from the profits that are made by the firm's operations.

CAPITAL AND INTEREST

The economic principle involved in a "free enterprise" system of economy is one in which capital or material worth is put to work to earn more capital. If capital is allowed to remain dormant, no risk is taken and no money is earned; however, uninvested wealth represents an actual loss since, potentially, it could be employed to earn additional money. A sum of money today is worth more next year by an amount equal to the interest it could earn if invested. This interest is referred to as the "time value of money."

One way that capital can be used to earn this interest is to invest it in a company that processes a raw material and produces a finished product. The product is worth more than the raw material by an amount equal to the costs of the processing. If the product can be sold for more than it costs to manufacture it, the excess is the profit or earnings made on all the capital involved in processing and selling the product. These earnings are the interest on the capital. (NOTE: This last statement is a matter of viewpoint. The economist considers profit as the excess of earnings above the going interest which money should earn; whereas the accountant ordinarily considers profit as the total earnings.) Further complications in defining the earnings arise from some accountants employing a basis "before profits tax" or "after profits tax" and others deducting actual interest paid out of earnings before calling the residue profit. In this text, the term *net profit* refers to the residue after deducting all costs, including profit taxes, from net sales; however, net profit includes any financial costs for borrowed money in accordance with usual accounting practice. If the financial costs for interest payments are deducted from the net profit, the residue is proprietary earning, or proprietary profit.

The amount of profit that a company makes depends upon the demand, or the desire, of the public for the product, the capacity of the plant to meet this demand, the availability of raw materials to supply the plant capacity, the competition from other companies making the same product, and the actual costs of making the product.

Profitability is computed as a percent of the total capital involved. Thus, the economic study of a company must consider both the dollars of profit and dollars of capital involved. A small company making $60,000 per year on a capital investment of $400,000 is making a 15 per cent profit, whereas a large company making $1 million annual profit with a capital investment of $20 million is earning only a 5 per cent interest rate on its capital.

The owner of the capital obviously wishes to earn as much interest as possible on the money he puts into a venture. He also wants to invest in ventures involving the least risk of losing his capital. Capital can be used to buy government bonds, but the interest rate is low (about 3 per cent) because the risk is not great. If the investor has a sufficient desire to increase substantially the earnings on his capital, he may invest it in ventures where the risk is greater. For example, he might invest in public utilities where the earnings are regulated by state and federal laws. They are regulated because the products (electricity, water, fuels) are necessary for the public welfare. Since these products are always in demand and the companies are protected somewhat from competition, the risk of capital loss is not too great, but neither are the earnings large.

If the owner of the capital is still more venturesome, he will invest in a private industry where the risk is greater. However, the opportunities for obtaining higher earning rates for his capital are also much greater. It is this "venture capital" that has produced the great economic development of the nation. In fact, the anticipation of high earnings has provided owners with the incentive to use their capital for new operations or processes which otherwise might not have been developed.

The accounting department in a company reports the status of the investors' capital in the form of a balance sheet, and the amount of company earnings is also stated periodically. These financial statements give only the results of past operations. A firm that shows no profit is valueless to the owners, since they could have kept their money in the bank at no interest, without running the risk of losing the capital. Thus, these periodic accountings enable the owners to follow a company's operations; if the indications are unfavorable, they can take proper steps to protect their capital before it is too late.

VALUE OF MONEY

The value of money (which is the interest that must be paid for its use) is not a fixed rate but varies with the general state of the national economy. For example, under conditions of an inflation economy, there is a rise in prices if the supply of goods cannot keep up with consumer demand for them or for other reasons, such as the payment of higher wages. Under such conditions, management attempts to increase its investment in order to enlarge production and take advantage of the increase in prices. This demand for investment funds by all management results in a decrease in the available supply of capital and money is said to be "tight," or the economic state is one of "hard money." Under these conditions, interest rates are high and credit is relatively difficult to obtain. Conversely, under a deflation economy money is "soft," or "easy," because low interest rates apply and credit is readily obtained for investment. Thus, the supply and demand of money in part controls its value. In addition, certain fiscal operations of the government may exert control over the value of money through the Federal Reserve Banking system, also known as the "Fed," or "Federal Reserve."

Briefly, the operation of the Federal Reserve tends to control interest rates in three main ways.[1]

(1) The limit of the legal reserve that banks must maintain to protect their depositors may be raised, causing a restriction in the money supply. On the other hand, it may be lowered, easing the money situation so that interest rates drop.

---

[1] The Federal Reserve System has other powers that affect the availability of money, but it is not the purpose of this text to consider all these refinements of the national economy.

(2) A source of income for a bank's funds is government bonds. If the bond prices go up, the bank prefers to loan more money instead of buying government bonds; thus, at such a time, business enterprises enjoy lower interest rates. Conversely, when bond prices go down, banks buy them; consequently, they have less money to loan to business, and interest rates go up. Thus, the Federal Reserve controls bond prices by *selling bonds* (their price decreases) or by *buying them* (the supply becomes less).

(3) The third control exercised by the Federal Reserve is the fixing of the rediscount rate, which is the interest rate charged to banks for borrowing money. In short, if banks must purchase money at a higher rate, the banks must charge their customers (business firms) at a higher rate also.

The preceding discussion and a consideration of the economic state provides the background that qualitatively determines the *manner* in which interest rates vary. The actual *numerical values,* however, may vary considerably. For example, in 1958 during a period of deflation, or recession, the Federal Reserve interest rate was about 1.75 per cent. The average earning of high grade bonds and securities was about 3.0 per cent. The rate of return, or interest, on money *varies* with the *risk* of the investment; that is, the likelihood of recovering the original capital accounts for the variation of interest rates for different types of enterprises. Investments in government securities, which entail less risk, yield less than 4 per cent; whereas, investment in more speculative economic ventures may yield 12 per cent or higher. However, these economic ventures do not include *highly* speculative ventures, such as wildcat oil operations or mining investments where the yields may vary from over 100 per cent to 0 per cent. These highly speculative ventures, while necessary in the development of natural resources, are considered to be pure gambles. Therefore, they must be treated as special cases in which the chances of losing the capital are compared with the possibilities of a very great return. Such a comparison is sometimes called *a calculated risk*.

INTEREST EQUATIONS

If $10,000 is put in a concrete box and stored in the ground, the original $10,000 can be dug up and recovered at any time. However, if that same $10,000 is put into a sound economic venture, not only the $10,000 is recovered but interest dollars as well. The total amount of interest dollars varies with the time during which the money is invested and also with the interest rate. From the above, it should be apparent that a given original $10,000 will be worth a varying amount in the future, depending upon what has been done with it.

The future value of the $10,000 also depends upon the kind of interest involved. In *simple interest* only the interest on the principal, $P,$ is computed each year. Thus, at 6 per cent interest on $10,000 the interest for each year is 0.06 × 10,000 = $600. If the time period involved is 10 years, the total interest is 10 × 600 = $6000. Thus, 10 years from now the $10,000 is worth 10,000 + 6000 = $16,000.

A second kind of interest is *compound interest* where any interest earned in one period is added to the principal, and the new total capital is used as the principal for the next period. Thus $10,000 at 6 per cent for the first year earns $600 to give a new capital of $10,600, which earns 0.06 × 10,600 = $636 for the second year to give a new capital of 10,600 + 636 = $11,236 for calculation of interest in the third year, etc. After 10 years the total value of the original capital is $17,908, which is larger than that for the previous calculation for simple interest.

These relationships will be explained in detail later in the chapter. They are expressed algebraically as a principal P, an interest rate i (expressed as a decimal), and the number of periods n. The pertinent relationships are as follows:

Future amount at simple interest $= P + Pin = P(1 + in)$

Future amount at compound interest $= P(1 + i)^n$

For most management economy decisions where the value of money is considered, the compound interest relations are employed, and they will be used in this text. Accordingly, most of the remainder of this chapter will be concerned with compound interest. However, many day to day banking transactions involving management employ simple interest because the time periods are usually less than one year. For this reason, several sections on commercial practice dealing with simple interest are included in this chapter.

THEORETICAL ECONOMY EQUATIONS

The term "theoretical economy equations" refers to calculations that include the value of money, that is, the interest that must be paid for the use of capital. In most cases compound interest is employed. The formulas for the mathematical relationships among time, interest rate, and worth of money are given in Table 2-1; their derivations can be found in standard mathematics or accounting handbooks. Before discussing these equations, two basic principles of (a) compound interest and (b) equivalence must be considered.

The basic premise in all these equations is the use of compound interest;

that is, any interest earned by the original capital is added to and becomes part of the capital at the end of the interest period so that, in succeeding intervals of each interest period, interest is earned on all previous interest payments as well as on the original capital.

The interest period is that time period for which the interest is calculated. Usually, it is annually, semiannually, quarterly, or monthly, but it can be any desired time interval. Unless otherwise stated, most bonds are considered to have *semiannual* interest periods and most other calculations to have *annual* interest periods, as is generally true in the real world. Although compound interest is usually computed on an annual basis, the formulas used must reflect interest for the actual time period considered. Thus, 6 per cent, quarterly, means 1.5 per cent for each of the four interest periods each year.

A second basic principle is equivalence, by which is meant that many different time series, or methods of paying back capital and interest, are *equivalent,* or mathematically equal for a given interest rate and the same interest periods. This will be explained by the examples later in the chapter.

TABLE 2-1. Theoretical Economy Equations

Nomenclature

s = compound interest factor equal to $(1 + i)^n$, or e^{in} for continuous compounding. (Sometimes s is called the amount of 1 at compound interest. The reciprocal of s is often called v^n, the present value of 1 at compound interest.)

i = interest rate per period

i_c = interest rate for capitalized cost

L = salvage value at some future date

n = number of interest periods

P = present sum of money

P_c = capitalized value

$a_{n/i}$ = present worth factor equal to $\dfrac{(1 + i)^n - 1}{i(1 + i)^n} = \dfrac{s - 1}{is}$, or $\dfrac{1 - e^{-in}}{i}$

($a_{n/i}$ is sometimes called the present value of annuity of 1 per period. The product, $sa_{n/i} = (s - 1)/i$, is sometimes called $s_{n/i}$, the final value of annuity of 1 per period as used in sinking fund calculations. The reciprocal $1/a_{n/i}$ is called the periodic rent of an annuity whose present value is one.)

R = end of period payment or annuity of uniform amount equivalent to P

R_s = sinking fund deposit

R_t = sum of all periodic annual charges including capital recovery

S = sum at future date at n periods

TABLE 2-1. (Continued)

Equation	Determines
(2-1) $\quad S = P(1 + i)^n = sP$	Single payment at end of nth period
(2-2) $\quad R = P\dfrac{i(1 + i)^n}{(1 + i)^n - 1} = P\dfrac{is}{s - 1} = P/a_{n/i}$	Uniform payment at end of period to recover P with interest, with no salvage value
(2-3) $\quad S = R\dfrac{(1 + i)^n - 1}{i} = R\dfrac{s - 1}{i} = Rsa_{n/i}$	Future worth of uniform payment R at end of nth period; recovers a sum S when S is numerically equal to P, as in a sinking fund. Also used to compute sinking fund deposit, R_s
(2-4) $\quad P = R\dfrac{(1 + i)^n - 1}{i(1 + i)^n} = Ra_{n/i}$	Present worth of a series of uniform payments with no salvage value
(2-5) $\quad n = \dfrac{-\log (1 - iP/R)}{\log (1 + i)}$	Payment time when L is zero or salvage value is neglected
(2-6) $\quad s = (1 + i)^n = \dfrac{1}{1 - i(P/R)}$	Used to calculate rate of return i when L is zero or salvage value is neglected
(2-7) $\quad R = \dfrac{P}{n} + (iP/2)\dfrac{n + 1}{n}$	Approximation of Eq. (2-2)
(2-8) $\quad R = (P - L)\dfrac{is}{s - 1} + iL$	Uniform payment allowing for salvage value
(2-9) $\quad R = \dfrac{(P - L)}{n} + (P - L)(i/2)\dfrac{n + 1}{n} + iL$	Approximation of Eq. (2-8)
(2-10) $\quad P_c = R_t/i_c$	Capitalized value of total annual payments, R_t

*Where these equations are to be used for any calculations in which a uniform periodic annuity due amount R_d is paid at the *start* of a period, then that payment must be multiplied by $(1 + i)$ to give the R to use in these equations; that is, $R = (1 + i)R_d$ where R_d is an annuity due uniform payment.

The equations in Table 2-1 constitute the ones most generally used in theoretical management economy studies. The use of these equations is illustrated in the following discussion, where an interest rate of 6 per cent is employed for a time period of 10 years, $n = 10$; a present sum of $1000

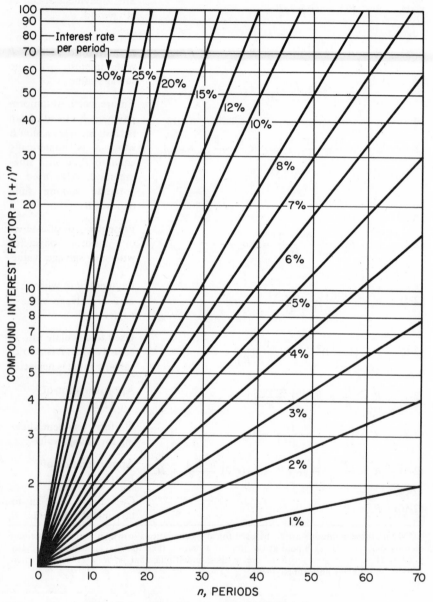

Fig. 2-1. Chart for compound interest factor.

EQUATION (2-1) 15

is applied in examples of the equations and to establish the concept of equivalence.

EQUATION (2-1)

What is the future worth after 10 years of a present sum of \$1000 if the annual interest rate is 6 per cent? This is the standard theoretical interest problem involving compound interest, which is solved by Eq. 2-1.

$$S = P(1 + i)^n = Ps, \text{ future worth of } P \qquad (2\text{-}1)$$

The term $(1 + i)^n$ will occur many times in the formulas, and for simplicity, it will be called the compound interest factor denoted by s. This factor as found in tables of interest factors is often called "the amount of 1 at compound interest." Its reciprocal $1/s$ is designated as v^n, "the present value of 1 at compound interest." For most calculations where an accuracy to three significant figures is sufficient, a log log slide rule may be employed. More precise figures for the factor are given in Appendix C. For approximations, the values for s may be read from Fig. 2-1. It will be noted that Eq. 2-1 is a simple exponential function which is linear on semilog coordinates as shown. Also Kulman[2] has nomographs for some of the relationships of Table 2-1.

Returning to the problem, one finds that the future worth of the \$1000 is:

$$S = 1000(1.06)^{10} = 1000 \times 1.791 = \$1791$$

The value for any other number of years or other interest rates is given by the appropriate substitution in the equation. The values of s (see Fig. 2-1) are the future worths of \$1 under various conditions of time and interest rates.

Thus, \$1791 in one lump sum is required 10 years hence to repay a sum of \$1000 borrowed at the present time; or \$1000 is the *present worth* of \$1791 paid 10 years from now, including interest. The \$1000 is also called the *discounted value* of \$1791 due 10 years from the present time with 6 per cent interest compounded annually. This establishes that over a 10-year period, considering the time value of money, the \$1000 will be worth \$1791 at the interest rate used. Any method of repaying the original \$1000, other than a lump sum, that results in the payments being worth \$1791 in 10 years is an equivalent time money series.

---

[2] C. A. Kulman, "Nomographic Charts," McGraw-Hill Book Co., Inc., New York, 1951.

EQUATION (2-2)

An example of a payment method equivalent to the lump sum procedure is one that pays a uniform amount at the end of every year. Thus, what uniform payment for 10 years at 6 per cent interest is necessary to pay off a present debt of $1000?

From Eq. (2-1) the future worth S is $P(1 + i)^n$. Therefore, the future worth of sum of uniform payments R in amount must also equal $P(1 + i)^n$. As shown in Table 2-2, the future worth of S of the R payments is:

$$S = \sum_{T=1}^{T=n} R(1 + i)^{T-1} = \frac{R[(1 + i)^n - 1]}{i}, \text{ future worth}$$

Equating this to $P(1 + i)^n$ and solving for R yields:

$$R = \frac{Pi(1 + i)^n}{(1 + i)^n - 1} = P\frac{is}{s - 1} = P/a_{n/i}, \text{ annual payment} \qquad (2-2)$$

The term $a_{n/i}$ as shown in Table 2-1 is a complex function of the interest rate and number of periods and also appears very often in economic calculations. It is the present value of 1 per period for n periods at i interest per period.

The reciprocal $1/a_{n/i}$ is called the periodic rent of an annuity whose present value is 1. Thus, as demonstrated in Eq. (2-2), the periodic payment, R, equivalent to an initial sum P is P times the periodic rent for 1. This annual payment, R, is at the *end* of the interest period, and the equation is called the *annuity* equation. Several alternative equations using the s and $a_{n/i}$ forms are given to simplify the arithmetic. The *present-worth factor* is derived from Eq. (2-4), which will be discussed below. Equation (2-2) is also called the *capital recovery equation* because it defines the annual uniform amount for recovery of a present sum plus interest on that sum.

Returning now to the actual problem—one finds:

$$R = 1000 \frac{0.06(1.791)}{1.791 - 1} = 1000 \times 0.1359 = \$135.90$$

or using the present worth factor, $a_{n/i}$, from Fig. 2-2 or Appendix C-2:

$$R = P/a_{10/0.06} = 1000/7.35 = \$135.90$$

Thus, paying $135.90 every year will repay the $1000 plus interest in 10 years, which is also equivalent to paying a lump sum of $1791 at the end

EQUATION (2-3)   17

of 10 years, even though the borrower actually pays on $1359 in 10 separate payments. The borrower pays $432 less than $1791 because he does not have use of the full $1000 for 10 years, since he reduces part of the principal each year. It is easy to show, however, by Eq. (2-3) that the lender who is receiving $135.90 each year for 10 years is receiving a time series of money which is worth $1791.

EQUATION (2-3)

What is the future worth of n uniform payments of $135.90 if interest is 6 per cent? The answer is given by Eq. 2-3.

$$S = R \frac{(1 + i)^n - 1}{i} = R \frac{(s - 1)}{i} = Rsa_{n/i}, \text{ future worth} \qquad (2\text{-}3)$$

The factor $(s - 1)/i$ is often called $s_{n/i}$ amount of annuity of 1 per period. Substituting in the equation gives:

$$S = 135.9 \frac{1.791 - 1}{0.06} = 135.9 \times 13.18 = \$1791$$

If the lender *reinvests these ten annual payments* at the same interest rate as before, they will actually produce $1791. Whether or not the lender does reinvest the payments is irrelevant; the payment of $135.90 each year for 10 years is still equivalent to the payment of $1791 in a lump sum. If the lender does not reinvest the payments, he merely loses the potential interest of $432, which he could have collected at a 6 per cent interest rate.

In Eq. (2-3) the R term is a constant and known as the *sinking-fund deposit*, R_s when it is computed from $Si/(s - 1)$, and the factor $i/(s - 1)$ is called the *sinking-fund factor*. This equation is used in depreciation accounting to determine what annual uniform deposit, R_s, should be made to equal a *future sum* of money, S, after n periods at a given interest rate. This future sum, S, in a sinking fund problem is numerically equal to the initial principal amount. The uniform payment should not be confused with the one in Eq. (2-2) where the principal amount is a *present* amount. In a sinking fund plan, a series of uniform annual deposits plus periodic interest on the accumulated amount will just repay an initial amount.

These relations may be clarified as follows: What sinking-fund deposit will provide a future sum of $1000; that is, the future worth of what series of payments must just equal $1000 in 10 years at 6 per cent? From Eq. (2-3):

$$R_s = \frac{1000(0.06)}{(1.06)^{10} - 1} = \$75.90$$

The present worth of this series is not $1000 (this is its future worth), but instead is:

$$P = 75.90 \times a_{10/0.06} = 75.90 \times 7.36 = \$558.40$$

This, of course, is also the present worth of the future $1000, or:

$$P = \frac{1000}{(1 + i)^{10}} = \frac{1000}{1.791} = \$558.40$$

If the annual interest on P of 0.06 (1000) = $60.00 were added to the $75.90, there would result the total annual capital recovery charge of 75.90 + 60.00 = $135.90. This is proven algebraically, since

$$R = R_s + Pi \quad \text{when} \quad \frac{i}{s-1} + i = \frac{is}{s-1} = 1/a_{n/i}$$

This equality states that annual uniform payment for capital recovery of a present sum is equal to a sinking-fund deposit for the present sum plus the annual interest on the present sum *when the interest rates are the same* for all calculations.

The annual charge of Eq. (2-2) is R, which recovers an initial amount plus interest and is a uniform payment; the annual sinking-fund *charge* (not deposit) is an increasing amount each year because of the interest on the accumulated amount. The charge for a Tth year is $R_s(1 + i)^{T-1}$, which can be demonstrated readily in a manner similar to that in Table 2-2. This nonuniform charge, if desired, may be converted to an equivalent present worth equal to $nR_s/(1 + i)$ by the methods shown in Table 2-2 and discussed in Table 3-2.

EQUATION (2-4)

This equation is used to determine the present worth, P, equivalent to a series of n uniform payments of amount R, if the interest rate is i. It can be obtained by algebraic manipulation of Eq. (2-2) or directly as the sum of the present worth of a series of future payments:[3]

---

[3] This is the same as the geometric series of $\frac{1}{(1+i)^n}[R + R(1 + i) + R(1 + i)^2 \cdots R(1 + i)^{n-1}]$ which has a value of $\frac{a}{(1+i)^n}\left(\frac{r^n - 1}{r - 1}\right)$ where $r = (1 + i)$ and $a = R$.

EQUATION (2-4) 19

$$P = \sum_{T=1}^{T=n} R/(1 + i)^T = R\frac{(1 + i)^n - 1}{i(1 + i)^n}$$

$$= R\frac{s - 1}{is} = Ra_{n/i}, \text{ present worth} \qquad (2\text{-}4)$$

What is the value at the *present time* of a series of 10 annual payments of \$135.90 with interest at 6 per cent?

$$P = 135.9 \frac{1.791 - 1}{0.06(1.791)} = 135.90 \times 7.360 = \$1000$$

This calculation will be recognized as the reverse of that used in the discussion of Eq. (2-2). The value of P is the *discounted value* of the uniform payments. The present worth factor is sometimes designated as $a_{n/i}$ called "present value of annuity of 1 per period."

The *present-worth factor*, $a_{n/i}$ is used quite frequently, and values for it are tabulated in Appendix C-2 and C-4. In addition, for approximation its value may be estimated from Fig. 2-2. This chart may also be used for trial and error calculations in solving for n or i as unknowns. The reciprocal of $a_{n/i}$ is sometimes called the *capital-recovery factor*. Note that

TABLE 2-2. Development of Certain Economic Equations[a]

End of Year	Single Payment, S Future Worth at End of Year		Uniform Annual Payment, R	
			Annual Deposit	Cumulative Total Future Worth at End of Year
0	P (start)		0	0 (start)
1	$P + iP = P(1 + i)$		R	R
2	$P(1 + i) + iP(1 + i)$ $= P(1 + i)(1 + i)$		R	$R + R(1 + i)$
3	$P(1 + i)^2 + iP(1 + i)^2$ $= P(1 + i)^2 (1 + i)$		R	$R + R(1 + i) + R(1 + i)^2$
T	$P(1 + i)^{T-1} + iP(1 + i)^{T-1}$ $= P(1 + i)^T$		R_T	$R[1 + (1 + 1) + (1 + i)^2 +$ $\cdots + (1 + i)^{T-1}]$ $= (R/i)[(1 + i)^T - 1]$[b]
$\vdots$	$\vdots$		$\vdots$	
n	$P(1 + i)^{n-1} + iP(1 + i)^{n-1}$ $= P(1 + i)^n$		R_n	$= (R/i)[(1 + i)^n - 1]$

[a] T is any number of periods; n is a specific number of periods.
[b] This is the sum of a geometric series with the nth term of form ar^{T-1} whose value
value is $a\dfrac{r^T - 1}{r - 1}$ where $r = (1 + i)$ and a is equal to R.

where there is no salvage value, $a_{n/i} = P/R$, which is a very useful relation for finding an unknown i or n from Fig. 2-2, or from Appendix C-2 or C-4 when P and R are known.

At this point it may be pertinent to review the theoretical basis for the equations previously discussed. This basis is illustrated in Table 2-2 in terms of the future worths of two independent methods of repaying an initial sum P.

$$S = P(1 + i)^n = R \frac{(1 + i)^n - 1}{i}$$

These equalities may be used to develop the relationships in Eqs. (2-1), (2-2), (2-3), or (2-4).

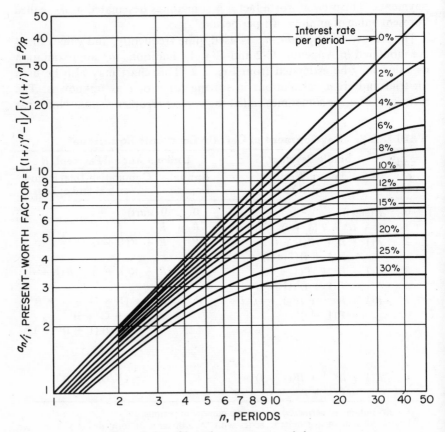

Fig. 2-2. Chart for present worth factor.

EQUATION (2-6) 21

In this discussion of theoretical economy equations, interest payments are considered as being made at the *end of the period;* however, one procedure (called annuity due) considers the payments as being made at the *start of the year;* however, this method is not usually employed for theoretical calculations. A uniform series with payments made at the start of the year differs from the standard annuity plan because a payment, R, at the end of a period has a value of $R/(1 + i)$ at the start of the period. Thus, for an annuity due series, the uniform payment R to be used in the equations of Table 2-1 to give the proper equivalent values should be $R = R_d(1 + i)$, where R_d is the annuity due payment at the start of the period. Expressed in another way the equivalence of an *annuity due series* and a *regular annuity series* for an *elapsed time* of n periods may be related through present worth as follows:

$$P = R_d + R_d \frac{(1 + i)^{n-1} - 1}{i(1 + i)^{n-1}} = R_d(1 + i) \frac{(1 + i)^n - 1}{i(1 + i)^n}$$

The validity of these two may be checked as an exercise.

EQUATION (2-5)

$$n = \frac{-\log (1 - iP/R)}{\log (1 + i)}$$

This equation is merely a rearrangement of Eq. (2-2) to permit solving for n when it is the unknown sought.

EQUATION (2-6)

$$s = (1 + i)^n = \frac{1}{1 - iP/R}$$

This equation is another form of Eq. (2-2) which is more convenient in solving for an unknown interest rate i when the values for n, P, and R are known. It may also be used instead of Eq. (2-5) to solve for an unknown n, if desired.

Another procedure for obtaining an unknown interest rate is to compute the factor, $a_{n/i} = P/R$, from the data given and then to use the charts in Fig. 2-2 or Appendix C-2 or C-4 to locate what value of i for the known periods n will give the same value of $a_{n/i}$. Thus, for twenty periodic payments of $1000 with an initial present worth P of $9818, the computed value of $a_{20/i}$ = 9818/1000 = 9.818. From the table or charts at n = 20 (or by trial and error calculation), for the $a_{20/i}$ factor at various interest rates it is found that i = 0.08, or 8 per cent is required for $a_{20/i}$ to equal 9.818.

EQUATION (2-7)

$$R = \frac{P}{n} + (Pi/2)\frac{n+1}{n}$$

This is an approximation of Eq. (2-2) where the term P/n is called *straight-line depreciation* and the term $(i/2)(n+1)/n$ is the *average interest rate* for the first and nth years.

CONTINUOUS COMPOUNDING

In the equations above the expression $s = (1 + i)^n$ is a discrete mathematical function where n is a discrete whole number. This is the historical procedure for all calculations involving the time value of money. However, modern management economy is beginning to employ more complex mathematical relations where continuous functions involving s are useful and where it becomes advantageous for s itself to be continuous. One form of s may be written as $(1 + i/m)^{mn}$, where i is the annual interest rate, m is the number of times per year interest is compounded, and n is the number of years. It will be observed that if n is one, then $(1 + i/m)^m$ will be greater than $(1 + i)$. For example, a 12 per cent annual rate when compounded monthly will yield $(1 + 0.12/12)^{12}$ = 1.127. The *effective* annual rate is i = 1.127 − 1.000 = 0.127, or 12.7 per cent, compared to the nominal annual rate of 12.0 per cent. Thus, the effective annual rate is $(1 + i/m)^m - 1$.

In continuous compounding where m approaches infinity, the value of $(1 + i/m)^{mn} \underset{m \to \infty}{\approx} e^{in}$, which permits the use of continuous equations containing fractional values of n. Tables are given in Appendix C for e^{in}, although slide rule values are usually sufficiently accurate. For continuous compounding, the effective annual rate is $e^i - 1$ when the nominal annual rate is i. For the nominal value of i = 0.12, the effective rate is:

$$e^{0.12} - 1 = 1.128 - 1 = 0.128 \text{ or } 12.8 \text{ per cent}$$

which is only slightly higher than for monthly compounding but appreciably larger than the nominal annual rate.

In a given situation where the discrete compounding of a nominal interest rate per period is the basis for calculations, the relations among present worth, future worth, etc., require the function $s = (1 + i)^n$. In order for the same numerical point to point results to be obtained with continuous compounding the compound interest factor, $s = e^{in}$, must be numerically equal to the s for discrete compounding. Thus, an *equivalent* interest rate, i, must be employed in the equations. It is evaluated from:

$$(1 + i)^n = e^{i'n}$$
$$e^{i'} = (1 + i)$$
$$i' = \ln(1 + i)$$

For example, if i is 12 per cent, the equivalent interest rate for use in continuous functions would be:

$$i' = \ln(1.12) = 0.113 = 11.3 \text{ per cent}$$

In most of this text, the presentation is in the discrete form, since this is the familiar form for management and banking. However, certain later chapters consider relations where time is considered as a continuous variable and where it is assumed that any sum can be reinvested immediately upon its receipt. The interest rate is expressed in the relation e^{in}. The equivalent *discrete* annual rate would be higher for any given point to point values for growth of capital, but the magnitude of the difference in interest rates would depend on the interest rate and number of periods involved.

SALVAGE VALUE

In many management decisions involving capital recovery, a portion of the original capital, P, is obtained after n periods by selling of the equipment, property, or assets. This salvage value, therefore, affects the economic equations in Table 2-1 with the corrections as shown in Eqs. (2-8) and (2-9).

From a practical standpoint, if a present investment has a salvage value, L, after n years, then the comparable cost at the present is really the initial cost, P, less the present worth of the salvage value which will be available n periods in the future. This present worth of the salvage value is $L/(1 + i)^n$ from Eq. 2-1.

CAPITALIZED COST EQ. (2-10)

Capitalized cost as a method of comparison is employed in economic studies, although it actually represents an abstract concept. The principle involved is that any uniform expenditure or payment represents interest payments in perpetuity (or for an indefinite future time). Therefore, this is equivalent to a certain sum, P_c, which must earn those interest payments at a selected interest rate, i_c. The amount, P_c, is called the capitalized cost of the uniform expenditures and is obtained by dividing the sum, R_t, of all annual costs involved by the assumed interest rate, or $P_c = R_t/i_c$. For example, if the annual cost for repairs and maintenance of a parking area is \$300 and money is considered to be worth 5 per cent, the capitalized cost for repairs is $300/0.05 = \$6000$. This is the capital equivalent to an annual charge of \$300 when the value of money is 5 per cent.

The total capitalized cost, R_t, for the parking area would be the sum of the capitalized costs for capital recovery, R, of the original investment plus those for the annual maintenance, R'. The annual capital recovery costs are (a) the annual capital recovery needed to *renew* the initial investment (allowing for salvage value) periodically and perpetually plus (b) the interest on the salvage value. A complete general equation including annual charges for maintenance R', therefore, would be:

$$P_c = \frac{R_t}{i_c} = \frac{1}{i_c} \left[(P - L) \frac{is}{s - 1} + iL + R' \right] \qquad (2\text{-}11)$$

Where i_c (the interest rate for capitalization) and i (the interest rate for capital recovery) are identical, this equation may be written in a variety of identical forms shown in Eqs. (2-11a, b, c). Thus, one form is:

$$P_c = \frac{1}{i} \left[iP + \frac{iP}{s - 1} - \frac{iL}{s - 1} + R' \right] \qquad (2\text{-}11a)$$

This equivalent form (see discussion of Eq. (2-3)) states that the annual capital recovery charges are (a) the interest on the original investment plus (b) the sinking fund deposit required for renewal (allowing for salvage). Other forms are:

$$P_c = P + \frac{1}{i} \left[(P - L) \frac{i}{s - 1} + R' \right] \qquad (2\text{-}11b)$$

$$P_c = \frac{Ps}{s - 1} - \frac{L}{s - 1} + \frac{R'}{i} \qquad (2\text{-}11c)$$

Note from Eq. (2-11c) that the capitalized cost of a present sum, P, is always equal to $P/(ia_{n/i})$, where $a_{n/i}$ is the present worth factor.

The remainder of this chapter applies these mathematical models to real types of problems where the value of money is considered. These same principles are utilized in a broader form in later chapters, and it is desirable for the reader to have an understanding of them. Problems provide a good test of this knowledge.

PRESENT WORTH

Problem

An accounting machine costing $1000 has yearly operating charges of $200. If it has a 10 year life and money is worth 6 per cent, what is the present worth of the cost for service provided by the machine?

Solution

From Eq. 2-4, the present worth for the operating charges is:

$$P = 200 \times 7.360 = \$1472$$

or, by Fig. 2-2, $\qquad P = 200a_{10/0.06} = 200 \times 7.36 = \$1472$

The total present cost for 10 years' machine service, therefore, is:

$$1000 + 1472 = \$2472$$

assuming that the equipment has no value at the end of 10 years.

Alternatively, the annual cost of the service may be considered as the sum of the operating cost plus the annual cost for capital recovery of the initial $1000, which must be recovered or paid for. Since $is/(s - 1)$ is 0.1359 for 10 years and 6 per cent interest, the capital recovery charges are $135.90 and the total annual cost is $135.90 + 200 = \$335.90$. From Eq. (2-4) it is found that this annual payment (or cost) has a present worth of $2472, which agrees with the above result.

Equation (2-8) relates to the case where all the original capital need not be repaid, since a portion is recovered as salvage or junk value. Thus, the machine purchased for $1000 may need replacement at the end of 10 years and may be worth $100 for parts and junk metal at the end of that time. The annual charge R for the recovery of capital is less here than for the above example using Eq. (2-2) because only

$$1000 - 100 = \$900$$

must be recovered. However, annual interest must be charged on the salvage value since this represents money in use. These points are included in Eq. (2-8):

$$R = 900 \times 0.1359 + 100 \times 0.06 = \$128.30$$

INTEREST RATE

Problem

Equation (2-6) is used to estimate the rate of return obtained when all factors but the interest rate are known. Thus, if the part-time services costing $135.90 per year for a man who periodically takes records in the time-keeping department can be eliminated by installing automatic recording devices (having a life of 10 years) at a cost of $1000, what is the rate of return earned on the investment, assuming that costs of servicing the instrument are negligible?

Solution

By Eq. (2-6) and trial and error, trying 6 per cent, one gets:

$$(1 + 0.06)^{10} = \frac{1}{1 - (1000/135.9)0.06}$$

$$1.791 = 1.791$$

The unknown interest being earned on the $1000 investment is 6 per cent. This problem may be solved[4] also by use of Fig. 2-2, since from Eq. (2-2)

$$a_{10/i} = \frac{P}{R} = \frac{1000}{135.9} = 7.36$$

This value of $a_{n/i}$ for $n = 10$ periods occurs at $i = 6$ per cent on Fig. 2-2, and from the Appendix C-2. Equation (2-5) or (2-6) or Fig. 2-2 also may be employed to estimate the payout time for the above recorder installation, where the interest rate that such investments must return is fixed. For this case, if such a rate is 6 per cent, then, by "solving," the payout time will be 10 years. (NOTE: In general practice, such investments must earn two or three times a 6 per cent return and, in addition, must be paid out in much less than 10 years for management to consider them seriously.)

---

[4] The relation $a_{n/i} = P/R$ is valid only for conditions of no salvage; where Eq. (2-8) applies, then $a_{n/i}$ as computed from observed data is:

$$a_{n/i} = (P - L/s)/R$$

THE APPROXIMATE EQUATIONS (2-7) AND (2-9)

For completeness, the remaining two approximate equations in Table 2-1 are illustrated by using the same data in the illustration for Eqs. (2-2) and (2-8). Thus, Eq. (2-7) can be employed to estimate the uniform annual payment required to repay $1000 in 10 years at 6 per cent:

$$R = \frac{1000}{10} + 1000 \, \frac{0.06}{2} \times \frac{10 + 1}{10} = \$133$$

The $133 approximates the $135.90 obtained by Eq. (2-2) to within 2.1 per cent. The first term on the right side ($100) is straight-line depreciation (see Chapter 3), and the second term ($33) is called average interest. Obviously, the using of straight-line depreciation plus 6 per cent annual interest of $60 would give a grossly incorrect answer ($160) for the annual payment that is to repay the $1000. This error is often made. The error would be due to the fact that by such a calculation the borrower is charged interest on the full amount each year, when actually he is reducing the principal continuously. The use of "average interest" on full principal is an approximate mathematical tool to give the same result as full interest on the gradually reducing principal. Such equations give low results for R, and the magnitude of the error increases with both time and interest rate. The user should appreciate these limitations in drawing conclusions in an economic study.

Similarly, Eq. (2-9) for a machine costing $1000 and having a salvage value of $100 at 10 years with interest at 6 per cent gives:

$$R = \frac{900}{10} + 900 \, \frac{0.06}{2} \times \frac{10 + 1}{10} + 100 \times 0.06 = \$125.70$$

The $125.70 approximates the $128.30 from Eq. (2-8) within 2 per cent.

CAPITALIZED COST

Problem

Compare the present worth and capitalized cost of an attachment for a textile weaving machine which costs $3000 and has an expected life of 3 years. The attachment has no salvage value, but requires $200 per year for maintenance and lubrication. Money is worth 8 per cent.

This problem might be written: What is the amount of capital equivalent to a present investment of $3000 plus an annual expenditure of $200 for 3 years at an interest rate of 8 per cent?

Solution

This permits a ready solution by the economic equations in Table 2-1.

The present worth of the capital invested is $3000 plus the present worth of $200 at the end of each of 3 years. The latter is computed from Eq. (2-4) or by using Fig. 2-2 for the $a_{n/i}$ factor.

		Dollars
P for capital invested		= 3000
P for annual expenditures	$\dfrac{200[(1.08)^3 - 1]}{0.08(1.08)^3}$	= 515
Total present worth		= 3515

The same result might be obtained by converting each future annual expenditure to its value at the present time by use of Eq. (2-1).

		Dollars
P for 1st-year expenditure	$200/(1.08)^1$	= 185
P for 2d-year expenditure	$200/(1.08)^2$	= 171
P for 3d-year expenditure	$200/(1.08)^3$	= 159
Total		= 515
P for capital invested		= 3000
Total present worth		= 3515

By use of Eq. (2-4), which is the sum of a series, the individual calculations of Eq. (2-1) are condensed. It must be noted that R in these equations is the equivalent *end-of-period* payment and that annual expenditures are usually considered as occurring at the end of the period, unless otherwise stated. Stepwise calculation would be required where annual expenditures were not the same each year.

The capitalized cost for the service rendered by the attachment is computed from any desired form of Eq. (2-11) where the salvage value, L, is zero. Using Eq. (2-11b), where the term $i/(s - 1)$ is the sinking fund factor, gives:

		Dollars
Sinking-fund deposit =	$3000 \dfrac{0.08}{(1.08)^3 - 1}$	= 924
Maintenance and operation		= 200
Total		= 1124

The present capitalized value of these annual expenditures is computed from Eq. (2-11b):

	Dollars
1124/0.08	= 14,050
Initial investment	= 3000
Total capitalized cost, P_c	= 17,050

The above method is preferred, but, alternately, the capitalized value could be computed by replacing the initial investment above by its equivalent which is the capitalized value of the annual interest on the investment, or $0.08 \times 3000 = \$240$, which is added to the other annual costs. Then, Eq. (2-11a) is used:

	Dollars
Sinking-fund deposit	= 924
Annual interest	= 240
Other annual expenditure	= 200
Total annual cost	= 1364

Capitalization of the $1364 results in:

$$P_c = \frac{\$1364}{0.08} = \$17,050$$

It will be noted that the $924 sinking-fund deposit plus the $240 annual interest on the initial investment is $1164, which is the same result as obtained by Eq. (2-2) for capital recovery with interest, or:

$$R = \frac{3000 \times 0.08(1.08)^3}{(1.08)^3 - 1} = \$1164$$

These results are the same, however, only when the capitalized interest rate and that for the sinking fund are identical.

Note that the present worth value is quite different numerically from the capitalized cost. This is true because the two are entirely different concepts. The present worth denotes the value of a service for a limited time. Capitalized cost represents a certain sum (but fictitious), the interest on which will provide a certain annual amount. The latter is equal to the sum of (1) the annual interest on the initial investment plus (2) an annual sinking-fund deposit which will provide for an infinite number of periodic renewals of the initial investment plus (3) an amount equivalent to annual operating charges.

CALCULATIONS WITH SALVAGE VALUE

Where the problem involves salvage value, the latter is a sum (after n years) having a certain present worth, which may be computed. Then, the amount is deducted from the present worth total of all other items, since the salvage value is a credit. For capitalized cost calculations involving salvage value the sinking-fund deposit is required only to recover $P - L$. The capitalized cost equivalent to the salvage value is included in the initial cost and is never credited, because for perpetual service it is always used over again for the periodic replacement.

Problem

If the salvage value in the preceding problem is $500, what is the capitalized cost for the service, assuming permanent operation?

Solution

The calculations are made from Eq. (2-11b) as follows:

		Dollars
Annual sinking-fund deposit (3000 − 500)0.308	=	770
Other annual expenditures	=	200
Total	=	970

Capitalized value of the annual expenditures 970/0.08	=	12,130
Initial Cost	=	3000
Total capitalized cost for the service	=	15,130

Alternatively, the initial cost could be replaced by its perpetual annual interest charge, or $3000 \times 0.08 = \$240$, which, when added to the other annual charges and capitalized, will give the same result.

$$\frac{240 + 970}{0.08} = \$15,130$$

The difference in capitalized cost for the two examples (that is, the ones with and without salvage value) is:

$$17,050 - 15,130 = \$1920$$

which is the equivalent capitalized value of the difference in the respective sinking-fund deposits; or it is the capitalized value of the sinking-fund deposit for the salvage value, which is:

$$\frac{500 \times 0.308}{0.08} = \$1920$$

The principle upon which the capitalization procedure is based is as follows: the annual charges are considered as the perpetual interest at the given interest rate for some unknown equivalent capital. The unknown equivalent capital is obtained by the subsequent capitalization calculation.

SPECIAL SUMMATION EQUATIONS

Where any first period item R' increases by a *constant increment, k,* at the end of each period the present worth of the annual payments for n periods can be computed for the combined arithmetic-geometric series:

$$
P = \frac{1}{1 + i} \left[R' + \frac{R' + k}{(1 + i)} + \frac{R' + 2k}{(1 + i)^2} + \cdots + \frac{R' + (n - 1)k}{(1 + i)n - 1} \right]
$$

$$
= \frac{R'[(1 + i)^n - 1]}{i(1 + i)^n} + \frac{k}{i^2(1 + i)^n} [(1 + i)^n - (1 + ni)]
$$

$$
= \frac{R'(s - 1)}{is} + \frac{k}{i^2 s} [s - (1 + ni)] \tag{2-12}
$$

Note that R' can be any constant figure such as all items in the brackets of Eq. (2-11). The capitalized value of these expenditures can also be computed by converting their present worth to an equivalent series of uniform annual expenditures and dividing by i_c to give (when $i_c = i$):

$$
P_c = \frac{R'}{i} + \frac{k}{i^2} \left[1 - \frac{ni}{(1 + i)n - 1} \right] = \frac{R'}{i} + \frac{k}{i^2} \left(1 - \frac{ni}{s - 1} \right) \tag{2-13}
$$

Similarly, where each succeeding annual expense increases a *constant fraction, f,* of the preceding period expense, the present worth for n periods is computed for the geometric series:

$$
P = \frac{R'}{1 + i} \left[1 + \frac{1 + f}{1 + i} + \frac{(1 + f)^2}{(1 + i)^2} + \cdots + \frac{(1 + f)^{n-1}}{(1 + i)^{n-1}} \right]
$$

$$
= R' \frac{(1 + i)^n - (1 + f)^n}{(i - f)(1 + i)^n} = R' [1 - (1 + f)^n/s]/(i - f) \tag{2-14}
$$

when $i = f$, then Eq. (2-14) becomes $nR'/(1 + i)$.

This present worth may also be converted to an equivalent series of uniform annual expenditures, which may then be capitalized by dividing by i_c to give (when $i_c = i$):

$$P_c = R' \frac{(1 + i)^n - (1 + f)^n}{(i - f)[(1 + i)^n - 1]} = R' \frac{s - (1 + f)^n}{(i - f)(s - 1)} \qquad (2\text{-}15)$$

THE BOND PROBLEM

The evaluation of bonds is an excellent example of the principles of capital recovery for which reason it is considered here, but it is not the intention to cover the broad field of stocks and bonds.

Bonds are issued for the purpose of raising capital and are sold as any commodity at a price determined by supply and demand. They are purchased by an investor for the purpose of earning a return on his capital. The bond is a specially printed piece of paper which states the value of the bond and the manner in which the interest will be paid. The company must pay the face value of the bond at its maturity date. By this method, the issuing company amortizes its debt, and the investor recovers his capital with interest. The bond is a promise of the company to pay the debt and is the legal document it gives the investor in return for the use of his capital.

A 10-year 6 per cent semiannual bond having a face value (principal) of $1000 is a promise to pay $30 every 6 months for a period of 10 years and $1000 at the end of 10 years. At any time from the date the bond is issued until it matures, the bond and its remaining interest payments have a value of $1000 if the earning rate (yield) desired by the investor for his capital is 6 per cent. This is true because the value of the bond is made up of two parts: (a) the present value of the future $1000 and (b) a series of interest payments which will also be made in the future. The dollar value of these two parts is computed by Eqs. (2-1) and (2-4), using the proper values of n and i. Employing the above bond as an example for a desired yield of 6 per cent, or 3 per cent every six months, gives the dollar value at the start of the first year as follows:

Dollars

Present worth of principal Eq. (2-1) $P = \dfrac{1000}{(1.03)^{20}} = 553.70$

Present worth of twenty $30 interest $P = \dfrac{30(1.03)^{20} - 1}{0.03(1.03)^{20}} = 446.30$
payments, Eq. (2-4)

Total $= 1000.00$

Thus, the present worth of this bond is $1000.

In a similar manner the present worth for a yield of 6 per cent at the start of the 4th, 6th, or any other year may be computed and will be found to be equal to $1000. In these cases the term "present worth" does not mean at the present time, but at a projected future date, after 3 years and 5 years, respectively, have elapsed from the time the bond was issued. The results are as follows:

At the start of:	1st year	4th year	6th year
Present worth of principal of $1000 to be paid at maturity date	$553.70	$661.10	$744.10
Present worth of remaining interest payments to be paid before maturity date	446.30	338.90	255.90
	$1000.00	$1000.00	$1000.00

In general, most bonds are understood to pay interest semiannually; however, any other series of periodic interest payments can be used, and would be so stated by the bond. Thus, a 6 per cent bond paying interest semiannually has a true, or effective, interest rate, computed from $(1 + i/m)^m - 1$ where m is 2; or, $(1.03)^2 - 1 = 0.0609$, that is, 6.09 per cent. The 6 per cent rate is called the *nominal* annual rate, and ordinarily the nominal annual rate is used to define the interest payments of a bond. Thus, in the problem following, the nominal 8 per cent rate is actually an 8.16 per cent yield to the investor on an annual basis.

All the above calculations are based on the investor's receiving a return on his investment equal to the interest rate stated on the bond. However, in many cases the investor desires a higher return and in some cases will be satisfied with a lesser return than the bond will pay. The interest dollars that the bond will pay in the future are fixed, but the purchase price paid by the investor is determined by the supply and demand of the bond. If the investor pays $1000 for the bond, the yield for his investment is 6 per cent; if he must pay more than $1000 for the $1000 bond, his yield will be less than 6 per cent (even though the bond is paying a rate of 6 per cent on the $1000 face value); and, if he can buy the bond at less than $1000, the investor's yield will be greater than 6 per cent. It is possible either to evaluate the yield for a given purchase price, or to determine the price to be paid for a bond to produce a given yield, by the capital recovery equations discussed in Table 2-1.

CALCULATION OF BOND YIELD

Problem

If a 10-year 6 per cent bond of $1000 face value has 4 more years before maturity, what price should be paid for it if an 8 per cent yield is desired?

Solution

The interest payments are $30 every 6 months, since this is a 6 per cent bond, but the *desired interest rate* used in the capital recovery formula is 4 per cent for two payments per year. A restatement of the problem might be: What is the total present worth of $1000 paid 4 years from now plus a series of eight semi-annual payments of $30 if money is worth 8 per cent? Note that the payments are at a nominal rate of 6 per cent because the bond states them to be, but the calculations are at a nominal rate of the desired 8 per cent.

$$\text{Dollars}$$

$$\text{Present worth of the principal } P = \frac{1000}{(1.04)^8} = 730.70$$

$$\text{Present worth of the interest } P = 30 \frac{(1.04)^8 - 1}{0.04(1.04)^8} = 201.90$$

$$\text{Total} = \overline{932.60}$$

Thus, $932.60 paid for the bond will yield 8 per cent during the 4 years it is held by the investor. If the purchase price of the bond is known, then the yield rate may be computed by a reverse calculation, where i is the unknown.

In a similar manner, the price paid for a bond at other yield rates or for other time periods may be computed. A summary of such calculations is shown in Fig. 2-3 for different conditions.

Problem

Calculate the yield on a $1000 bond which pays 4 per cent, semiannually, and has 6 years to go to maturity. The buyer pays $900.

Solution

The present worth of the purchase is known at $900 and the interest rate i is the unknown for semiannual interest payments of $20 each.

$$\text{Present worth of face value } + \text{ present worth of interest } = \$900$$

$$\frac{1000}{(1 + i)^{12}} + 20 \, a_{12/i} = \$900$$

This is a trial and error solution for i. Trying i = 6 per cent annually, or i = 0.03, gives:

$$\frac{1000}{(1.03)^{12}} + 20(9.95) = \$900$$

This value of i gives a result for the left side that is sufficiently close to $900 to be an acceptable solution for the expected yield on the purchase to be 6 per cent.

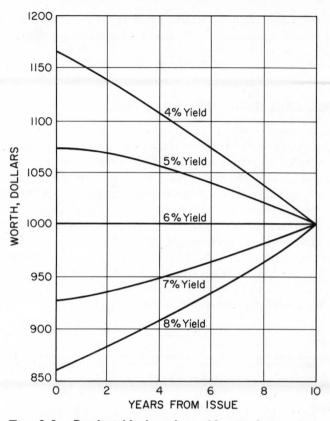

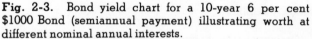

Fig. 2-3. Bond yield chart for a 10-year 6 per cent $1000 Bond (semiannual payment) illustrating worth at different nominal annual interests.

A study of the evaluation of bonds shows that the methods are based on determining the equivalent time-money series that applies in order to place the results on a comparable basis. It should be noted further that if the bondholder actually wishes to attain the yields indicated, it is his responsibility to collect the interest when it becomes due and to reinvest the interest dollars immediately at the yield rate of interest.

When a bond is purchased for an amount V greater than its face value or redeemable value F in the future, the excess amount is called a premium. From the foregoing calculations it is possible to derive an equation for this premium of $V - F$ where the bond is redeemed at its face value.

$$V - F = Fa_{n/i}(i' - i) \qquad (2\text{-}16)$$

where i is the *yield* rate for the purchaser and i' is the stated interest payment rate.

In a similar manner the discount which is the excess of the face value over the purchase price, or $F - V$, may be obtained directly as:

$$F - V = Fa_{n/i}(i - i') \qquad (2\text{-}17)$$

COMMERCIAL BANKING PRACTICE

In the section on interest equations it was pointed out that for short period operations (one year or less) most management dealings with banks are based on simple interest for economic problems. Accordingly, the nominal interest referring to the annual rate is often reduced to a daily simple rate, and the interest may then be computed from the number of days involved. There are certain common practices observed, however, which should be known.

On a daily basis *ordinary* simple interest is 1/360 of the annual rate, but *exact* simple interest is 1/365 of the annual rate. In determining the days to use, the first day is not counted, but the last day is included. Thus, from August 10th to October 1st constitutes 21 days in August plus 30 days in September plus one day in October, or 52 days.

Problem

If the annual interest rate were 6 per cent, what would the ordinary and exact simple interests be for this period on $2000?

Solution

The solution is as follows:

		Dollars
Ordinary simple interest	$= 2000(0.06/360)(52) =$	17.33
Exact simple interest	$= 2000(0.06/365)(52) =$	17.10

The dollar difference for the two rates varies of course with the amount of the principal.

COMMERCIAL DISCOUNT

Management needs to have cash available for its operations. At times this cash or working capital may be decreased to such an extent as to force management to curtail its activities. To avoid this it may call on a bank to extend cash on the basis of a loan or a discount of other commercial papers (such as bills payable from the company's customers) which are not due until some time in the future. The bank will make a loan on these future items but will deduct a discount or the interest on the amount of the loan; that is, the bank will pay the simple discount present value of a future sum. This is very much the same as a present worth calculation with compound interest but differs in two respects. First, simple interest is used; and second, the interest charge is deducted from the maturity value of the future sum. The calculations will be demonstrated by a real problem.

At simple interest the equation for interest is:

$$\text{Interest} = Pin \qquad (2\text{-}18)$$

where P = a present sum
    i = the interest rate (expressed as a fraction)
        for one period (a day, a month, or a year)
and n = the number of periods
    Then the future amount is:

$$S = P + Pin = P(1 + in) \qquad (2\text{-}19)$$

and the present value, P, of a future sum, S, would be:

$$P = S - Pin = \frac{S}{1 + in} \qquad (2\text{-}20)$$

At the bank, a loan of $100 to be paid in one year at simple interest of 6 per cent will mean a future amount after one year of $106 to be paid. However, bank procedure is to deduct the $6 from the $100 loan to give the borrower $94 immediately as proceeds of the loan. He pays back $100 one year from now. Thus, the bank has discounted the maturity value of a sum at an interest rate slightly higher than the nominal rate of 6 per cent as shown in the following comparison.

Problem

(a) borrower receives $100 and repays $106 one year from now. What annual interest does he pay? (b) The borrower receives $94 now as the proceeds of the loan, and repays $100 one year from now. What annual interest does he pay?

Solution

From Eq. (2-19)
  (a) $in = S/P - 1$ $\qquad (2\text{-}21)$
    for n = one year, the simple interest rate is
      i = $(106/100) - 1 = 0.06$, or 6 per cent
  (b) for n = one year, the *bank discount rate* is
      i = $(100/94) - 1 = 0.064$, or 6.4 per cent

The proceeds from a bank loan due in the future are readily computed from a formula based on a 360 day year.

$$P_b = S - Si_d n_d = S(1 - i_d n_d) \tag{2-22}$$

where P_b = the proceeds received by borrower
S = the maturity value (value at some future date)
i_d = the daily interest rate (discount rate) based on a 360 day year
n_d = the number of days involved

In bank discount calculations the proceeds of a loan are smaller in proportion to the interest paid when the discount period is increased. Thus, as the discount period decreases, the simple interest rate approaches the bank discount rate.

Note that the maturity value may be any type of negotiable financial paper which the bank cares to discount. For example, it might be a note bearing interest at some rate different from the bank discount rate, a bank draft, or any other valuable paper having a future specific value which the bank will accept as security that the money paid today will be regained at some future maturity date.

CASH DISCOUNTS AND INTEREST EARNED

Sellers of goods often give a discount to buyers who pay promptly when bills are rendered on the first of the month. For example, a policy of one per cent discount in 10 days or net in 30 days means the following: if paid in 10 days, a $100 charge for merchandise may be paid off for $99; whereas, if paid in from 10 to 30 days, the full amount is due the seller. In effect, payment in 10 days means that the buyer loans his money (which he could otherwise use) in an amount of $99 for the month and earns interest of one dollar. This is equivalent to a monthly rate of $(100 - 99)/99 = 1.01$ per cent at simple interest; the annual interest rate for 30 day periods would be:

$$i = (360/30)\left(\frac{i'}{1 - i'}\right) = 12(0.01/0.99) = 12.12 \text{ per cent}$$

where i' is the discount rate allowed per period.

This high effective simple interest rate that the buyer earns is at the expense of the seller. The latter encourages the buyer to pay quickly to avoid losses from bad debts in amounts greater than the discounts offered. The rate also assumes that the buyer can repeat the transaction continuously every period, which in the real world he may not do. Note that the effective compound rate would be:

$$i_e = (1 + 0.0101)^{12} - 1 = 0.128 = 12.8 \text{ per cent}$$

MORTGAGE CALCULATIONS

A mortgage loan is similar to any other bank loan except that it usually extends over a long period of time, the interest on it is compounded, and it requires some tangible asset as security. This asset may be a home, land, or other property that in the future will always be likely to have a value equal to or greater than the amount remaining to be paid on the loan. Usually, a uniform periodic payment, R, is made once a month from which the interest deduction is made, the remainder of the payment being used to reduce the principal of the loan. Since the interest is computed separately at each period on the unpaid amount of principal, the interest rate represents the true rate. A mortgage of an initial amount, P, represents the present worth of an annuity series of payments of amount, R, and n in number. This is demonstrated by the following.

Problem

What are the monthly payments on a $7000 house mortgage to be paid off in 15 years at 6 per cent annual interest.

Solution

The monthly interest is 0.5 per cent and 15 years is 180 periods; thus, from Eq. (2-2) the monthly payment is:

$$R = 7000 \frac{(0.005)(1.005)^{180}}{(1.005)^{180} - 1}$$

$$= 7000 \frac{(0.005)(2.454)}{1.454} = \$59$$

The borrower pays $59 each month, and his current balance is figured by computing the interest for one month, adding it to the unpaid balance, and then subtracting the $59 payment. Thus, at the end of the first month the interest would be $(7000)(0.005) = \$35$; the total due is $7035, but the remaining balance is $7035 - 59 = \$6976$, which earns interest starting at the beginning of the second month.

From the basic compound interest relations the amount still owed on a mortgage after n payments is:

Equivalent amount of an initial P after n periods, Eq. (2-1), is Ps

Equivalent amount paid after n periods, Eq. (2-3), is $R(s - 1)/i$

$$\text{Amount owed} = Ps - R(s - 1)/i \qquad (2\text{-}23)$$

Problem

After paying $59 a month for 10 years on the preceding mortgage problem, how much is still owed?

Solution

The solution is as follows:

$$\text{Amount owed} = 7000\,(1.005)^{120} - 59\,[(1.005)^{120} - 1]/0.005$$
$$= 7000\,(1.8194) - 59\,(0.8194)/0.005 = \$3060$$

Note that in these problems the *effective annual compound interest rate* for a nominal rate of 6 per cent is:

$$i_e = (1.005)^{12} - 1 = 0.0617, \text{ or } 6.17 \text{ per cent}$$

INSTALLMENT AND LOAN CALCULATIONS

It is interesting to compute the interest charges on certain installment and loan procedures commonly used.

One common practice on financing automobiles is to charge 6 per cent (per year) on the amount of the loan. For example, a car purchased for $3000 with $1200 down payment leaves an unpaid balance of $1800 to be financed. Thus, the charge is 6 per cent of 1800 or $108 for each year of the loan to be paid off in equal installments. If the loan were for 15 months, the total contract would be:

	Dollars
Initial amount	1800
Interest for 15 months 0.06(15/12)(1800)	135
Total contract	1935
Equal monthly payments of 1935/15	129

Thus, a series of 15 uniform payments of $129 is used to pay off a debt of $1800.

Problem

With $R = \$129$, $P = \$1800$, and $n = 15$, what is the annual effective compound interest rate?

Solution

The rate per period may be computed from Eq. (2-4) as:

$$1800 = 129 \, a_{n/i}$$

$$a_{n/i} = 13.953 = \frac{(1 + i)^n - 1}{i(1 + i)^n}$$

The tables in this text do not show $a_{15/i}$ values for the small interest rate needed, so a trial and error solution is necessary. Assume i equal to 0.0094; the right hand side of the equation is:

$$\frac{(1.0094)^{15} - 1}{(0.0094)(1.0094)^{15}} = \frac{1.1502 - 1}{0.0094(1.1502)} = 13.90$$

or, trying i equal to 0.0090, the answer is about 13.95.

Thus the interest rate for one period is 0.90 per cent. The annual effective rate, therefore, is:

$$(1.0090)^{12} - 1 = 11.30 \text{ per cent}$$

Unlike the mortgage calculation, the rate on interest and loan calculations is much greater than the implied nominal annual rate of 6 per cent. The effective rate is higher because the borrower receives no reduction in interest for the periodic reduction of the principal. The interest was originally computed for an elapsed time period, assuming no payments. Thus, the borrower is denied the use of his own capital, but at the same time is paying interest on it. This can be stated in another way as follows: the financing agency has its money repaid and can reloan it at the same time it is getting interest on it from the first borrower.

In some states the law permits a 3 per cent monthly interest rate on personal loans. This is equivalent to an effective annual compound rate of $(1.03)^{12} - 1 = 0.425$, or 42.5 per cent.

SUMMARY

The essential points of this chapter are: (a) capital when put to work earns more capital, (b) the amount earned varies with the time it is at work, and (c) different methods of payment, resulting in an actual difference in dollars paid, may be equivalent. The application of the equations has been illustrated, and charts have been presented for simplifying the calculations. These calculations are tools that management employs in setting up models to aid in decision making.

PROBLEMS

2-1. Certain United States Savings Bonds cost $75 and pay the bondholder $100 at the end of 10 years. What is the interest rate?

2-2. A present investment, P, requires how many years at 4 per cent to double its value?

2-3. What is the difference in dollars after 10 years for a $12,000 investment at the present time if interest is 10 per cent when (a) the interest payments are annual and (b) the payments are quarterly, based on the future worth of the original investment for the 10-year period?

2-4. What interest rate is earned by spending $1200 now for renovation of a pump which will eliminate buying a new pump at a cost of $2000 at the end of 4 years?

2-5. What annual payment for 8 years is equivalent to investing $9000 at 7 per cent interest? Show that the future worth of both is the same.

2-6. A man is paying $75 per month rent. What amount can he spend to buy a house having a 20-year life if money is worth 5 per cent and taxes and repairs on the new house will amount to an average of $15 per month?

2-7. What is the value at the present time of an annual cost of $2400 for repairs on a distillation column which has a life of 10 years if money is worth 8 per cent? What is the capitalized cost of such an expenditure for an indefinite future time?

2-8. A company making plastic toys purchases molding machinery with an expected life of 14 years for a price of $75,000. (a) How much money must be placed annually in a 5 per cent sinking fund to replace the machinery when it is worn out if it has no salvage value? (b) If the equipment has a salvage value of $2800, what must the annual deposit be?

2-9. What is the present worth of a series of expenditures made as follows: $500 at the end of 1 year, $1000 after 2 years, and $2800 at the end of 5 years if money is worth 6 per cent? Prove the answer by determining the future worth of the present sum in a single payment after 10 years and comparing it with the total of the future worths of the separate expenditures after 10 years.

2-10. A student contemplating graduate work estimates the cost as $5000 for the next 2 years. If after 2 years the difference in income will amount to an average of $300 per year for 30 years with money worth 6 per cent, would it be economical for him to continue his studies? (Assume that the $5000 is the present worth 2 years from now.) If the cost for the next 2 years were $3000, would it pay? What factors other than the financial return must be considered?

2-11. Because of corrosion, equipment for heating water requires replacement every year at a cost of $1000 with interest at 12 per cent. Calculate the present worth of five heaters for 5 years' service (a) by the present worth formula and (b) by determining the total sum of the individual present worths of each of the five payments, which are 1 year apart and provide 5 years' service. HINT: This problem is of an annuity due form.

2-12. A man retiring at the age of 65 has a life expectancy of 16 years and $50,000 in the bank, drawing 2 per cent interest. How much can he take out annually in equal installments before the fund is used up in 16 years? This problem might be stated as follows: What annual payment for 16 years is equivalent to the value of a present sum of $50,000 at 2 per cent?

2-13. What is the effective interest rate on a mortgage paid monthly if the nominal rate is 5 per cent?

2-14. If a finance company charges 5 per cent on a bill to be paid in 12 equal installments, what is the effective interest rate?

2-15. Calculate the value of a 10-year 6 per cent bond with semiannual interest and 7 years to go if the investor desires a yield of 4 per cent. Compare your result with that shown in Fig. 2-3.

2-16. Compute the worth of a 10-year $1000 bond just issued which pays 6 per cent, semiannually, if the desired yield rate is 7 per cent. What is the value when it has 5 years to go before maturity at a 7 per cent yield? Compare your answers with Fig. 2-3.

2-17. A 20-year $1000 bond paying 4.5 per cent annually is selling for $900 with 10 years to go. What is the yield rate?

2-18. A 10-year bond paying 6 per cent, semiannually, costs $1356, and the yield rate is 8 per cent with 6 years to go to maturity. What is the face value of the bond?

2-19. Repeat even numbered problems using continuous compounding.

2-20. Repeat all odd numbered problems using continuous compounding.

2-21. Prove Eqs. 2-12 and 2-13 by numbers.

2-22. Prove Eqs. 2-14 and 2-15 by numbers.

2-23. (a) With money worth 4.5 per cent annual interest, what amount does a business man receive on a 90 day loan of $10,000 when borrowed from the bank? (b) What effective simple interest rate does he pay? (c) If the interest were compounded four times per year, what would be the effective annual compound interest rate?

2-24. A purchaser of goods obtains terms of one per cent for payment in 10 days after receipt of bill. (a) What does he pay on a $12,000 order? (b) What effective annual simple interest would he earn by paying his bills promptly? (c) What effective annual compound interest would he earn by paying his bills promptly?

2-25. The maturity value of a financial paper six months from now is $40,000. The bank will discount it to yield $39,000. (a) What is the bank discount rate? (b) What is the effective simple annual interest rate? (c) What is the effective compound annual interest rate based on a semiannual period.

AMORTIZATION: DEPRECIATION ACCOUNTING 3

Among the numerous uses for interest calculations is their utilization in any model selected for *amortizing* a debt or initial capital outlay. Amortization as employed in this text is a generic term describing the equivalence of a capital sum over a period of time, although in accounting it may have a more restrictive meaning. In an industrial company amortization may mean a program or policy whereby the owners (stockholders) of the company have their investment of depreciable capital partly protected against loss. In such a context, the invested capital may be employed as working capital, or be used for the construction of buildings, the buying of machinery, merchandise, or other assets.

From the foregoing it is clear that one must make a distinction between the original invested capital and the working capital. The *working capital* is employed to pay salaries, buy raw materials, and to pay other expenses in connection with operations. The working capital is normally replaced by the sales dollars that the company receives for its products as fast as the working capital is spent, and therefore this capital is always available for return to the owners. However, the *original invested capital* utilized for equipment and buildings (fixed assets) cannot be directly converted to the original capital when these physical properties have decreased in value. They decrease in value because they *depreciate,* that is, they wear out or become obsolete. Thus, some provision must be made to return that part of the original capital that the owners would lose because of depreciation. In addition to their recovery of this depreciating capital, the owners expect the total capital to earn interest, which is the profit on the investment. In Chapter 2 mathematical tools were described for making the calculations involved in the recovery of investment capital. The purpose of the present chapter is to illustrate the application of these tools to certain problems.

DEPRECIATION

Depreciation has many meanings, but only two will be discussed here. The first meaning is descriptive and refers to the loss of value of capital with time when equipment (capital) wears out or becomes obsolete (becomes less valuable because something better is available). The second meaning refers to the systematic allocation of the costs of an asset that produces income from operations using the asset. This allocation is recorded in an accounting book and is the *estimated* dollars charged as a cost in making the product during the time period considered. This is an estimate because no one can predict with certainty when a piece of equipment will wear out or become totally obsolete. For tax accounting purposes, governmental agencies[1] establish the estimated life of industrial equipment as a guide. For the purpose of this book, these estimated lives will be considered as the actual life of the equipment, which will permit a simplification in comparing theoretical accounting with income tax accounting. In certain types of management economy studies, such as the selection of alternatives for investment, the estimated life of the equipment is often much less than that used for accounting, in order to minimize the risk of obsolescence. In other types of studies, depreciation data as available from the accounting department are employed as that part of capital recovery represented by the first term in Eq. (2-7) or (2-9), which is straight-line depreciation.

Since depreciation is such an important factor in the economics of an enterprise, it is important to have an understanding of the methods utilized by the accountant for offsetting it. In short, depreciation may be considered as a cost for protection of the depreciating capital *without interest* over the period (the minimum time being established by governmental agencies) during which the capital is used. When the selling price for the product is equal to all costs including depreciation, the loss in value of the machine (or asset) producing the product is being recovered without profit. Any excess of the selling price over the total cost of producing the product is profit (or interest) on the operation.

If an air conditioner cost $1100 and at the end of 10 years has a salvage value of $100, the $(P - L) = \$1000$ and $N = 10$ years. Any procedure may be employed to record a certain amount each year as the cost of the air conditioner that should be allocated (as a cost) for that year of its operation. This will be recorded in a depreciation account, and at the end of 10 years the cumulative amount (plus the salvage value) will be equal

---

[1] U.S. Treasury Dept., Bur. Internal Revenue, Bull. F; Supt. of Documents, Washington, D. C. (See Appendix B).

to the initial cost of the asset (air conditioner). In this case the amount will be $1100, which will then pay for replacing the wornout equipment so that operations can continue, if provision has been made to have $1100 available in cash. This might be accomplished in one of the following ways:

1. Straight-line Method (SL)

If N is the years of expected life, then the fraction, f_1 of the $(P - L)$ that must be recovered each year is $1/N$, or

$$f_1 (P - L) = \frac{1}{N}(P - L) = \frac{1}{10}(1000) = \$100 \qquad (3\text{-}1)$$

The straight line rate to recover $(P - L)$ is 0.10; the cumulative sum of these deposits, on which no interest is charged, is called depreciation re-

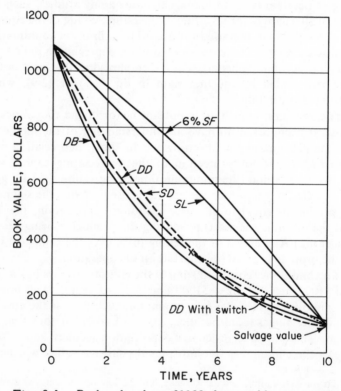

Fig. 3-1. Book value for a $1100 depreciable asset with $100 salvage value at 10 years.

serve. (NOTE: The term *reserve* as used herein refers to a bookkeeping account and does not necessarily mean a sum of cash money. It is that portion of the total cost of an asset that has been allocated in the past as a cost of operation.) This procedure, called straight-line depreciation, makes use of the first term in Eq. (2-9).

That part of the original cost of the asset which is in excess of the accumulated depreciation at any time is called the *book value* of the asset. The manner in which the book value decreases for the air conditioner is shown in Fig. 3-1 for a variety of depreciation methods. A summary of depreciation equations is given in Table 3-1, and the manner in which the depreciation account accumulates is shown in Fig. 3-2.

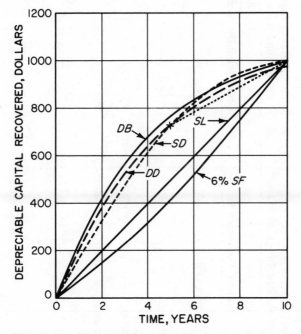

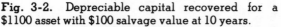

Fig. 3-2. Depreciable capital recovered for a $1100 asset with $100 salvage value at 10 years.

2. Sinking-fund Method (*SF*)

In this method an annual *deposit,* as computed from Eq. (2-3), is made. Then, each year interest at rate i on the previous accumulated sum is also added to give a depreciation *charge* for the year. After N years, an amount equal to the original value of the asset less salvage value will be accumulated. The factor f_2 for the annual deposit is a function of N and the interest rate as shown in Table 3-1. Since the previously accumulated

TABLE 3-1. Special Equations for Depreciation

Nomenclature

s' = compound interest factor equal to $(1 + i)^N$
f = factor for computing yearly depreciation charge; see discussion of each method
i = interest rate as a fraction
L = salvage value of asset at end of Nth year
M = estimated units or production during life of asset
M_n = units of production serviced in any one year
N = estimated life in years
n = end of year age for which depreciation is computed
P = initial value of asset
R_s = sinking fund deposit as computed from Eq. 2-3

Method	Annual Depreciation	After n Years		
		Accumulated Depreciation	Book Value[a]	
1. Straight Line, (SL)	$f_1(P - L) = (1/N)(P - L)$	$(n/N)(P - L)$	$P - (n/N)(P - L)$	
2. Sinking Fund (SF)[b]	$R_s = f_2(P - L) = \dfrac{i}{s' - 1}(P - L)$	$R_s\dfrac{(1 + i)^n - 1}{i}$	$P - R_s\dfrac{(1 + i)^n - 1}{i}$	
3. Declining Balance (DB)	$f_3(\text{book value})_{n-1} = (1 - \sqrt[N]{L/P})(\text{book value})_{n-1}$	$P[1 - (1 - f_3)^n]$	$P(1 - f_3)^n$	
4. Double Declining (DD)	$f_4(\text{book value})_{n-1} = (2/N)(\text{book value})_{n-1}$	$P[1 - (1 - f_4)^n]$	$P(1 - f_4)^n$	
5. Sum of Digits (SD)	$f_5(P - L) = \dfrac{2(N - n + 1)}{N(N + 1)}(P - L)$	$\dfrac{n(2N - n + 1)}{N(N + 1)}(P - L)$	$P - \dfrac{n(2N - n + 1)}{N(N + 1)}(P - L)$	
6. Units of Production (UP)	$f_6(P - L) = (M_n/M)(P - L)$	$(P - L)\dfrac{\sum\limits_{k=1}^{k=n} M_k}{M}$	$P - (P - L)\dfrac{\sum\limits_{k=1}^{k=n} M_k}{M}$	

[a] Book value is P minus accumulated depreciation. [b] For sinking fund annual depreciation deposit is $\dfrac{i}{s' - 1}(P - L)$; annual charge requires an additional amount equal to interest charge for the year on all previously accumulated amounts.

sums increase each year, the interest charges increase each year, and the total depreciation charge for each year increases as a progression. The accumulated depreciation for n years is computed readily. For the air conditioner where $(P - L) = \$1000$, $i = 6$ per cent, and $N = 10$ years, the annual deposit is \$75.90, and the book value decreases as shown in Fig. 3-1.

The sinking-fund method is employed by some governmental agencies for depreciation, but is not widely used by industrial operations for reasons that will be discussed at the end of the chapter.

3. Declining-Balance Method (DB)

In this method a fixed percentage f_3 of the remaining book value of the asset for the preceding year is used for the annual depreciation charge. The factor, f_3, is a function of the life of the asset and salvage value. In fact, salvage value must be known or estimated to utilize this method, since the factor is computed from Eq. (3-2) as used for the air conditioner problem.

$$f_3 = 1 - \sqrt[N]{L/P} = 1 - \sqrt[10]{0.0909} = 0.214 \qquad (3\text{-}2)$$

After 10 years, for P of \$1100 there will be a residual value of \$100. This residual is the salvage value. The equations and results for the air conditioner are shown in Table 3-1, Fig. 3-1, and Fig. 3-2. This method is used in some European countries.

4. Double-Declining-Balance Method (DD)

As will be explained later, any method of computing depreciation that gives large amounts in the early years is desired by management because it permits a more rapid recovery of part of the original investment. The Federal government regulates depreciation rates for income tax purposes, and one allowable procedure is the double-declining-balance method (DD). This method is of the same type as the DB method, but the factor, f_4, is established differently by assuming it arbitrarily with the limitation that the rate cannot exceed double the straight-line rate. Therefore, as a maximum, $f_4 = (2/N)$, regardless of the salvage value. This causes a peculiar result in that after N years a specific book value results which bears no relation to the salvage value. For example, with the air conditioner the straight-line rate here is $1/N$, or 0.1, and f_4 is $2 \times 0.1 = 0.2$, so the book value after N years from Table 3-1 is $P(1 - f_4)^N$. For $N = 10$ years, the value is:

$$P(1 - f_4)^N = 1100(0.8)^{10} = 1100(0.108) = \$119 \qquad (3\text{-}3)$$

Values for the *DD* method are shown in Figs. 3-1 and 3-2 with the book value terminating at $119 in Fig. 3-1. Since the salvage value is only $100, there is an unamortized amount of $19 which is not recovered by this method. Furthermore, the law prohibits the undepreciated amount after *n* years to be less than the salvage value. If the salvage value were actually $200 for the air conditioner, there would be a violation of the law. In this case, more than the original cost would be recovered since 200 + (1100 − 119) = $1181.

In order to correct for these inadequacies, the law permits a switch to straight-line depreciation for the last part of the service life, so that the terminal book value can be the same as the salvage value. This correction is shown in Fig. 3-1 as a straight long dashed line for the latter years of the *DD* curve. The straight line connects the *DD* curve at the year of switching with the salvage value at the terminal year. The switch point for best management policy occurs very close to the midlife of the asset if no salvage value is involved. If salvage value is available, the optimum time for switching to straight line occurs beyond the midlife and depends on the ratio of salvage value to initial cost. In certain situations, the double-declining-balance-with-switch method has only a minor advantage over other methods.

5. Sum-of-Digits Method (*SD*)

This depreciation procedure is permitted by Federal law and consists of an annual charge which is in proportion to the ratio of the digit for the years of life remaining divided by the sum of the digits for the total life. For example, for a 10 year life, the sum of the digits of the life is 1 + 2 + 3 + ... + 10 = 55. The depreciation factor, f_s, for the first year is 10/55, for the second year, it is 9/55, etc. This factor expressed algebraically is:

$$f_s = \frac{N - n + 1}{1 + 2 + 3 + \ldots + N} = \frac{(N - n + 1)}{\tfrac{1}{2}N(N + 1)} \tag{3-4}$$

where the symbols are defined in Table 3-1. The accumulated depreciation for a given *n* represents a series which can be evaluated as shown in Table 3-1. Results by this method for the air conditioner problem are plotted in Figs. 3-1 and 3-2.

In Fig. 3-3 a nomograph is given to assist in sum-of-digits calculations. For example, what is the amount of depreciation at end of the sixth year for a machine which cost $12,000, has no salvage, and has a life of 8 years? From Fig. 3-3, for an 8 year life, with $n = 6$, the fraction depreciated should be 0.083 or 0.083 × 12,000 = $1000. From Table 3-1, line 5, the calculated value is:

$$\frac{2(N - n + 1)}{N(N + 1)} P = \frac{2(8 - 6 + 1)}{8(8 + 1)} 12,000 = \$1000$$

6. Units-of-Production Method (*UP*)

This method follows the same concept as the straight-line procedure. The factor, f_6, for depreciation is based on the estimated number of units to be processed or serviced by the asset rather than the estimated number of years of service. The depreciation for any one year, therefore, is computed for the relative proportion of production units for the year compared to the estimated total. This is a linear or straight line relationship, with production as the variable, and would give a plot similar to *SL* in Figs. 3-1 and 3-2 if the base line were production units. The equations are given in Table 3-1.

THE DEPRECIATION MODEL

The relations in Table 3-1 plus the charts in Figs. 3-1 and 3-2 constitute the depreciation model. They provide management with some very

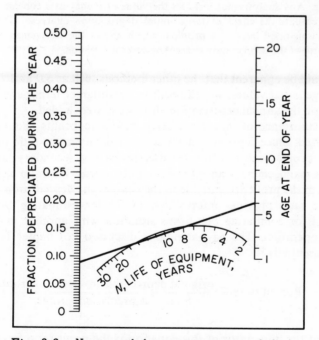

Fig. 3-3. Nomograph for computing annual depreciation by sum-of-digits method (After D. S. Davis, *Petroleum Refiner, 36* (3): 164 (1957)).

powerful visual and quantitative aids in making decisions regarding depreciation. For example, depreciation is an allowable item of cost which can be deducted from a firm's income before calculating profits taxes. Calculations consider the profits tax to be t (expressed as a fraction) and the tax credit to be tD dollars, where D is the annual depreciation. This tax credit is actually the dollars that are deducted from the tax where the tax is computed on profits before depreciation is taken. Management desires to make D as large as possible and to recover it as soon as possible during the life of the asset for several reasons.

(1) One author has likened the accelerated depreciation writeoffs to an "interest free loan" to industry. If the tax rate is lower after the initial period, the loan is underpaid because fewer taxes are paid. However, if the rate is higher, the loan is overpaid because a larger total amount is paid in taxes over the entire period.

(2) It provides a source of capital early in the history of the asset when the earnings are most likely to be relatively large, that is, before obsolescence sets in or before competition reduces the market for the firm's services.

(3) In the comments in (2) above money is considered to be at a zero interest rate. Any analysis that includes the value of money as a consideration will accentuate the effect of large initial depreciation charges. This will be demonstrated later in a problem which shows the differences in present worth of the depreciation charges by several methods.

It should be apparent that the other methods such as SD and DD which have large initial values, would result in much larger terminal amounts if interest on the annual depreciation allowances were added.

The importance of the depreciation model to profits should also be emphasized. Since depreciation is a cost of operation, the ratio of depreciation to profits (either before or after taxes) is a secondary model that provides management with information. In words, the ratio defines that fraction of the profit attributable to the charges for depreciation. In numbers, the ratio provides information useful for estimating the payout time and for comparing one firm's situation with another in the same type of operation. The payout time is discussed in Chapter 11, but the model essentially is:

$$\text{Payout time} = \frac{\text{original depreciable investment/profit}}{(1 + \text{depreciation/profit})}$$

Note that the numerator of this equation is the profit ratio based on depreciable investment. This also is discussed in a later chapter.

COMPARISON OF METHODS

The variety of depreciation methods available to management causes some confusion as to which is best. However, no method is an absolute first choice, since each has certain advantages under certain conditions. The service life, salvage value, type of service rendered, the value of money, purpose for computation, basis for computation, etc., all affect a decision in any particular situation. For those who are interested, a list of major companies, according to the method of depreciation they use, is available.[2]

Although each individual case must be considered separately, certain generalizations can be made about the various methods. However, these generalizations are limited by the factors mentioned above, and should be applied judiciously.

The sinking-fund method (*SF*) allows for the value of money by including interest considerations, which the others do not. The annual depreciation charge (deposit plus interest on depreciation account) plus the interest on the undepreciated value of the asset is a constant, as discussed under Eq. (2-3) in the previous chapter. This is a theoretically correct procedure, which apparently is its principal merit.

Where simplicity of calculation is desired, the straight-line method (*SL*) and units-of-production method (*UP*) obviously are the best. They are zero interest rate procedures for amortizing the value of an asset. For the *UP* method the annual charge will vary from year to year, depending upon the units produced or serviced each year.

The other three methods are similar in that they permit amortization at an accelerated rate in the early life of the asset, with lower annual charges near the end of service life. For physical assets which require repairs and service, charges for maintenance usually are small in the early life and tend to increase in the terminal stages. Thus, for any of the accelerated procedures, the total charges for depreciation plus service are approximately uniform throughout the asset life.

The choice of an accelerated method depends upon conditions. As shown in Figs. 3-1 and 3-2 for the air conditioner, all three accelerated methods give curves similar in shape and position. In two reports[3] an analysis of this problem was made under a specific set of conditions. The results may be summarized as follows: Using as a base the sum-of-digits

---

[2] Taken from "Accounting Trends and Techniques," 10th ed., American Institute of Accountants, 1956.

[3] Anonymous, *Capital Goods Review*, Nos. 27 and 28 (1956), Machinery and Allied Products Institute, (MAPI), Washington 6, D.C.

method (SD) with an after profits tax return of 10 per cent and no salvage, the analyses showed that the double-declining methods (DD) with and without switch were quite similar. Above a five year life, the sum-of-digits method was slightly better (with 0.5 per cent increase in return); however, below a five year life, both DD methods were superior, with up to a 3.2 per cent improvement in return.

Where salvage value was considered, the second report gives a chart indicating the quantitative relations among service life and ratio of salvage value to initial cost; this is for the specific case of a 10 per cent return after taxes. For these conditions, over a band of salvage-value ratios of 6 to 13 per cent, the DD method with switch was better than DD without switch. However, the difference was not great; and when either was compared with the sum-of-digits method (SD) a simple rule was proposed. This rule implies that the SD method should be used if the salvage value is less than 10 per cent of cost; if above 10 per cent, then the DD method without switch should be used.

Based on these data, on simplicity, and on the assumption that salvage value usually does not exceed 10 to 15 per cent in practical situations, the sum-of-digits method appears to be the most generally useful. However, Grant and Norton[4] recommend a declining balance method as the most generally feasible; their recommendations are based on theoretical and practical accounting considerations and general economic theory.

In 1962 the U. S. Treasury Department permitted a modification of its depreciation schedules. For new procedures under the Internal Revenue Service revenue procedure 62-61 guidelines are set up for broad classes of assets, as shown in Appendix B. A comparison of the depreciation allowances in the United States with those in some foreign countries is shown in the following tabulation. The double-declining-balance method is utilized for the U. S. data; the percentages recorded indicate the approximate proportion of the asset cost that may be charged off in the first five years.

	Per Cent		Per Cent
Canada	71	Italy	100
United Kingdom	64	Sweden	100
France	76	Japan	68

	Per Cent
New Procedures with 12 year life	60
Old Procedures (prior to 1962)	43

---

[4] E. L. Grant and P. T. Norton, "Depreciation," Rev. Ed., The Ronald Press Co., New York, 1955.

Certain other tax credits are allowed for capital expenditures, and in addition, the tax laws are being revised at the time of the writing of this text. Thus, for exact numbers to use in models employing depreciation (and depletion), the most up-to-date government procedures should be consulted. The basic principles given in this text are not altered by the particular procedures employed. In general, the tax credit consists of an initial direct reduction in taxes for a new capital expenditure, but the depreciable capital must then be based on the initial capital less the initial tax credit. For example, a 7 per cent tax credit on an investment of $10,000 would permit a reduction of $700 in the tax bill the first year. However, the subsequent depreciation would be based on the initial cost of $10,000 less the first year tax credit of $700, that is, a basis of $9300.

In addition to this tax credit and the regular depreciation, 20 per cent of the actual cost of certain capital expenditures having a life of more than six years is allowed in the initial year. This special 20 per cent allowance must be deducted from the cost before the regular depreciation is taken. In the example above if the flat 20 per cent special depreciation is taken on $10,000, then the depreciation would be figured on $10,000 − $700 − $2000, or on a residual of $7300. Thus, the total real dollars available the first year are (1) $700 special allowance directly on the tax bill, (2) $2000 special first year depreciation, plus (3) the regular depreciation by a selected method based on $7300 (with no salvage value).

PRESENT WORTH OF DEPRECIATION PAYMENTS

In certain management economy calculations, it may be desirable to compute the present worth of the depreciation payments. At *zero* interest, these payments (including salvage value) are worth exactly the investment at the present time. However, where the time value of money must be considered, each future payment must be reduced to the present time by dividing by the factor $(1 + i)^n$, where n is the end-of-year time for which depreciation payment will be made. The sum of all payments is determined. For uniform payments, the present-worth summation is quite simple, but for nonuniform payments more complicated summation expressions result. These are summarized in Table 3-2 for future reference. Note that the total present worth of these payments (including salvage value) will be less than the initial investment, and the *DB* method has the highest value, as shown by the following example.

Problem

A company purchases a piece of machinery for $1100 with a salvage value of $100 after 10 years; money is worth 6 per cent. What is the present worth of the depreciation charges, omitting the salvage value and using methods 1, 2, 3, and 5 in Table 3-2?

TABLE 3-2. Present Worth of Depreciation Payments[a]

Nomenclature

s = compound interest factor equal to $(1+i)^n$

f = factor for computing depreciation, see Table 3-1

i = interest rate as a fraction

L = salvage value at Nth year

M = estimated units of production during life of asset

M_n = units of production serviced in any one year

N = estimated total life in years

n = end of year age at time of computation; it is the number of terms in the summation

P = initial value of asset

$a_{n/i}$ = present worth factor = $\dfrac{(1+i)^n - 1}{i(1+i)^n}$

R_s = sinking fund deposit

Method	Present Worth for n Years	
1. Straight Line (SL)	$\dfrac{P/N}{1+i} + \dfrac{P/N}{(1+i)^2} + \cdots + \dfrac{P/N}{(1+i)^n}$ $= (P/N)\,a_{n/i}$	(3-5)
2. Sinking Fund (SF)[b]	$\dfrac{R_s}{1+i} + \dfrac{R_s(1+i)}{(1+i)^2} + \cdots + \dfrac{R_s(1+i)^{n-1}}{(1+i)^n}$ $= nR_s/(1+i)$	(3-6)
3. Declining Balance (DB)[c]	$(1 - \sqrt[N]{L/P})(P)\left[\dfrac{1}{1+i} + \dfrac{(1-2/N)}{(1+i)^2} + \cdots + \dfrac{(\sqrt[N]{L/p})^{n-1}}{(1+i)^n}\right]$ $= \dfrac{P(1\sqrt[N]{L/P})[1 - (\sqrt[N]{L/P})^n/s]}{(1+i) - \sqrt[N]{L/P}}$	(3-7)
4. Double Declining (DD)[c]	$2P/N\left[\dfrac{1}{1+i} + \dfrac{(1-2/N)}{(1+i)^2} + \cdots + \dfrac{(1-2/N)^{n-1}}{(1+i)^n}\right]$ $= \dfrac{2P}{N}\,\dfrac{1 - [(1-2/N)/(1+i)]^n}{i+2/N}$	(3-8)
5. Sum of Digits (SD)	$\dfrac{2P}{N(N+1)}\left[\dfrac{N}{1+i} + \dfrac{N-1}{(1+i)^2} + \cdots + \dfrac{N-n+1}{(1+i)^n}\right]$ $= \dfrac{2P}{Ni(N+1)}(n - a_{n/i})$	(3-9)
6. Units of Production (UP)	$P\left[\dfrac{M_1/M}{(1+i)} + \dfrac{M_2/M}{(1+i)^2} + \cdots + \dfrac{M_n/M}{(1+i)^n}\right]$ $= (P/M)\displaystyle\sum_1^n M_n/(1+i)^n$	(3-10)

[a] When salvage value is to be considered at a terminal point after N years, replace P in Eqs. (3-5), (3-6), (3-9) and (3-10) by $(P - L)$ and add a term $L/(1 + i)^N$ to all equations to get total present worth for N years. Care must be taken here to explain the salvage value term, since it may vary with n as in the case of turn in value, or it may be considered a nondepreciating value such as working capital or land as occurs in some calculations, in which case it may be considered constant for any value of n or N. Only P is used in Eqs. (3-7) and (3-8).

[b] Note that from the definition of a sinking fund the *future* worth of an external sinking fund deposit must be exactly equal to the initial sum to be recovered. Equation (3-6) applies for an internal sinking fund where a nonuniform annual charge equal to $R_s(1 + i)^{n-1}$ is paid annually as discussed in the text. If an external plan with a uniform annual deposit of R_s was used, the present worth would be $R_s a_{n/i}$ like in Eq. (3-5).

[c] These two methods always have a residual value at the Nth year. For method 3, this is the salvage value; for method 4, it must be computed as $L = P(1 - 2/N)^N$ to obtain the total present worth.

TABLE 3-3. Comparison of Results for Problem on Present Worth of Depreciation

Method	Initial Annual Charge[a] and Remarks	Present Worth[b]
1. Straight Line (SL)	$1000/10 = \$100$, uniformly	$100(7.36) = \$736$
2. Sinking Fund (SF)	Internal account, (see text, page 00)	$10(75.90)/(1.06) = \$715$
	External account, $1000 \dfrac{0.060}{0.791} = \$75.90$ uniformly	$75.90(7.36) = \$558$
3. Declining Balance, (DB)[c]	$1100(0.214) = \$235.40$, and 21.4 per cent of book value each succeeding year	$\dfrac{1100(0.214)(1 - 0.0508)}{1.06 - 0.786} = \$815$
5. Sum of Digits (SD)	$\dfrac{(2)(10)(1000)}{10(10 + 1)} = \$181.81$ and progressively smaller each year	$\dfrac{2(1000)}{110(0.06)} (10 - 7.36) = \$804$

[a] See Table 3-1 for methods of computation.

[b] The present worth of salvage value $= 100/1.791 = \$56$, which should be added to get total present worth. For SF method see comment in text on internal vs. external account (page 00).

[c] The factor for this method is $1 - \sqrt[10]{100/1100} = 0.214$.

Solution

The numerical results are shown in Table 3-3 which the student can check for an exercise. The s factor is equal to 1.791, and for this problem $n = N$. Table 3-3 indicates that the greatest variation is for the sinking fund method, although the others differ appreciably.

In Chapter 2 the sinking fund deposit was discussed, and it was shown that a uniform annual deposit provides a future sum of money. This deposit plus interest on the previously accumulated fund results in an annual nonuniform quantity that is equivalent to the future sum. Numerically, the annual charge in the Tth year was shown to be $R_s(1 + i)^{T-1}$. The present worth, therefore, for this annual charge in any Tth year is $R_s(1 + i)^{T-1}/(1 + i)^T = R_s/(1 + i)$. Then, the total present worth for an n series of these charges if actually paid out each year is:

$$\sum_{T=1}^{T=n} R_s/(1 + i) = nR_s/(1 + i)$$

This is the present worth of this particular series of annual charges for a company that arranged internally to provide for amortization of capital assets. If a company amortized such assets by an external sinking fund where only the annual uniform deposit R_s was paid out, the present worth would be $R_s a_{n/i}$ as for any uniform payment series. Thus, the equivalent present worth of an internal plan is larger because more of the company's funds must be utilized to pay the accumulated interest than for the external plan where only the deposit must be paid out each year. The interest is earned externally for such plans.

The variations in present worth are a function of both interest rate and life period with the DB method in general giving the highest present worth values. The explanation lies in the fact that the accelerated methods return greater amounts in the initial periods, and the interest on these amounts (plus the interest on the interest) is available for a longer period of time. Therefore, their future worths (or their equivalent present worths) are greater for the accelerated procedures. When short lives and salvage value considerations are involved, this conclusion may be altered for specific interest rates, which accounts for the variations in conclusions of MAPI and Grant and Norton, as previously mentioned.

COMMENT ON DEPRECIATION

As has been indicated earlier, the sum of the book value and depreciation account by any method is always just equal to the original investment.

Thus, if each year a company records in a reserve account the amount indicated by a chosen method and if the air conditioner depreciates in exactly the manner predicted by the method selected, the book value is a measure of the actual value at any time during the 10-year period. If the air conditioner does not last as long as predicted, the portion of the capital represented by the remaining book value has vanished, and nothing can be done about it except to write it off as a loss. If the air conditioner lasts longer than predicted, any value after 10 years represents an error in estimating its life. The depreciation accounts can be corrected periodically to make the predicted life conform to the actual life, which is necessary for accurate cost accounting. This is sometimes called multiple-straight-line depreciation when straight-line procedures are used for each part of the total life.

In Table 3-1, all depreciation methods serve an identical purpose, namely, each provides a means for estimating an annual charge which company management must make to protect the depreciable capital and to recover the investment. At any one year, the book values by any one method vary considerably; similarly, therefore, the amount in the depreciation account varies. However, these differences are reconciled at the tenth year, when each method gives the same final result. Thus, each method attains the same objective based on a predicted life, and the protection of the capital is based on the accuracy of the predicted service life. In this connection a detailed discussion of the principles involved in predicting the life of industrial property is given by Marston, Winfrey, and Hempstead.[5]

The discussion of depreciation up to this point has considered replacement of the *original dollars invested*. This does not mean that a machine asset can necessarily be replaced for the same amount of money. During an inflation period the replacement of a worn-out physical asset will require more *current dollars* than was paid for the original asset if the real value of the replacement is the same as the original asset. Such variations in replacement costs are not allowed by the government in depreciation accounting. However, the allowable accelerated depreciation methods provide for some relief in this respect. One simple correction is to consider the initial investment, P, as fluctuating with the dollar in such a manner that the term P should be replaced in all equations by $P(1 + k)^n$ where k is a percentage factor and n is the years involved. For example, the factor, k, in the long run is often considered as equal to 0.03; that is, on the average, the national phase of the U.S. economy has proceeded in

---

[5] A. Marston, R. Winfrey, and J. C. Hempstead, "Engineering Valuation and Depreciation," McGraw-Hill Book Company, Inc., New York, 1953.

such a manner that the annual increase in the cost of material goods is about 3 per cent.

DEPRECIATION ACCOUNTING

In most industries, the amount shown in the depreciation account does not appear as money unless a special fund is set aside specifically for this purpose. Chapter 4 will explain that an amount equal to the depreciation account appears as other assets, such as working capital, and that the physical form of such assets may be raw materials or finished products in storage. When it is necessary to buy a replacement, management must convert such physical assets into cash unless sufficient cash is available in the bank to pay for the new equipment.

Sometimes where the sinking-fund depreciation method is employed, a depreciation reserve fund may actually be invested outside the company to earn the interest rate used in sinking-fund depreciation calculations. However, since such interest rates are ordinarily less than the earning rate of the company, it is more prudent for the company to employ the money for its own operations, assuming of course that it will continue to earn more than a conservative sinking-fund rate of interest. Therefore, some question arises as to what interest rate should be used for sinking-fund depreciations. This rate can be a conservative low risk rate. At such a rate, the annual charges are higher, thus making the remaining book value less in any one year than would be the case for sinking-fund depreciation rates based on the earning power of the company.

A customary convention in bookkeeping assumes that any new asset dates its life from the middle of the year in which acquired. Thus, one half year depreciation is charged for the year acquired. However, in this text all assets are considered to be acquired on the first day of the year in question, since this simplifies calculations in the demonstration of principles.

One other point worth mentioning is that the portion of original capital employed to purchase land is not depreciable. (If mineral rights are considered, their initial value is recoverable through a depletion account). Land values remain more or less constant or appreciate in value, except in a depression. Therefore, no reserve is needed.

DEPLETION

The exhaustion of certain natural resources such as petroleum, metal deposits, orchards, vineyards, timber reserves, fisheries, etc., which disappear with use is termed depletion. In some cases, nature may renew these resources over a period of years, but in other instances they disappear forever. Since the value of a property becomes less as its resource is utilized in the process industries, accounting must make some provision to recover the initial investment. This is done by a procedure similar to that employed in allowing for depreciation. A depletion account is set up to recover that portion of the initial asset used. This account will accumulate in the same manner as a depreciation account. Thus, a depletion account can be considered as part of the lump sum paid to the owners when operations cease as a result of the exhausting of a natural resource. Alternatively, a depletion allowance can be paid annually to the owners where the life of the resource can be estimated with a fair degree of accuracy. For metallic ores, coal, etc., which are not renewable, one method of repayment is in the form of a sinking-fund deposit at one interest rate for the depletable capital plus interest (or profit) on the investment at another rate.

Where physical equipment for processing the raw material is installed, the cost for depreciation of this equipment must be considered in addition to the cost of depletion, proper allowance being made for the land value and salvage value of any equipment moved to another site. However, some of the invested capital is recovered in the usual manner through depreciation, and other parts of the capital are regained through the depletion allowance. The methods employed ordinarily follow the federal regulations. For example, in the case of oil and gas properties, the development costs (as represented by money spent for exploration and operations preliminary to the actual obtaining of the oil or gas for sale) must be recovered by writing such costs off in the year they occur or they must be regained through the depletion allowance. However, such development costs cannot be charged as depreciation.

The depletion allowance may be computed as a fixed percentage (27½ per cent for oil and gas wells, 23 per cent for most minerals and non-minerals—although for some the allowance is 15 per cent—and 5 per cent for common building stones) of net sales, but in no case can it exceed 50 per cent of the net income before the deduction of depletion as allowed by the federal government. The depletion allowance may also be computed on a cost per unit basis. (NOTE: The terms net sales and net income are discussed in detail in Chapter 5). For the time being, the term net sales may be considered as the annual dollars taken in by the company; when all the costs for doing business are subtracted from the net sales, the result is net income. The cost per unit basis is similar to the units of production method (*UP*) for depreciation. Cost per unit is computed by determining the unit depletion cost (ratio of intangible development cost plus other depletable costs divided by the estimated total units potentially recoverable). The total units recoverable may be estimated if the number of years of service and the annual production rates can be estimated.

Problem

An oil company has an oil lease on a field with an estimated reserve of 1.2 million barrels which the engineers consider to be 60 per cent recoverable in 3 years. The intangible development costs (excluding a $20,000 bonus to the land-owner which is recovered through depletion) all occur in the first year. They are estimated at $160,000 and are considered as an expense. The depreciable capital (casing, machinery, etc.) is $900,000 with a life of 9 years. If 300,000 barrels of oil are sold the first year at $2 per barrel and the operating and other expenses are $50,000 for the year, what is the depletion charge for the year (a) using the fixed percentage method and (b) employing the unit cost and a 3-year basis? (c) What are the total allowable first-year charges for capital recovery in parts (a) and (b)? (d) What are the first-year net incomes by both methods? (e) What are the second-year net incomes by both methods if 300,000 barrels are sold and the second-year expense is $40,000?

Solution

A preliminary accounting of the cost items for the year is required which may be tabulated as follows:

		Dollars
Net sales, 300,000 × 2	=	600,000
Annual depreciation, 900,000/9	=	100,000
First-year expense	=	50,000

Part a. On a fixed percentage rate of 27.5 per cent of net sales the depletion would be:

$$0.275 \times 6000,000 = \$165,000$$

If this does not exceed 50 per cent of net income before allowing for depletion, the allowance would be $165,000. The net income must be computed. Therefore, where the $20,000 bonus is recovered as part of the depletion charge, the net income must be computed as follows:

	Dollars
First-year expense	= 50,000
Development expense	= 160,000
Annual depreciation	= 100,000
Total annual cost	= 310,000
Tentative net income, 600,000 − 310,000	= 290,000
50 per cent of net income	= 145,000

The maximum allowable depletion is $145,000, and not $165,000.

Part b. The unit depletion costs for the lease and development are:

$$\frac{160,000 + 20,000}{0.60 \times 1,200,000} = \$0.25 \text{ per barrel}$$

The allowable depletion based on unit costs for the first year is:

$$0.25 \times 300,000 = \$75,000$$

This amount would be allowed even if it exceeded the amount permitted by the percentage method, but the unit cost method must be used each year.

Part c. The total allowable first-year charges for capital recovery by the above alternate methods may be summarized as follows:

Basis	Percentage depletion, Dollars	Unit depletion, Dollars
Annual depreciation (equipment)	100,000	100,000
Development expense	160,000	
First-year depletion	145,000	75,000
Total first-year charges for capital recovery	405,000	175,000

Part d. The net income (before taxes) would be the net sales minus the allowable deductible costs; for the first year net income would be:

	Percentage depletion, Dollars		Unit depletion, Dollars	
Net sales		600,000		600,000
First-year expense	50,000		50,000	
Development (100% in year incurred)	160,000			
Depreciation	100,000		100,000	
Depletion	145,000	455,000	75,000	225,000
Net income		145,000		375,000

Part e. For the second year, net income would be:

	Percentage depletion Dollars		Unit depletion, Dollars	
Net sales		600,000		600,000
Second-year expense	40,000		40,000	
Depreciation	100,000		100,000	
Depletion	165,000[a]	305,000	75,000	215,000
Net income		295,000		385,000

[a] In this case, 50 per cent of the net income before depletion is $230,000 so that the full 27.5 per cent of net sales is allowed.

The percentage depletion method permits recovery of a greater amount of capital the first year, but there are no further development charges unless new ones are incurred. In subsequent years, however, the depletion may still be computed as 27.5 per cent of the net sales. In the unit depletion method the depletion charges for subsequent years will be based on the number of barrels produced. Additional development costs or changes in estimated recoverable oil will require calculation of new adjusted unit costs which are then used for depletion in subsequent years. Although the net income by the unit method is greater than by the percentage method in the above problem, the total net income plus capital recovery by the latter method is greater and the profits tax is lower. Sometimes the unit method has an economic advantage where the mineral rights are purchased outright at a relatively high price.

Depletion accounting for wasting assets involves considerable complications because of the uncertainties of future development costs, of actual recoverable material, of future value of the deposit, and of the scale of operations. Variations in these factors may result in changes in the depletable value, necessitating separate calculations each year for the depletion charge.

It is evident that the term depletion, like depreciation, may have one of several meanings, and the accounting department must be consulted for exact data, since different companies may allow for depletion in a variety of ways. Furthermore, the individual states may have special regulations concerning such allowances in regard to state taxes.

The term accretion applies to the increase in value of certain types of assets such as the growth of timber, etc. When such increase in value results, an allowance for it must be made on the accounts of the company.

SUMMARY

Amortization is a procedure for allocating the cost of an asset over a period of service life for the asset. Amortization is necessary because of obsolescene of the asset or because it wears out and must be replaced or discarded.

The use of the depreciation model, such as presented here, permits an over-all viewpoint of the effects of different procedures. This is much more useful than certain isolated numbers, which when compared out of context from the whole picture might lead to erroneous decisions. Such a model serves as a good example of the objective of this text, which is to illustrate how charts, equations, and diagrams can assist management to operate more effectively.

PROBLEMS

3-1. Special machinery for a furniture factory is installed at a cost of $50,000 and has an estimated life of 8 years with a salvage value of $2000. Make a plot showing the book value and depreciation reserve for any three methods of allowing for depreciation. Use 5 per cent interest for the sinking fund.

3-2. Eight years ago a ceramics company constructed several drying kilns at a cost of $120,000. (a) What is the book value of the installation today by sinking-fund and sum-of-digits depreciation methods if the interest rate is 4 per cent and estimated life for depreciation is 10 years with a salvage value of $12,000? (b) Compare the future worths of the 10 annual charges for both methods, using the given interest rates.

3-3. A soup manufacturing company has an annual production of 100,000 cases of 24 cans each. If a packing machine costing $32,000 is purchased with an estimated life of 4 years, what is the unit depreciation charge per can of product for the packing operation?

3-4. A coal mining property estimated to contain 2 million tons of recoverable coal is purchased for $1.2 million; $800,000 is the estimated mineral value and the land is valued at $75,000, exclusive of mining equipment and of the coal. (a) If 400,000 tons are mined each year and development costs are $20,000 annually, what is the unit depletion charge and what is the annual depletion charge? (b) If the mining equipment has no salvage value, what are the total annual capital charges for each year?

3-5. A textile manufacturer installs weaving machinery at a cost of $150,000 with an estimated life of 16 years and a salvage value of $10,000. Make a plot of book value against service life for the sum-of-digits method and declining-balance method. After 12 years, what is the book value by each method?

3-6. A paper products manufacturer purchases cutting and folding machinery at a price of $186,000. The products being made are specialties which may not be popular after an estimated 8 years. Develop a suitable depreciation model, which in your judgement would apply for this problem. Assume that the salvage value is $12,000.

3-7. In order to replace a distillation column in 10 years at a cost of $68,000, how much must be placed in a sinking fund annually at 4 per cent? If after 8 years the column is junked and the salvage value equals the dismantling cost, what additional sum is required to provide the new column? If a sum-of-digits depreciation had been used, what would the additional sum be?

3-8. In the depletion problem of the text (page 62), if 120,000 barrels of oil is sold the third year with an expense of $20,000, (a) what are the net incomes by the percentage and unit methods? (b) What are the total net incomes for 3 years? (c) What are the total capital recoveries for the 3 years by both depletion methods?

3-9. Compute the present worth of the depreciation payments (including salvage value) for a high speed computing machine costing $10,000 with an estimated life of 10 years and money worth 12 per cent for these calculations. Take salvage value as $2000 and use the (a) straight-line, (b) declining-balance, and (c) sum-of-digits methods.

3-10. For problem 3-9, check the calculations by the summation formulas against a step by step calculation, assuming a 3-year life and $2000 salvage value.

3-11. Compare the present worth of the depreciation payments on $100,000 depreciable capital for money valued at 14 per cent after 3 and 9 years of a 15-year life when (a) external sinking-fund, (b) straight-line and (c) the double-declining methods without switch are used for depreciation. (d) Discuss the future worth of the payments for the entire life of 15 years. NOTE: Consider salvage value at 15 years to be equal to the value found for (c).

3-12. Using both the straight-line and sum-of-digits depreciation payments, what are the present worths for $250,000 of depreciable capital at 10 per cent interest for a twenty year life? Discuss the results.

3-13. (a) What is the total present worth of the sum-of-digits depreciation for a $15,000 investment with $940 salvage value and a 4-year life if money is worth 10 per cent? (b) What is the present worth when double declining balance is employed at a zero interest rate?

3-14. (a) Develop analytically the expression for the present worth of the annual charge (deposit plus interest on accumulated account) for sinking-fund depreciation payments: (b) what is the uniform annual payment equivalent to the answer in (a)? Note that this is a different series from the one in Table 2-2 because the latter is the sum of a series to give a *future* worth equal to $P(1 + i)^n$ in terms of an annual payment, R (which is sought). The present problem seeks the present worth of a series of increasing annual *charges* to give a future worth equal to P; it differs from the present worth, P, for a series of uniform *deposits* of the sinking fund.

THE ECONOMIC STATE
AND INVESTMENT

4

Management decisions must take into account a complex of many factors. First these factors are found to affect only an individual, next the firm, then the industry, and finally the national economy. This chapter emphasizes the aggregate of all of these factors, especially their total effect on the national economy. The *economic state* may be defined as the over-all situation at any instant of time under the specific conditions applying at that particular time. The economist describes this economic state as the general level of the economy, or the performance level.

The economic state can be better understood by the aid of a simple diagram that demonstrates the economic relations in a capitalistic economy. Such a diagram, being actually a model providing for a quantitative analysis, is useful to management as it considers the firm's immediate problems in connection with the over-all economy.

THE ECONOMIC STATE

The economic state, as illustrated in Fig. 4-1, may be thought of as representing typical operations of either the national economy or any individual firm. The firm may be a manufacturing plant, a service facility, a farming operation, or any other enterprise whose function is the production of economic goods or services. However, at any one instant of time conditions may be such that a particular firm or individual operates at a loss, whereas the gross aggregate of all operations in the nation may be at a profit. Actually, the sum total of operations control the economic state. Thus, as a representation of national income, the diagram applies to the aggregate of all of the firms in the nation.

If the total demand for economic goods in the nation is just being met by the supply, the economic state is one of equilibrium. Although such a state is considered an ideal goal for the national economy, the conditions for obtaining equilibrium are continually changing with time. Manage-

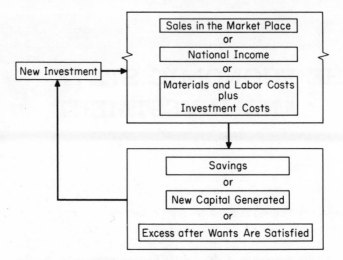

Fig. 4-1. Basis of the economic state. Savings must be used as new investment in order to generate more new savings. This repeats in each succeeding cycle for an expanding economy.

ment is concerned with the over-all economic state because the national economy's relative nearness to equilibrium may influence a firm's policy decisions and determine the economic feasibility of operations. This is true because certain changing conditions can be controlled to some extent by management decisions. For example, financing decisions (based on interest rates available) and production schedules (based on projected sale, etc.) are amenable to control.

In the diagram of Fig. 4-1, the numbers are supplied by economists who collect the figures and analyze them for classification purposes and for the avoidance of duplication. The data can be obtained from "Survey of Current Business" published monthly by the U. S. Department of Commerce which supplies a wide range of information indicative of the economic state of the nation. This information is classified by type of economic activity, each type often being referred to as an economic indicator. A discussion of the analysis and use of such data is given by Samuelson[1]. In Fig. 4-1 "sales in the market place" refers to all business operations, and includes rents, interest payments, individual professional expenses, direct taxes, etc. The sum of all of the costs of producing the collective net product plus the sum of all of the profits is the national income. The national income then is also equal to the total sales value either as physical products or as intangible services (such as washing

---

[1] P. A. Samuelson, "Economics," 3rd ed., McGraw-Hill Book Co., Inc., New York, 1955.

TABLE 4-1. Magnitude of *GNP*[a]

Year	*GNP* (Current Dollars) $ billion	*GNP* (Constant 1954 Dollars) $ billion	Price Index (1957–59 = 100)
1950	285	318	
1954	363	363	
1955	398	393	
1958	444	401	
1960	503	440	103.1
1961	519	448	104.2
1962	554	472	105.4
1963	579 (est)		106 (est)
1964	595 (est)		108 (est)

[a] U.S. Bureau of the Census, Statistical Abstract of the United States, Washington, D.C., 1963; *Survey of Current Business,* 43, 4:5–7 (1963).

laundry and cleaning clothes), since collectively these provide the income for all operations. As shown in Fig. 4-1, the dollar value of the National Income (*NI*) is made up of two parts. One is the costs associated with the productive operations and the remainder is the earnings or profits associated with the operations.

The dollar value of the *NI* when compared for various times is ordinarily adjusted for price changes in order to make valid comparisons. The adjustments are made by employing the price or cost indices discussed in the next section. It should be stressed that the *net* national production or income[2] is utilized where the costs of intermediate products is not counted twice. For example, if the sales value of a suit of clothes is $50, which is considered as one item of *NI*, then none of the value of the thread, the fabric, the buttons, etc., should be added as other items of *NI*, since they would be counted twice in totalizing the *NI*. The economist avoids this error in computing *NI* by adding together only the *increases in value* for each step or operation involved in making the fabric, thread, and buttons and in selling the suit to the consumer. Thus, only the net over-all increase is considered for the final figure.

Problem

A gallon of gasoline sells for 33 cents at the gas pump. Excise taxes are 10 cents; the wholesaler sells it for 19 cents; and it costs him 17.8 cents. The refinery buys crude oil at 6 cents a gallon (of which a portion is gasoline) and produces and delivers the gasoline to the wholesaler at 17.8 cents a gallon. Assume that the

---

[2]Income and net national production are used loosely here as synonymous. Strictly the economist distinguishes between the two by calling the national income equal to the value of net production minus indirect taxes.

amount of crude oil in the ground required to make one gallon of gasoline has a sales value of 5 cents as a resource. What is the contribution of one gallon of gasoline to National Income?

Solution

By listing the sales value at each step, one arrives at a total value of 70.8 cents, which obviously is incorrect, since there is a duplication of steps.

	Cost, Dollars	Sales Value, Dollars	Added Value, Dollars
In the ground[a]	0	5	5
Produced and delivered[a]	5	6	1
At the refinery and delivered	6	17.8	11.8
At the wholesaler and delivered	17.8	19	1.2
At the gas station	19	23[b]	4
Total	47.8	70.8	23

[a] Assuming equal value for all components of crude oil regardless of yield of gasoline, that is, one gallon of gasoline in crude form is worth 5 cents.

[b] Note that indirect business taxes are not included by government statisticians, although personal taxes are included.

By employing only the added value of 23.0 cents, one obtains that part of the *NI* contributed by one gallon of gasoline.

For a comparison of the *NI* at different time periods, changes in price' must be adjusted to a common level; otherwise, a false value of *NI* will result. For example, if the price of bread was 15 cents in 1950 and 25 cents in 1960, the portion of the *NI* represented by bread would be a much greater amount (by the ratio of 25/15) in 1960 than in 1950 for the same quantity of bread sold in each year. Similarly, this holds for other products and services sold on the market place. If prices on all products had advanced, even though the quantities remained the same, one would conclude erroneously that the *NI* had increased (if not adjusted for price change) where actually it had remained constant. When the data are given in terms of current dollars, corrections may be applied for price changes so that the indicator can be on a basis of constant dollars.

One of the most popular economic indicators to describe the national economic condition is the Gross National Product (referred to as *GNP*). It is a measure of the total output of goods and services at market prices. *GNP* embraces all government and consumer purchases, gross private investment, and net exports, as well as sales and excise taxes. However, it differs from the National Income by including allowances for depreciation. Table 4-1 shows its magnitude in terms of current dollars and constant dollars on a 1954 basis of 100. The tabulation also shows the consumers price index which is based on 1957–59 prices as being 100. This index has only recently been adopted as a revision of previously used indexes.

INVESTMENT

In Fig. 4-1 there will be found investment costs listed as a part of the production costs. Thus, they are in the same category as wages or raw materials. This is a very important item, since it is the critical factor in a capitalistic system. The national income dollars representing products or services sold or consumed are used to pay the costs of operation plus the earnings on the capital invested in the firm. That portion of the national product which is in the form of earnings and is not consumed is savings or new capital. It has been created and in itself must earn more new capital. Thus, a diagram such as Fig. 4-1 with numbers on it represents only an instantaneous situation which must be growing in magnitude; otherwise, a stationary state develops in which profits decline percentagewise. This stationary state could result if, as the capital supply increased, the unconsumed portion of the national income was not utilized for new investment. Some unconsumed income in the form of savings may be deposited in banks. These banks in turn can create new capital by making loans to entrepeneurs who have a good credit rating.

As the over-all economy expands, the new capital derived from savings increases, and the national income of Fig. 4-1 also increases, which means more consumption by customers, and so on in an endless cycle. If the products in the market place are not consumed, the cycle stops. Therefore, certain factors such as increase in the standard of living (more people enjoy more of the luxuries), increase in population, the advent of new inventions and discoveries, etc., are needed to increase consumption in order to use up[3] the amount of wealth created in the form of earnings. If this does not occur, the economic cycle will be stopped because the new capital cannot be absorbed, that is, there is no market for new investment equivalent to the earnings or savings portion of the national income. These earnings may be in the form of inventories, cash, new buildings, etc., where they represent additions to the supply of capital.

THE EQUILIBRIUM STATE

An equilibrium exists when the many variables of the economic state are in an essentially stable condition. This equilibrium state varies with the amount of national income as is demonstrated in Fig. 4-2. The dashed line represents the situation when sales in the market place are just equal

---

[3] A very basic economic concept is involved here. Since the national income is the sum of consumption (consumed goods) and savings (capital generated), an increase in consumption lowers the proportion available for new investment unless the total is increasing with increased consumption.

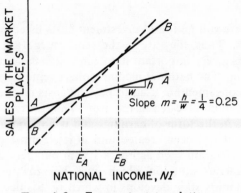

Fig. 4-2. Economic state relations.

to the national income at any level of national production (income). This dashed line is necessarily drawn at 45 degrees to represent this case, which is mathematically expressed as:

$$S' = m(NI), \text{ annual dollars} \tag{4-1}$$

where S' = sales in the market place

(NI) = national income or national production dollars as specifically defined by the economists

and m = slope of the line which is equal to one for the dashed line since it is drawn at a 45 degree angle.

Consider next the line AA on Fig. 4-2 which represents a proposed relationship between national spending (sales in the market place) and national production (national income). Wherever operations on line AA are below the dashed line, sales in the market place are less than national production; thus the costs for the latter cannot be met, and production is slowed down. Similarly, where the line AA is above the dashed line, the sales exceed the production and business increases production. At the intersection E_A of line AA and the dashed line, the equilibrium state is reached, since both sales and production are equal. Thus, there is an automatic corrective effect that tends to stabilize the economic state at E_A, that is, it approachs an equilibrium level.

The line BB compared to line AA represents a set of circumstances where the rate of change of sales with respect to the national income is different for AA and BB. This difference is determined by the slope of the respective lines, or for the general case;

$$\text{slope } m = \frac{\text{change in sales}}{\text{change in production}} \qquad (4\text{-}2)$$

Thus, the line AA intersects the dashed line at a lower value than that for BB, and equilibrium is attained at a lower national income as shown by comparing E_A and E_B. Note that when the actual relation between sales and national income coincides with the 45 degree line, the production and sales are equal.

At any equilibrium point, the excess of sales dollars over costs is the profits, which equal the savings when costs and consumption are equal. The new investment corresponding to the equilibrium national income is at E_A or E_B. If not, the equilibrium state will not hold, and then an adjustment will occur in national production (income) until the new investment and savings are in balance. Under these instantaneous conditions of equilibrium, the national income grows with time by the amount of the new investment made during each time period considered. For example, should full employment exist and the population remain the same, people will consume more and more, thus raising their standard of living.

MATHEMATICAL RELATIONS

The position of the lines in Fig. 4-2 is defined by the simple linear equation. This is really the locus of all points on the dashed line.[4]

$$S = m(NI) + b, \text{ sales dollars} \qquad (4\text{-}3)$$

where m = the slope of the line (see Eq. 4-2)
and b = the intercept on the sales axis.

From this mathematical relation, it should be apparent that the numerical value of the slope m is a measure of how fast the economic state is moving toward, or away from, the equilibrium point. As demonstrated in Fig. 4-2, the absolute value of the equilibrium point depends upon both the slope m and the intercept b of Eq. (3-3).

These diagrams and equations constitute the model which describe an economic state of the national economy. They give to management a quantitative value for those factors which it must consider in making

---

[4] Note that the lines are not necessarily linear but will be considered straight lines for the purposes of this discussion.

decisions, since management may evaluate those factors in its own firm, which may be moving toward or away from equilibrium.

MANAGEMENT AND THE ECONOMIC STATE

Management is concerned with Fig. 4-2 because the chart demonstrates visually whether or not it is advisable to increase production, depending upon whether the national income is at the equilibrium state for the prevailing economic conditions. Thus, using line AA, if sales are indicated to be above production, the condition to the left of the E_A point is prevailing and production should be increased. Conversely, if sales are less than production as determined by line AA to the right of E_A, consideration should be given to decreases in production.

The exposition in the preceding paragraph is an oversimplification of the real problem because many other factors, tangible and intangible, also must be considered in a policy decision of whether to increase or decrease production. However, the application of the principles of economics to the over-all economy as illustrated in Fig. 4-2 is quite helpful to management in understanding how the individual firm may be affected by changes in the national economy.

A very definite limit to the oversimplified discussion, which should be mentioned, is illustrated in Fig. 4-3. This diagram is based on one by Samuelson[5], and clearly shows the situation applying for inflation and deflation. In Fig. 4-3 the national income at a represents what it might be at

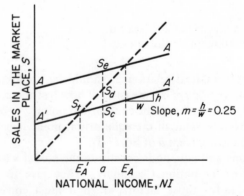

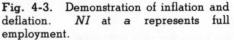

Fig. 4-3. Demonstration of inflation and deflation. NI at a represents full employment.

---

[5] P. A. Samuelson, op. cit.

full employment. If the line AA represents the general economic relation, it is expected that an increase in production to the right toward the equilibrium value, E_A, is the proper policy. However, production facilities are already at their maximum at a. The result is inflation because demand in the market place being in excess of all possible production causes a rise in prices.

If the general economic state is defined by the line $A'A'$, then at point a the economic state is above the equilibrium point, and operations can and will move to the left, resulting in a deflation. The movement here is possible, whereas the indicated movement to equilibrium for inflation is impossible.

For any economic situation where an increase in production is possible and indicated (such as line AA on Fig. 4-3 below a), it is interesting to consider how new investment results in the additional value of production.

Consider that for every $100 of a new consumption activity, 0.75 of it or $75 is respent in purchases of capital goods or services and then 0.75 of the second investment of $75, etc., is spent, and so forth, in a cumulative series. This means that each dollar of new productivity has a cumulative effect because 0.75 of each dollar is consumed additionally, and then 0.75 of the $0.75, etc.

Thus, the total cumulative effect is the sum of a geometric series, or:

$$100 + 100r + 100r^2 + \ldots + 100r^{n-1} = \frac{100}{1-r} \qquad (4\text{-}4)$$

where r is the fraction of each additional consumption that is used over and over again for additional productivity. The value of r in this case is 0.75 and is known as the *marginal propensity to consume*. The term $\dfrac{1}{1-r}$ (in this case it equals $1/0.25$, or 4) is known as the multiplier, and defines the effective dollars of new income (production dollars) resulting from each actual dollar of consumption at a given marginal propensity to consume, (that is, it is a quantitative measure of the proportion of each dollar received that is respent to cover costs of operations and of desired goals and services). The larger the value of r the greater is the over-all cumulative effect as demonstrated by smaller and smaller values of the denominator of Eq. (4-4). This is also logical because the equation states that the greater the proportion of income consumed the greater the over-all cumulative effect from the respending.

Returning now to Fig. 4-2, one finds that the value of m and r may be reconciled, since a vertical change of sales dollars may be considered as an addition to investment (the sale of capital goods). However, it was demonstrated above that such a vertical change, h, in investment is equivalent to an effective horizontal change, w, in production in an amount equal to the multiplier, or $1/(1-r)$, times the vertical change, that is, change in production dollars $= 1/(1-r)$(change in sales dollars).

$$w = [1/(1-r)]h \qquad (4\text{-}5)$$

But from Fig. 4-2 and Eq. (4-3) (which defines the value of m in terms of national income, that is, dollars worth of production), one obtains:

change in production dollars $= (1/m)$(change in sales dollars)

$$w = (1/m)h$$

Hence, $m = 1 - r$, or the slope of the line, m, is the reciprocal of the multiplier. This means that for a given line such as AA, relating sales dollars to production dollars, the multiplier represents the increase in the number of production dollars to produce one dollar of new investment at a given marginal propensity to consume.

Thus, where $1 - r$ represents the marginal propensity to save (that is, it is a form of potential investment, since it is an excess of income available above consumption), then, for that investment, the available savings resulting from one additional dollar of production is also $1 - r$. The implication also works in reverse, since a loss of one dollar of savings represents a decrease in income dollars of $1/(1-r)$. Thus, a loss of one dollar means the dollars of income have decreased $4 if r is 0.75.

REAL CAPITAL INVESTMENT

After recognizing the economic concept of the role that capital exerts in the over-all economy, one must have a real world viewpoint to understand the forms in which capital investment is utilized. If the new investment is to pay for its use, it must provide a profit. To provide a profit in a going enterprise usually requires tangible and material facilities such as land, buildings, and equipment. Such items are referred to as fixed assets. In addition, however, capital is required as cash to purchase raw materials and supplies, to pay salaries, and to meet expenses until such time as the venture pays its own way.

The amount of capital required for buildings and equipment varies with the type of operation, the size of the plant, and the time at which the investment is made. Other factors are operating to control the amount of the investment in a proposed venture; among these are the economic condition of the world and the changes in the market supply and demand for the products or services provided by the company. Many of these factors are irreducible (cannot be evaluated in terms of dollars) and require consideration by management before a decision is made to proceed with the venture. However, once the decision is made, it is the function of managers and engineers to select the equipment items and determine their costs. The equipment for a process plant may be classified into two types—the equipment required to process the material and the equipment and facilities necessary for providing the services to the process and for the efficient operation of the plant. Typical lists of equipment for these process and nonprocess categories are given in Table 4-2.

Since it obviously is impossible to compile a detailed cost list of all kinds of equipment and to maintain each item's cost up to date, various methods are employed to simplify such lists, thus providing a practical means for estimating equipment costs at any time with reasonable accuracy. These methods usually involve two procedures: (a) the use of price indices whereby prices known at some given date are corrected to the current time, and (b) the establishment of a summary of price data on individual items where the price is related to some size or capacity factor.

TABLE 4-2. Typical Fixed Asset Items

Process Equipment:
  Bins and tanks for storage of raw materials
  Equipment for manufacturing and converting raw materials into finished
    products
  Product finishing, forming, storage, and packing equipment
  Automation and process control facilities
  Materials-handling equipment; cranes, lift trucks, trucks and cars, etc.
  Buildings for housing process equipment

Service or Nonprocess Facilities:
  Power generation and distribution; water distribution
  Waste disposal
  Office buildings, furniture
  Laboratory and research facilities
  Safety, infirmary, and first aid installations
  Warehouses, shops, maintenance facilities
  Roadways, parking areas, fences, sidings, docks, etc.

PRICE INDICES

A price or cost index tabulation, such as is given in Table 4-3, gives a relative comparison of price from year to year referred to some datum

TABLE 4-3. Price and Cost Indices

Year	Consumers Price, Base Year: 1957-59 = 100	Marshall-Stevens All-Industry Equipment, Base Year: 1926 = 100	Engineering News-Record Building Costs, Base Year: 1947-49 = 100
1926		100	
1940		87	60
1950		168	112
1952		180	124
1954		185	133
1955		191	140
1956		209	146
1957		225	151
1958	100	229	156
1959		234	163
1960	103.1	238	166
1961	104.2	237	169
1962	105.4	238	170
1963	107 (est.)	239 (est.)	171 (est.)
1964	108 (est.)	240 (est.)	172 (est.)

year, which in the tabulation is 100. Ordinarily, the index is based on a composite figure made up of a number of related items; these items may be weighted individually, according to their relative importance in affecting the index.

The index provides for comparing the price in any one year with the price in some other year, since the relative prices vary in the same manner as the ratio of the indices.

Problem

If a textile weaving machine cost $78,000 in 1954, what is the expected cost for a similar machine in 196X?

Solution

Any price-index tabulation applicable to such machines is suitable for this estimation. However, from Table 4-3, using the Marshall-Stevens all-industry equipment price index, assume the 196X index is 231; then the 196X price would be estimated as:

$$\text{New price} = \text{old price} \times \frac{\text{new index}}{\text{old index}} \qquad (4\text{-}7)$$

$$196\text{X price} = \$78,000 \times \frac{231}{185} = \$97,000$$

Each area of economic activity has its generally accepted special index for studies in a given field. Examples of several various types are given in Table 4-3. There are numerous other indices, including separate ones for labor costs, material costs, stock market prices, mineral prices, etc. Each particular field of interest has certain periodic publications which carry the index values for that field. In many cases, the indices relate to the magnitude of an activity instead of prices. For example, the yearly relative construction and building activity may be expressed in terms of an index rather than total dollars.

The base for the indices is usually taken as 100 in some base year. However, over a long period of time, changes in an index may be so great as to make it desirable to change the base year and establish a new datum base of 100. The old and new indices are related by their values in any one year.

The components of the indices shown in Table 4-3 will now be summarized; a plot of each index is demonstrated in Figs. 4-4 and 4-5.

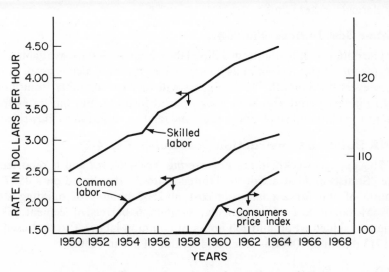

Fig. 4-4. Consumers price index and cost of labor.

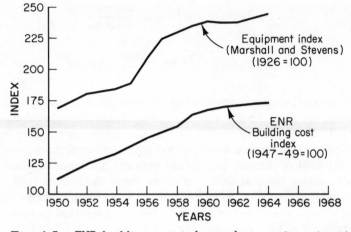

Fig. 4-5.   ENR building cost index and an equipment cost index.

Consumers Price Index (Fig. 4-4)

This index, published monthly in the *Monthly Labor Review* and in "Statistical Abstract of the United States," is based on the approximate price for 300 goods and services taken in 46 cities by sampling according to a specified time schedule. The index measures the effect of price change only—not the relative amounts spent for living by city wage earners and by clerical worker families.

Labor Cost Indices (Fig. 4-4)

The data for skilled and unskilled labor wage scales are available from *Engineering News-Record* in its quarterly cost report, which is given in one issue every third month. These figures will vary considerably from industry to industry and also with geographical location, but any rates from year to year should show about the same relative variation as illustrated.

ENR Building Construction Cost Index (Fig. 4-5)

This appears weekly in the *Engineering News-Record* and is reported in the "Statistical Abstract of the United States." It is based on a 1947–49 datum of 100 for a composite cost of 2500 pounds of building steel, 1088M board feet of a particular lumber, 6 barrels of cement, 68.38 hours of skilled labor, and 200 hours of common labor. (NOTE: Based on a 1913 datum of 100, this index had a value of 376 in 1950.)

Marshall-Stevens Equipment Cost Index (Fig. 4-6)

This index is prepared by Marshall and Stevens, Inc., Chicago, Illinois. The data for different process and related industries are compiled separately. The average of all industries is given in Table 4-3. These figures are from *Engineering News-Record* which reports the figures in its quarterly cost report issue every third month.

FIXED ASSET INVESTMENT

Designers and engineers concerned with calculations involving equipment for plants have devised certain relations to assist them in estimating the value, or cost, of fixed assets. Obviously, keeping to date on the cost of all items in which estimators are interested is impossible. However, if the value of an item of a specific size is known at a certain date, its value at some other year can be estimated from the ratio of price indices for the two years. A price index pertinent to the subject should be applied such as the Marshall and Stevens Index in Fig. 4-5.

Where the value of an asset is sought but the size is different from the size at a known cost, it is possible to estimate the unknown value. For many physical assets the cost does not increase linearly with size, that is, an electric motor of 50 horsepower does not cost twice as much as a 25 horsepower one. Instead of the larger one costing double ($50/25 = 2$) the smaller one, the larger one costs about 2^a times as much, where a is some power less than unity. In general, for many types of equipment, a averages about 0.6, and this relation is known as the *"sixth-tenths factor."*

Mathematically, the relation is expressed as:

$$p_2 = p_1 (Q_2/Q_1)^{0.6} \qquad (4\text{-}8)$$

where p_2 = the price sought for a size Q_2
and p_1 = a known price for a size Q_1

The validity of the price relation can be checked by plotting price and size data on log-log paper, since the basic relation for Eq. (4-8), where the price p varies as some power of the size or quantity Q with b a constant, is as follows:

$$p = bQ^a \qquad (4\text{-}9)$$

When price and size data are plotted on log-log coordinates as illustrated in Fig. 4-6 for storage tanks, one gets:

$$\log p = a \log Q + \log b \qquad (4\text{-}10)$$

Here, the power factor, a, is the slope of the plot as shown in Fig. 4-6. Note that in using Eq. (4-9) for comparing two prices at two different sizes, the ratio is taken and the constant, b, drops out as was demonstrated

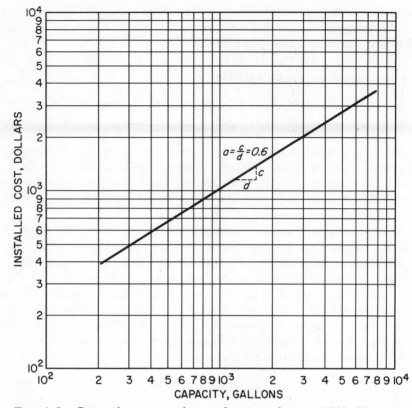

Fig. 4-6. Price of storage tanks as a function of size in 1960: Marshall-Stevens all-industry equipment index = 234.

in Eq. (4-8). The parameter used for Q may be any desired criterion of size, such as weight, area, diameter, number of bobbins on a textile machine, etc. An example of the use of these relations is as follows:

Problem

Management has decided to triple the production of one type of woven rug. In 1957 the machinery cost $64,000. What is the 196X estimated cost for the investment?

Solution

This problem consists of two parts: (a) a change in price index and (b) a change in capacity. Both these changes may be incorporated into the price estimation. Using the Marshall and Stevens all-industry price indices for the two years (225 in 1957 and 231 in 196X) and Eqs. (4-7) and (4-8) results in:

$$p_2 = \$64{,}000 \, (3/1)^{0.6} \, (231/225) = \$64{,}000 \, (1.93) \, (1.03) = \$127{,}000$$

From the form of the equation, it is apparent that the correction for size or price change can be made in any order.

The total fixed-asset investment for an enterprise consists of more than the cost for equipment. Some capital is employed in the planning stage before a decision is reached to proceed with the proposed operation. Additional capital is required for architects and engineers to design the physical plant, to procure equipment, to construct facilities, and to install the equipment. Finally, management incurs numerous preoperation charges, such as test runs, off quality production, and the training of operators. All of these items must be included in the determination of the fixed-asset investment. In general, the costs for these components of the fixed investment can be evaluated as some proportion of the purchase cost of the equipment. Thus, if a list of the needed equipment is available, a factor based on experience can be utilized to estimate other investment costs that may be involved for the design, the direct materials, the direct labor, the construction, the installation, the buildings, the costs of delivery to plant site, etc. Indirect expenses for purchasing, planning, and inspection can also be expressed as a factor related to the purchase price of the equipment. However, these indirect expenses are usually part of the design cost.

It is not the object of this text to give these factors in detail, since they vary widely from industry to industry, depending on the size of the enterprise. However, certain figures are given in Table 4-4 for operations involving processing of a raw material to a finished product. From this table, a nominal figure is a fixed-asset investment of about three times the cost of the equipment. This number is not a firm value for all installations, however, since it might be six or eight times the equipment cost of certain installations.

TABLE 4-4. Certain Factors Relating Components of Fixed-Asset Investment to Cost of Equipment

Cost of equipment (assume equal to C dollars)	$1.0C$
Erection cost	$0.2C$
Complete installation cost including all facilities	$1.5C$
Design and engineering costs	$0.3C$
Contingency allowance	$0.1C$
Preoperation charges	variable charges
Total	$3.1C$ plus variable charges

Obviously, the more detailed information available the more accurate the cost estimate. Thus, where extremely firm values for investment are desired, reasonably complete information is required from the designers. However, management in reaching a decision on a proposed operation must balance the possibility and economics of an error in the investment estimate against the effect of delay and additional cost for more detailed plans and design.

In estimating equipment costs by utilizing data from a variety of sources, the estimator must be careful to allow for differences in the bases employed for fixing costs by the different sources. In general, there are three main points of difference: (1) the year the costs were taken may vary; (2) in some cases, the costs of installation, foundations, and field work are already included in the cost listed; however, in other instances, these costs must be added (the authors of the sources being used ordinarily provide the necessary data for estimating the installed cost, if it is not included in the cost figures); and (3) the cost of auxiliaries such as motors, switch boxes, and service equipment (ejectors, condensers, gear reducers, etc.) is sometimes included and sometimes not.

In selecting an investment, management should be aware that the optimum investment is not always the one calling for the smallest capital outlay. The optimum size for a unit requires consideration of factors other than the initial price since, as will be explained in Chapter 9, the optimum cost for producing a pound of product is a function of the operating costs as well as the fixed charges for depreciation of the equipment. Thus, the highest price for equipment may produce the lowest over-all processing costs. On the other hand, the use of multiple small units may be the most economical when the production rate varies. This is true where it is more efficient to operate a small unit at full capacity than a large unit at part capacity. If the same labor crew is required for high and low production rates, the operating cost per unit of product will vary directly with the efficiency, and multiple small units may be more economical.

A summary of actual cost data to be practical for cost-estimation purposes must be concise and reasonably accurate. The availability of such information for design and estimation purposes provides a quick and efficient means for determining costs that otherwise would require considerable correspondence with equipment manufacturers regarding specifications and bids. Furthermore, if any considerable time elapsed before design and estimation or if major changes in equipment requirements were made, most of the cost data would be out of date, necessitating a repeating of the correspondence.

The fixed-asset investment is not the total plant investment, since capital is needed for other purposes in a financial enterprise. This is discussed in the next section.

TOTAL CAPITAL INVESTMENT

The total capital investment comprises certain items other than the fixed-asset investment. These items may be listed in a sample tabulation such as Table 4-5.

TABLE 4-5.   Total Capital Investment Schedule

1. Total fixed assets (Table 4-4)	_____
2. Cost for land and site preparations	_____
3. Construction interest charges[a]	_____
4. Working capital[b]	_____
Total capital investment	_____

[a] Estimated at about one-half the going rate for money for the total elapsed time based on progressive demands for capital outlay as construction proceeds.
[b] Estimated at about three times the value of one month's production to include inventories, semifinished goods, capital for salaries, accounts payable, and cash in bank. Usually working capital amounts to from 15 to 50 per cent of annual sales dollars.

In Table 4-5, the item for cost of land and site preparation is separated from fixed assets for two reasons. First, it may vary considerably with the selection of a plant location; in fact, the site-preparation cost can vary from very little to a very expensive operation where dredging, installation of docks, special water facilities, etc., may be required. By listing this item separately, management can evaluate readily the effect on total capital requirements of alternative sites. Second, land is not considered depreciable; thus, land requires a different treatment in the bookkeeping than does the first item in the tabulation. Fixed assets are subject to yearly decreases in value.

The construction interest charge is the cost for capital tied up in the project from the start until the project is operating and bringing in some return. Consider half of the total capital for items 1 and 2, or its equivalent, as being tied up for the full time before operation. By using half the interest rate on the full capital, one gets an approximate cost equal to that obtained by applying the going rate to half the capital. For example, suppose that the sum of the cost for items 1 and 2 is $1 million. Also, assume that an estimated two years total time is involved from the start of expenditures until the construction is completed and that the plant is in normal operation. Further, assume that the expenditures are made in a uniform manner (that is, at a rate of about $250,000 every six months) and that the going rate for money is taken as 6 per cent for this analysis. Thus, for the first year $500,000 is spent, but all of this capital is not needed for the full year. As an approximation, half the interest rate (or 3 per cent) can be applied to the $500,000 to give an interest charge of $15,000. For the second year, the full interest rate on the first $500,000 is charged plus half the interest rate on the second $500,000. The total interest charges, therefore, are:

		Dollars
FIRST YEAR:	$0.03 \times 500,000 =$	**15,000**
SECOND YEAR:	$0.06 \times 500,000 =$	30,000
	$0.03 \times 500,000 =$	15,000
TOTAL:		**45,000**
TOTAL INTEREST CHARGES:		**60,000**

This total of $60,000 is the same result obtained by using an annual interest rate of one half the going rate on the full capital. Thus, by this method for two years the total interest charge is:

$$\text{Total interest charge} = (2)(0.06/2)(1,000,000) = \$60,000$$
$$\text{(two years)}$$

In addition to the capital equal to the costs incurred in getting a plant to the operating stage, another important type of investment is necessary. This is the working capital and is discussed further in a later part of this chapter. It represents (a) cash in the bank to cover salaries and bills that must be paid over a short period of operation—usually the cost of one month's operation (equivalent to the value of one month's production); (b) the value of products sold for which payment has not been received (the value of one month's production); (c) the value of the inventory of semifinished and finished products for one month's operation; and (d) the value of consumable raw materials and other supplies sufficient for one month's operation. As an estimation, the working capital approximates three times the value of one month's production, since the product at different stages has a different value. This value is the cost value or book value of the product, whichever is lower for conservative estimations; the actual dollars involved obviously will vary with the dollar value of the product. Where known values of nonconsumable materials such as catalysts, recycled solvents, carrier materials, etc., are used, their values are included as part of working capital. However make-up costs for lost material constitute an item of annual operating cost, and are not considered here. For very high-valued products such as gold or pharmaceuticals, it is probable that the working capital is not equivalent to three times the monthly production rate because rapid processing and quick turnover of inventories are desired. Another special case exists in some food processing industries. In the manufacture of potable spirits where aging of stocks ties up considerable capital as inventory, the working capital is relatively high. Other estimates for working capital consider it to approximate between 15 and 50 per cent of the value of the annual sales, the average being about 25 per cent. Here, again, however, the type of plant and the nature of the product are determining factors.

The above approximations are generalizations, and each plant requires separate consideration of all factors involved. The estimations are just as reliable as the accuracy with which the detailed data are assembled for the cost estimations. In round figures, process facilities vary up to 50 per cent of the total investment, whereas service facilities including buildings approximate 30 per cent. Working capital is usually less than 50 per cent of the fixed assets (land, buildings, service facilities, and process equipment) unless the product has a very high unit value and may be taken generally as about 20 per cent of the total investment.

The magnitude of total investment with respect to the number of employees often influences decisions where labor production is involved; that is, in those industries where automatic production equipment is installed at a high investment of capital, the ratio of investment per worker is high. For the same amount of production, however, the productivity per worker is also high. Thus, higher investment costs are balanced by decreased operating costs.

Certain available data in 1959 indicate the average investment in manufacturing was about $18,000 per worker. Other figures for some specific groups or industries were as shown in the following tabulation:

Industry	Plant Investment per Worker, Dollars
Leather and textiles	5000
Metal products, paper, ceramics and foods	13,000
Chemical	27,000
Petroleum	40,000

CAPITAL TURNOVER

The capital turnover is a useful measure for comparing certain aspects of investment capital with other financial data. The capital turnover, T, is defined as the ratio of annual sales to investment; that is, capital turnover is the number of times per year that the annual sales equals the investment. Mathematically, it is expressed as:

$$T = S/I = pQ/I \qquad (4\text{-}11)$$

where T = the capital turnover
$\quad S$ = annual sales dollars
$\quad I$ = total investment capital including working capital
$\quad p$ = unit price of the product
$\quad Q$ = annual units of product sold

(NOTE: the capital ratio, $R = I/S$, which is the reciprocal of the turnover ratio is often used as a financial guide in discussions of the type under consideration here).

Where the magnitude of the annual sales is known, the expected investment for those sales can be estimated within a given industry. As a starting point for discussion, it can be stated that a distribution of values for T will show that they distribute around a value 0.5 in 1958. Those industries, such as the heavy metal industries, which are of a stable type and form a major background for the economy show a relative low turnover ratio of about 0.3; whereas, industries based on operations which are subject to obsolescence, such as the pharmaceuticals and drug industries, have high turnovers ranging up to 1.0.

Further generalizations can be made with respect to the turnover ratio. For example, the percentage of working capital increases with businesses having high turnovers because in general less value is added by manufacture in such operations. This is explained by more capital being tied up in raw materials which constitute a larger proportion of the sales price. Thus, completely integrated companies that start with basic raw commodities will have lower turnovers with smaller working capital. Where the product is a large volume one, the turnover tends to be low. Similarly, where the product made is further from the consumer (such as heavy industry), the turnover is low.

Profit on high turnover operations tends to be higher, but there is no implied relationship that high profit ratios on investment necessarily follow high turnovers when all things are considered for comparative purposes.

SOURCES OF CAPITAL

The capital required for a new financial venture may be obtained in a variety of ways. It may be acquired by converting certain assets to cash by selling stock, or by selling bonds. Although some information on financing a venture is included in Chapter 16, a detailed discussion is not attempted here, since the financing of a venture involves many considerations covered by books on finance. However, certain implications of these methods of financing and several comments on the subject are presented at this time to show the differences among them.

The assets owned by a company may be utilized to expand a company's operation if the assets are in an available form, such as cash from a surplus account. (The definition of surplus is considered in the next section).

The selling of additional stock, however, is quite different from the use of surplus funds. In the former case, the number of stockholders is increased and the equity, or share of the company's assets held by each original stockholder, is decreased proportionately. There are many types of stock issued, but only common stock and preferred stock are considered here.

Capital stock (common or preferred) is the term for the certificate that designates that its owner has a definite share in the assets of the company. The profits of the company are distributed as dividends in terms of so much a share on an equal basis. The stockholder may or may not receive equal dividends every year, since the rate is not fixed.

Preferred stock, as its name connotes, indicates a special privilege described by the company's charter. Usually, a guaranteed dividend payment rate, and only that rate, must be paid to preferred stockholders before any dividends are distributed to the common stockholders. (If no profits are earned in any one year, no dividends are paid; however, the stock may include an obligation to pay all back dividends when, and if, future profits are made). Any excess dividends are then allocated to the common stock, which may receive a higher rate in some years than the preferred stock. There are many variations in the types of preferred stock, regarding dividends, voting power, ownership, and other matters.

The par value of common stock is the estimated worth at the time of issue, particularly in the case of new stock. The actual selling price of stock on the market may be much higher or lower than the par value. In many cases, stock is issued with no par value. Its worth will be determined by what the public will pay for it. The par value of a preferred stock has a definite significance, since it represents the value upon which the guaranteed dividend rate is based.

Bonds are entirely different from stocks in that the bondholders usually do not control the company's operations. There are many varieties of bonds, all of which may be considered as a legal obligation of the company to repay borrowed capital to the bondholders at some future date. In addition, the company agrees to pay interest to the bondholders at a stipulated rate. The bondholder in reality is loaning his money on the basis of collateral (company assets) put up by the company to ensure that the bonds will be paid off. If the company is unable to pay off its bonded indebtedness, the bondholders have a prior claim to the assets of the company, and the stockholders' equity is only the remaining assets.

The proportion of capital raised by selling stocks or bonds varies with the type of industry and the individual company. The selling of additional stock reduces the share (equity) the original stockholders have in the company, but the selling of bonds entails an annual obligation (to be paid whether or not a profit is earned) for the company.

In general, banks are used for financing short term loans up to about 3 years, where the monies are utilized to get the enterprise started.

Other sources of capital are venture capital groups, private estates, and other established companies. However, financing of this type usually requires sacrifice of some share or equity of the company to the sources of capital. This sacrifice of equity also means granting those providing the capital some control in the company.

THE BALANCE SHEET

After a company has been operating for a period, the various items for which the original capital was employed change in value. In order that the owners may periodically have an accounting of the condition and distribution of their capital, a *balance sheet* is prepared to show that distribution. An understanding of balance sheet terminology and relations is necessary in order to interpret intelligently the effects of improvements or developments made by management on the over-all economic success of the company. Thus, a balance sheet is a model. Although this section does not discuss the details of balance-sheet accounting, a summary is given of the accounting classifications employed for distribution of the capital involved in a going concern.

The simplified balance sheet shown in Fig. 4-7 is a typical one for a petroleum oil company. However, reference should be made to accounting texts for more detailed information. The balance sheet is the central point to which are directed the capital items, or assets, owned by the company and from which radiate those items owed by the company. These two sets of items must be equal, or, in other words, in balance. Since the stockholders own the company, management is accountable to them for invested capital plus earnings or profits on operations retained by the company. This accountability to the stockholders is analogous to a liability of the company in the same way as are the amounts owed to suppliers of materials in the accounts-payable item.

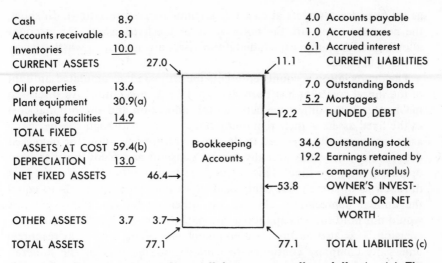

Cash	8.9		4.0	Accounts payable
Accounts receivable	8.1		1.0	Accrued taxes
Inventories	10.0		6.1	Accrued interest
CURRENT ASSETS		27.0	11.1	CURRENT LIABILITIES
Oil properties	13.6		7.0	Outstanding Bonds
Plant equipment	30.9(a)		5.2	Mortgages
Marketing facilities	14.9		←12.2	FUNDED DEBT
TOTAL FIXED				
ASSETS AT COST	59.4(b)	Bookkeeping	34.6	Outstanding stock
DEPRECIATION	13.0	Accounts	19.2	Earnings retained by
NET FIXED ASSETS		46.4→	——	company (surplus)
			←53.8	OWNER'S INVEST-
				MENT OR NET
OTHER ASSETS	3.7	3.7→		WORTH
TOTAL ASSETS		77.1	77.1	TOTAL LIABILITIES (c)

Fig. 4-7. Typical balance sheet (all figures are million dollars). (a) The process investment is $30 million. (b) This is the original cost and, omitting the value of real estate, is the depreciable property. When the accrued depreciation is subtracted, the resulting figure is the book value for this depreciable property plus the value of the land. Current total investment is sum of assets (net) plus working capital, or: 46.4 + 3.7 + (27.0 – 11.1) = $66 million. This is the same as total assets at net value minus the current liabilities. (c) Sometimes called total liabilities plus net worth.

For purposes of clarification, a brief summary of the terms employed in the balance sheet is given here.

Current Assets

Capital represented by items essentially the same as cash (such as raw materials, materials in process, finished products, and bills that customers owe the company) are called current assets. Cash and accounts receivable are called quick assets because they can in a short time be converted into real dollars. In connection with current assets, all materials are evaluated either at their actual cost or at the market value, whichever is lower.

Fixed Assets and Depreciation Account

The fixed assets are capital represented by plant equipment, trucks, machinery, land, and buildings, which, while of a definite dollar value, would require some time (or the purchase by a party interested in acquiring the company) before they could be converted into real dollars. Since the fixed assets represent depreciable capital (except for land), their book value decreases with time and use. The fixed assets are ordinarily shown as *net fixed assets* (assets at cost less accumulated depreciation). In short, the net fixed assets are the book value of the fixed assets. An account allowing for depletion is included if the assets are subject to wasting away or being used up.

The depreciation account represents the estimated accumulated amount of the initial cost that has been allocated for depreciation. As discussed in Chapter 3, the difference between this allocation and the original cost of the fixed assets is their remaining book value. Although the net fixed assets are normally expected to decrease with time to allow for increased depreciation, they may actually remain constant in a plant that has been operating for some time. This is true in such a situation because the cost of renewals of some plant equipment at various intervals tends to equal the annual depreciation charges on other equipment. However, this would not hold for an expanding company where the fixed assets are increasing. It is also untrue for a company processing a mineral resource, unless the company continues to exploit new deposits of the mineral. Where such continuous exploitation occurs, the depreciation and depletion accounts may build up in the form of other assets to repay the investors upon completion of operations.

Some balance sheets show the fixed assets at cost on the asset side. If fixed assets were shown this way in Fig. 4-7, the total assets would be increased by an amount equal to the depreciation account; by the same token, the depreciation account would be shown on the liability side so that total liabilities would be equal to the total assets. As previously mentioned, the depreciation account represents an allocated cost. An amount equal to this cost is retained from sales dollars each year as an item of cost or expense. This retained amount protects the owners' depreciable capital. However, the asset may be employed as part of the working capital in the running of the company.

Other Assets

This section of the balance sheet shows the intangible capital represented by "good will," patents, franchises, trade-marks, etc., that are owned or controlled by the company. Great care must be exercised to avoid overevaluating these other assets, since this can make a company appear to be in a sound financial condition when actually it is not.

Current Liabilities

These are obligations that the company has currently outstanding such as bills for materials. Also included are interest owed and taxes prorated for that portion of time elapsed since they were last paid. However, since these items are not yet due, they are called accrued liabilities.

Funded Debt

Funded debt includes obligations that must be amortized over a long period of time. It represents capital obtained from sources other than from the owners of the company.

Owners' Investment or Net Worth

This item, sometimes called proprietorship, is confusing to the student unless he recognizes that the company is accountable to the stockholders (owners of the company) for the capital that they originally put into the company. The amount shown on the balance sheet for the stock is generally the par value of the amount received for all shares when first sold. In addition, the company is accountable to the stockholders for any earnings retained in the company for purposes of expansion or for other contingencies. These retained earnings are called surplus because they represent that part of the cumulative total profits in excess of the amount distributed to the stockholders. The sum of the stockholders' investment plus the surplus is the net worth. However, if there is no funded debt, the excess of total assets above the current liabilities is the owners' investment, or net worth. In setting up the balance sheet the sum of the assets must equal the sum of the liabilities (including net worth). If the total assets are less than the liabilities, the surplus account is actually a deficit. Moreover, if unprofitable company operations prevent erasing the deficit, the capital put up by the stockholders gradually disappears into an increasing deficit and may eventually be completely lost.

Working Capital

As mentioned previously, a certain amount of capital is required to operate a plant in addition to that required for its construction. Thus, the balance sheet not only shows the distribution of this working capital but also that it is not necessarily cash in the bank. Working capital may appear in the form of any asset that can be converted readily to any other asset without disturbing the operation of the company. For example, the excess of current assets over current liabilities is working capital because a conversion of current assets to cash permits paying off the current liabilities, leaving the excess available for the purchase of other assets. Fixed assets that are being used for operations cannot be converted directly to cash because the operation would have to stop making its product. If a company is operating at a profit and all the profit is not distributed to the stockholders, the working capital, the assets, and the surplus account increase by an equal amount. If, however, the increase is used to purchase equipment, the working capital remains the same, but the fixed assets and the surplus account increase. Under this condition of undistributed profit, the total capital invested also increases.

As has been pointed out, management retains from sales dollars an amount equal to the annual depreciation of depreciable capital. Consequently, the depreciation account *increases* by this amount. However, since the fixed assets have *decreased* by this amount, the total assets remain the same if no profits are retained by the company. In fact, since net fixed assets have decreased, current assets and working capital must increase if other assets and current liabilities remain constant.

Working capital is a very important consideration in a plant operation, especially for new plants or processes which do not have established markets for the products. An ample supply of working capital provides flexibility to management for dealing with unpredictable conditions, such as strikes, fires, recessions, or delays in getting a plant into an efficient state of operation. Such delays include the initial period when a plant is obtaining acceptance of its product by the consumer. In addition, availability of working capital permits a plant to expand its operations, particularly during a period of rising costs, and also to purchase raw materials and other supplies at low prices when it is advantageous to do so.

INVESTED CAPITAL

In analyzing the balance sheets of a company, one should bear in mind that the invested capital can be selected from a variety of viewpoints. Accordingly, any ratios based on the investment must state clearly what investment is involved. For example, the *current* total investment at any time might be the net worth (NW) plus the funded debt (FD) if any. However, such a basis includes all changes in surplus and in fixed or other assets from the start up of operations. The *initial* total investment is the sum of initial net worth—excluding earned surplus, $(NW)_{\text{ex surplus}}$, plus the initial funded debt, $(FD)_{\text{initial}}$.

The initial total investment represents the capital raised by selling stock and floating bonds. This capital was used to provide the equipment and working capital (WC). Since the current assets increase each year by the cash equivalent to the depreciation, D, and the added surplus, S, then, other things remaining the same, the current investment will increase unless there is a compensating increase in current liabilities.

Where the total investment is computed from the assets, one must recognize that year to year changes affect the result. The *current* total investment is the total assets at net value, $(TA)_{\text{net}}$, minus the current liabilities (CL). The initial investment is the current investment minus the surplus, if other things have not changed. The total assets at net value, $(TA)_{net}$, are the total assets with the fixed assets at their net value, not at cost.

The current total investment will remain constant if there is no change in surplus or current liabilities, since the decrease in fixed assets because of depreciation is just equal to the gain in current assets because of the depreciation deposit.

Expressed in equation form the above relations are as follows:

$$\text{Current total investment}$$
$$= (NW) + (FD) = (TA)_{net} - (CL) \tag{4-12}$$

$$\text{Initial total investment}$$
$$= (NW)_{\text{ex surplus}} + (FD) = (TA)_{net} - (CL) - S \tag{4-13}$$

The initial total investment is often used in evaluating management economy models because it is a fixed number. However, in evaluating the current financial situation it is more realistic to consider the current total investment as a basis for study. The book value is ordinarily considered to be the tangible portion of the total investment; thus intangible assets are neglected.

The working capital increases with time because of depreciation deposits and surpluses, if any. However, the working capital can remain constant when combinations of depreciation and changes in surplus and liabilities are mutually compensating. If the plant operates at a profit which is all distributed to the stockholders, the total assets and the sum of net worth and funded debt remain constant, as they do when there is no change in current liabilities. The form in which the assets appear changes when the depreciation is deducted, but the current assets increase by a like amount. As noted previously, however, some balance sheets show the fixed assets at cost with an account for depreciation shown on the liability side; to maintain the assets and liabilities in balance, the current assets and total assets must increase. This leads to a false conclusion when total investment is based on total assets, since the fixed assets are carried at a cost value, whereas they actually have decreased in value. For this reason, the form of the balance sheet as given in Fig. 4-7 showing fixed assets at book value is preferred. However, both procedures (and others) are used in balance sheets.

When some of the profits from operation are retained in the company, the surplus account and the assets will increase by like amounts, and the total investment will also become greater. Thus, if the same annual profits are made, the earning rate based on total investment will decrease from year to year. Accordingly, the earning-rate base must be clearly stated in an economic evaluation. For example, if the current liabilities are included as part of the investment, the term *total capital in use* (implying capital in addition to debt capital and the owners' capital) is sometimes used as a base for earning rate.

In certain companies pension and insurance liabilities are assumed by the company itself; an account for such eventual costs may be accumulated or be set out of surplus and appear as a separate liability item on the balance sheet. Where such accounts are shown as a separate entry on the liability side, they may be deducted from the total investment to give the *invested capital*. Invested capital may be used in place of total investment for calculating any of the ratios considered later, where it is desirable that pension and insurance liabilities be handled as not part of the investment capital. This method of calculation is sometimes employed in tabulations of financial data for different companies.

When one wishes to compare data where the basis is cost of the productive facilities only, then the *process investment* is employed. This includes only the capital used for the equipment and services directly related to a given process as listed in Table 4-2. Such comparisons often are valuable where alternative procedures are being compared to determine relative economic advantages. It is assumed that the *investment* (such as service facilities, office buildings, etc.,) would be the same for all alternatives.

BALANCE SHEET RATIOS

The balance sheet provides data for analyzing the financial condition of a company by means of certain ratios listed in Table 4-6 where the sample calculations are based on the data in Fig. 4-7. Numerous other ratios are used by accountants. These can be found in any standard accounting book, and the reader should refer to such volumes for the detailed significance of the ratios. *The purpose of presenting this material here is to provide the basic information needed to understand the accounting terminology.* The accountant's job is to assemble such information, since his knowledge is required to interpret the significance of the detailed bookkeeping accounts and their effect on the individual ratios.

Returning to Table 4-6, one finds that the current ratio (item 1) and worth-debt ratio (item 2) are standard ratios used to estimate the credit of a company. For item 1, the ratio should be greater than 200. The values of the current ratios for several types of firms are in the following ranges:

Railroads and steel industry   200 to 250

Merchandizing, chemicals, tobacco   340 to 360

The worth-debt ratio measures the relative amount of borrowed and owners (proprietary) capital in use. Another criterion often employed by financial analysts is that the working capital should equal or exceed the amount of bonds and preferred stock. In general, an industrial company should show less than 25 per cent of its capitalization in the form of bonds.

Item 3 is a measure of the company's ability to pay current obligations; it should exceed 100 per cent. Items 4 to 8 show the distribution of the various items of capital as they appear in the accounts of the company.

TABLE 4-6. Schedule of Ratios for Studying the Balance Sheet (Based on Data in Fig. 4-7, All Figures in Millions of Dollars)

1. Current ratio $= \dfrac{\text{current assets}}{\text{current liabilities}} \times 100 = \dfrac{27.0}{11.1}100 = 244\%$

2. Worth-debt ratio $= \dfrac{\text{net worth}}{\text{total liabilities}} \times 100 = \dfrac{53.8}{77.1}100 = 69.5\%$

3. Acid test $= \dfrac{\text{quick assets}}{\text{current liabilities}} \times 100 = \dfrac{17.0}{11.1}\ 100 = 153\%$

4. Fixed asset-worth ratio $= \dfrac{\text{fixed assets (net)}}{\text{net worth}} \times 100 = \dfrac{46.4}{53.8}100 = 86.1\%$

5. Inventory ratio $= \dfrac{\text{inventory}}{\text{total investment}^{a}} \times 100 = \dfrac{10}{66.0}\ 100 = 15.1\%$

6. Fixed asset ratio $= \dfrac{\text{fixed assets (net)}}{\text{total investment}^{a}} \times 100 = \dfrac{46.4}{66.0}100 = 70.2\%$

7. Working capital ratio $= \dfrac{\text{working capital}}{\text{total investment}^{a}} \times 100 = \dfrac{15.9}{66.0}100 = 24.1\%$

8. Process investment ratio[b] $= \dfrac{\text{process investment (cost)}}{\text{total investment}} \times 100 = \dfrac{30.0}{66.0}100 = 45.4\%$

[a] Invested capital may be used as a basis in lieu of total investment if desired.

[b] This can be computed only where process investment can be separated from the total investment. The process investment includes the investment in productive facilities only as listed in Table 4-2.

The actual ratios shown in Table 4-6 will vary considerably with the type and size of processing plants, and, accordingly, it is difficult to establish what exact values should apply for any one plant. Current financial literature and data obtainable from the Securities and Exchange Commission, Washington, D.C., will provide a means for comparing the financial operation of any given company to that of others.

SUMMARY

It may be stated as an economic principle that, at any instant, the national economy tends toward an equilibrium state, where sales in the market place equal the dollar value of production. Furthermore, the amount by which total savings produced exceeds consumption equals the new investment. The pressure of this excess exerts a cumulative effect to produce an expanding economy.

In considering the state of the national economy, management is influenced in its decisions by the magnitude and direction of any departure from the equilibrium state. Thus, the relations between the equilibrium state and conditions at full employment indicate to management whether inflation or deflation is the prevailing condition. This chapter has presented certain mathematical expressions for evaluating these relations in a quantitative manner. These mathematical models assist management in arriving at investment-policy decisions based on the current economic state.

Capital investment is an important managerial function, and top-level personnel are guided by many factors of a technical, economic, and psychological nature. Decisions on capital investments and their magnitude, which are of the wrong kind or which are made at the wrong time, can be disastrous for a company. Certain technical aspects of investment capital and its accounting have been covered in this chapter.

PROBLEMS

4-1. Select some commercial product with a known selling price. From available data determine the added value for each step of manufacture going back to as basic a starting material as possible. HINT: see "Statistical Abstract of the United States" as a source of data.

4-2. From the following data on marketing orange juice, estimate the value added to the annual national income for each step in the operation: selling price in the store $2.40 per dozen cans of 6 ounce size; wholesale price $2.00 per dozen cans; price on the tree $2.20 per box. One box may be considered to yield 1.3 gallons of frozen orange concentrate. Assume that an annual crop of 60 million boxes is used for frozen orange juice.

4-3. The slope of a plot of sales versus national income such as in Fig. 4-2 is 0.20. What is the marginal propensity to consume? What is the value of the multiplier? What is the change in national income for a $10 billion change in sales dollars?

4-4. A company is planning to purchase a large quantity of office furniture which in 1954 was estimated at $14,000. What is the current estimated cost for this material?

4-5. A gear manufacturing company is planning to increase the power on certain stock-forming operations now using five-horsepower motors which cost $300 each in 1957. What would be the current total estimated cost for 10 new motors, each of 10 horsepower?

4-6. Management is considering extension of operations for which the purchase price of equipment delivered is estimated at $228,000. Determine the individual cost components and estimated total cost for the installation based on the relations in Table 4-4. Preoperating charges are estimated at $32,000.

4-7. If the plant in problem 4-6 is to be placed on land where the cost including site preparation is $80,000, what is the total capital investment if annual sales are estimated at $400,000 and construction time is estimated at 9 months, the annual interest rate being 4 per cent.

4-8. From the reported balance sheet for three companies in different industries, determine the turnover values for any year for which data are available.

4-9. The following data are available for a company with no funded debt: current assets $250 million, current liabilities $40 million, stock $8 million shares with par value of $50 per share, quick assets $150 million, surplus $610 million, fixed assets at cost $1200 million, and "other" assets none. Construct a balance sheet and compute the ratios of Table 4-6 for those possible from the available data.

4-10. A company shows the following accounting data: working capital ratio 40 per cent, total investment $30 million, current liabilities $3.2 million, with no funded debt, a depreciation account of $3 million, and an inventory of $5 million. Neglecting "other" assets, construct a balance sheet, and compute the balance-sheet ratios in Table 4-6.

4-11. A composite balance sheet for a group of oil companies shows the following: inventories $1.5 billion, "other" assets $0.2 billion, total assets $11 billion, stock $4.5 billion, long term debt $1.1 billion, surplus $4 billion, and fixed asset-worth ratio 63 per cent. Calculate the balance-sheet figures and Table 4-6 ratios as completely as possible.

4-12. An analysis of a company's operations indicates the following: current ratio 288 per cent, acid test ratio 154 per cent, current liabilities $7 million, total assets $44 million, fixed assets-worth ratio 70 per cent, and depreciation reserve $8 million. There are no "other" assets. Construct a balance sheet and compute the ratios of Table 4-6.

4-13. The data available from the balance sheet of a pharmaceutical company are as follows: current liabilities $18 million, current assets $46 million, total liabilities $63 million, funded debt $2 million, inventories $12 million, and depreciation reserve $8 million. What are the balance-sheet ratios of Table 4-6, assuming no "other" assets?

4-14. For a company fabricating rubber products the following data are available: long term debt $3 million, inventories $5 million, cash and accounts receivable $3.6 million, fixed assets (cost) $13 million, depreciation reserve $7 million, current liabilities $2.8 million, surplus $6 million, and total liabilities $14.6 million. Compute the ratios of Table 4-6, assuming no "other" assets and omitting item 8.

COSTS AND EARNINGS: THE CASH FLOW MODEL

5

In Chapter 4, investment capital was discussed with respect to its source, its use, and its form of distribution after a prolonged period of financial operation. However, the present chapter considers the flow of capital or cash into the company and how it is paid out. There are only *two* main cash flows. The first is income (or revenues) from sales of products, for services rendered, or for earnings upon outside investment; these are *cash inflows*. The second is the expense (that is, costs) involved for the venture in performing its function; these are *cash outflows*.

The detailed recording of company sales and expenditures is handled by bookkeepers in the accounting department. When these records are classified, they provide the source data for determining actual costs of individual operations in a company. Thus, cost accounting involves allocation of such data to individual operations and detailed analysis of costs. Although production personnel should be familiar with cost accounting terminology and the relative importance of various costs affecting operations in a plant, it is beyond the scope of this book to consider cost accounting techniques in detail. Students should refer to other texts for more complete discussion.

The terminology of cost accounting is confusing because many words are used synonymously and interchangeably. This is perhaps not true for any one company or accounting system. However, because there are no absolutely fixed definitions for accounting terms, presenting a survey of the subject that applies to all cases is difficult. Furthermore, certain accounting terms (such as earnings) may be used differently by the accountant and economist. Therefore, in this chapter, restricted terminology is employed in the interest of simplification. In fact, no matter how comprehensively the subject is presented, it could hardly cover all possible modifications of accountants and different authors. Thus, for actual plant operations a production person must determine the precise meanings of the terms utilized by the accounting department of his employer.

NOMENCLATURE	
C_f = annual financial costs	P = an initial sum of money
C_T = annual total cost or expense	$a_{n/i}$ = the present worth factor of
D = annual depreciation, dollars	Chapter 2
F = annual fixed costs, dollars	R = an annual return
I = investment, dollars	R_s = the annual sinking fund de-
i = an annual interest rate expressed	posit
as a decimal	S = annual sales dollars
i_c = capitalized earning rate, ex-	T = annual profits tax, dollars
pressed as a decimal	t = tax rate expressed as a decimal
n = the annual number of produc-	V = variable costs, dollars per unit
tion units, or the number of	Y = annual net profit
years for capital recovery	Z = annual gross profit
n_c = capitalized payout time, years	Z_t = annual taxable profit

The general cash flow principle is shown as a simplified diagram in Fig. 5-1.

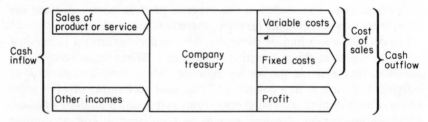

Fig. 5-1. Diagram of cash flow in an economic venture.

TYPES OF COSTS

In any kind of economic venture, a cost (or expense) may be defined as the monetary value of materials and supplies consumed or the equivalent of effort expended. Accordingly, costs appear in a variety of forms. However, since they may be classified into types common to many different kinds of ventures, they are discussed in general terms.

As demonstrated in Fig. 5-1, the cash outflow is made up of two major types of costs, which when added to the profit dollars constitute the total cash outflow that must equal the cash inflow. This may be stated in another way as follows: the profits or earnings of a company are determined by the difference between what the customer pays for the company's product and what it costs the company to manufacture, package, store, and sell it to the customer. It is these aggregate company costs relative to the sales dollars that determine whether or not the company is a profitable venture. Moreover, if there is a possibility of reducing costs, an analysis of aggregate company costs will show management which plant operations should be studied most intensively to obtain the maximum dollar advantage. For the purpose of this book, all costs are considered to fall in either of two, and only two, groups: *variable costs* are those annual costs considered to vary directly, more or less, with the annual production; on the other hand, *fixed costs* remain constant, more or less, for the year regardless of the production rate. The annual profit that a company makes, therefore, is simply the difference between the income from sales and the sum of the variable costs and fixed costs; the latter sum is called the cost of sales. However, from the cost accountant's viewpoint, such a simplification is not possible, since many costs are neither fixed nor do they vary linearly with production. Thus such other existing costs are important; allowance must be made for them.

Figure 5-2 may be used to demonstrate the relations among costs. Thus, the annual fixed costs, F, are shown as a horizontal straight line because they are considered *constant, regardless of the annual production*. Note, however, that the fixed cost per unit of production, F/n, varies with production. However, the term fixed costs may always be considered to refer to the annual constant value unless otherwise stated.

The variable costs are shown as a rising straight line in Fig. 5-2, the annual amount varying with the annual production; V is the average variable cost and is considered to be *constant per unit of production*. Thus, the product of V and the number of annual units is nV, which is a straight line starting at the origin in Fig. 5-2.

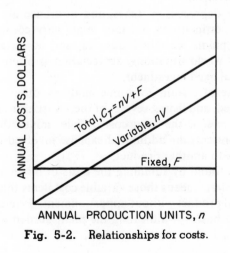

ANNUAL PRODUCTION UNITS, n

Fig. 5-2. Relationships for costs.

If the annual variable and fixed costs are added together, the result is the total cost, C_T,

$$C_T = nV + F \qquad (5\text{-}1)$$

This is a straight line in Fig. 5-2 if the conditions of constant unit variable cost and constant annual fixed costs, F, are assumed applicable. When these simplifying assumptions are made, the model in Fig. 5-2 can serve as a useful tool in discussing costs and earnings in the real world. Later, when principles are understood, sophisticated analysis can be applied to plots like Fig. 5-2 if the real world relations show curved lines. This subject is discussed in Chapter 7.

The details that make up the annual cost and expense items in the operation of an economic venture are numerous and varied in character. They also may vary considerably in relative importance from industry to industry. Even within a given plant, which might be producing the *same* product by *different* methods, the relative costs can show variation.

A typical cost schedule and its relationships as shown in Fig. 5-3, may be followed for estimates of cost distribution. Figure 5-3 includes most of the cost items likely to be encountered; in addition, it classifies them into either of the two basic cost groupings. Many other variations of such a cost schedule are used in industry. The percentage figures shown in the figure are a guide only. They represent the approximate distribution of the total cost of sales on a yearly basis at 100 per cent capacity, except for those costs referring to the plant investment. However, at lower plant capacities, the variable costs represent a proportionately lower fraction of total cost of sales, since under such conditions the annual fixed charges remain relatively constant.

The figures for depreciation, taxes, and insurance are those normally used in plant cost estimations, but they might vary considerably for federal-government plants, certain industries, and at certain geographical locations. In real world situations, depreciation is often only 5 to 8 per cent of the original fixed investment.

For the purpose of certain economic analyses, other classifications of costs are often used, and listing several of these variations is worthwhile.

The *conversion cost* is the total cost of sales minus the raw materials cost. Thus, it represents the additional expense involved in converting an original raw stock to another product.

Prime costs are found by summing the costs of direct labor and direct materials, where *direct* means those variable cost items that can be directly related to the final product without apportionment. Prime costs are sometimes utilized as a basis for allocating the plant burden in the making of different products.

Annual
Cost
of Sales

Annual
Variable
Costs,
60 to 70
per cent

A. Raw materials used in making the product, make-up solvent, or catalyst, etc.
B. Direct labor (operators and helpers); add 15 per cent for social security, pensions, vacations, etc.
C. Process services (steam, power, water, refrigeration)
D. Maintenance (labor and materials), about 4 to 6 per cent of fixed plant investment
E. Miscellaneous supplies and others at about 0.5 per cent of fixed plant investment
F. Direct supervision (foremen), about 10 per cent of direct labor
G. Laboratory charges (process control)
H. Royalty (charges for using some other company's process)
I. Packaging and storage charges
J. Credit for by-products
K. Spoilage and other losses
(Items I and K can be estimated at 5 per cent of total raw materials plus labor plus services)

Annual Fixed
Costs,
40 to 30
per cent

A. Indirect plant cost, or plant burden, 15 per cent
   1. Investment costs, 5 per cent
      a. Depreciation, 10 per cent of fixed plant investment
      b. Taxes, 2 per cent of fixed plant investment
      c. Insurance, 1 per cent of fixed plant investment
      d. Financial costs (interest on inventories, plant equipment, total capital where allowed)[a]
      e. Other assessments, penalties, etc., for certain items of inventory)
   2. Overhead, 10 per cent
      a. Technical (engineering)
      b. Nontechnical (office force, plant protection, etc.)
      c. Supplies (those not chargeable to direct costs)
      d. Rent (included where equipment, buildings, land, or a service is rented)
      e. Others
B. Management expense, 5 per cent
   1. Executives
   2. Legal
   3. Research (technical research and market research)
C. Selling expense, or distribution expense, 10 to 20 per cent
   1. Cost of selling
   2. Delivery and warehouse costs
   3. Technical service, 3 to 5 per cent

[a] Interest is a form of profit unless borrowed capital is used. This item normally is not included but may be important where valuable material is in inventories and a comparative study of alternative plants is being made.

Fig. 5-3. Detail cost schedule for annual cost of sales in one plant.

Incremental costs may be employed as a synonym for variable costs or for certain portions of variable costs such as the sum of materials costs and all direct costs. At constant annual fixed costs, marginal costs are equal to incremental costs. In this text, the terms "variable costs" (which includes costs of materials) and "direct costs" are considered synonymous unless stated otherwise.

The sum of the variable cost and indirect plant cost, or burden, (see Fig. 5-3) is sometimes called the *factory cost*, or the *manufacturing cost of sales*.

The cost and expense data are assembled into a diagram or table such as Fig. 5-3 by referring to past accounting records or by making estimates of future amounts according to the best information available. A basic time unit of one year is usually assumed to obtain the most representative amount, since some expenses such as taxes, insurance, vacations, etc., occur irregularly or only once during the year. Thus, a shorter time period may give unduly high values or omit some items entirely unless the total for the year was prorated back over a shorter time period. The cost-of-sales schedule is ordinarily made at some per cent (80, 90, or 100) of capacity and is based on annual production and annual costs, or it is based on some unit of production (per pound, per gallon, or per ton) which is then multiplied by the annual production in the proper units.

VARIABLE COSTS

Variable costs are essentially constant per unit; however, their annual total varies more or less linearly with production. In this book, variable costs are considered to be zero at no production and to increase directly with the number of product units made, unless otherwise stated.

As shown in Fig. 5-3, certain items of variable costs may actually vary nonlinearly. However, for the purpose of simplification, these items of variable cost are considered to be directly proportional to the rate of production. Examples of variable costs that vary nonlinearly are easily enumerated. Thus, in controlling the process, the laboratory testing charges may be smaller per unit at high production rates than at low capacity operation. Also, power costs per unit may decrease at higher production rates. However, the effect of these variations on the over-all variable cost ordinarily is small, and only with detailed cost accounting is it necessary to consider such variations.

Certain other costs applying directly to a specific operation or product, such as direct maintenance, may not actually be constant per unit of production; however, for each year's operation, their totals may average out as essentially constant per unit of production. Such costs are often called direct costs because they apply to a specific product or operation. Here, unless otherwise indicated, the total of all direct costs for all operations is considered to vary linearly with production and to be synonymous with variable costs.

The self-explanatory list of variable costs in Fig. 5-3 requires little elaboration. Usually, the accounting department has each item of variable cost available in the form of the dollars per unit produced, the equivalent number of man-hours per unit, and the quantities of items (such as services per unit of production). Thus, variable costs either are known or can be estimated with considerable reliability. When individual cost items are known, one can ascertain the amount per unit of product that they contribute to the total cost of sales. Thus, in making economic analyses, one can evaluate the influence of individual cost items, directing attention to those that affect total cost to the greatest extent. However, it is impossible to assign a rigid classification to all of the many cost items encountered in manufacturing and servicing operations. The most important point is to omit no cost items where such exclusion would lead to erroneous results. Use of check lists in cost accounting is highly recommended to obviate such ommissions. For example, federal old-age and security taxes, vacations, sick leave, retirement, etc., are more or less proportional to direct labor, and amount to about 10 to 15 per cent of the nominal man-hour wage rate.

Values for process services per unit for any plant are ordinarily available in company files. However, for cost estimates or for comparative purposes data can be found in other publications.

Under variable costs the item of maintenance (repairs to equipment) can be ascertained from actual accounting records. For estimating purposes, an approximation of 5 per cent of the fixed investment can be employed as the annual cost for this item (at 100 per cent capacity); this approximation is divided by the annual production to get the unit cost. However, any major improvement in equipment is not considered an annual cost, but is treated as a capital investment recovered by depreciation. An alternative procedure to using a rough 5 per cent of investment for annual maintenance is to employ 1 per cent of the accumulated investment times its age for all plant items. Year by year, this cumulative total is corrected for plant additions and removals. Other methods of allocating maintenance costs prorate the total maintenance costs over each operation in proportion to the utilities consumed; this is feasible since repair work should be in proportion to activity.

Cost of direct supervision varies considerably with type of operation. For example, a nontechnical skilled laborer may be in charge of a group of mining operations, whereas a highly technical pharmaceutical process may require a graduate engineer.

Laboratory charges vary with the number of samples required daily and the complexity of the analysis. Direct charges usually are about $10 per man hour of laboratory work.

Variable costs for miscellaneous supplies and packaging vary greatly with the type of product, but are not ordinarily a large proportion of the total variable costs, except for those products made for human consumption, that is, pharmaceuticals, cosmetics, and food products. In such cases, the packaging costs require major consideration because of governmental requirements (such as the pure food and drug regulations) and sales-appeal factors.

Royalty charges generally vary directly with the rate of production; a range of 1 to 5 per cent of the sales price per unit is reasonable.

Credit for by-products requires careful consideration in estimating costs. However, the simplest procedure is to apply all costs to the major production item and then to credit the total variable costs with the value of the by-products. If two or more valuable products result from a given series of processing operations where no single product is of major importance, the term "joint products" may be employed, all plant and material costs being apportioned to each product on a basis of relative tonnage, relative cost value, relative market value, or some other selected basis.

Detailed cost analyses are recommended where a variety of products are being made, where the value of the by-product may exceed that of the major product, and where all variable-cost items are apportioned to each product on the basis of relative unit value per pound (or on some other basis for apportionment as discussed under fixed costs below).

Spoilage is the cost of a poor-quality product minus the actual value of it that can be reprocessed.

FIXED COSTS

The second major group of costs are the expenditures necessary regardless of production rate. The cost items in this group are more or less "fixed" in amount for each year and accordingly are called fixed costs. These fixed costs must be added to the variable costs to obtain the total cost (cost of sales) of producing the product. Figure 5-3 gives a typical list of items making up fixed costs. These items may vary considerably in number and actual dollars for any one plant, depending upon the magnitude of operations and the company's particular system of accounting. Thus, production personnel must become familiar with a company's methods in order to have an intelligent understanding of its costs, particularly since terminology regarding overhead and fixed costs is subject to variations in interpretation. Generally, the total fixed costs are equal to about one-half the total variable costs at 100 per cent capacity operation. However, since these relative values are subject to considerable variation in different types of plants, actual cost data must be used whenever available.

The major cost items falling into the category of fixed costs are listed in Fig. 5-3. In this book, three groups have been employed: (A) indirect plant costs, (B) management expense, and (C) selling expense. There is no one way for grouping such cost items, since each company may have its own special manner of handling them. Any grouping, however, should be simple and logical. It is believed that Fig. 5-3 meets these requirements without involving the fine points of cost accounting found in more extensive texts on the subject. The items under fixed costs apply to any one product made at a given plant. If but a single product is manufactured, the sum of the indirect costs, the management expense, and the selling expense is charged against that one product. However, if the plant produces several products, the fixed costs may be prorated to each joint product on some basis selected by the cost accountants.

Where a company has several plants making only one product, the management-expense and selling-expense items shown in Fig. 5-3 are prorated to each plant on a definite basis, such as the capacity of each plant. The indirect plant costs for each individual plant involved are added to these prorated costs to obtain the total fixed costs for the product at each plant. If the company has several plants making a variety of products, the management and selling expense are prorated as before and added to the indirect costs at any one plant to determine the total fixed costs for that plant; this latter total is then apportioned to each product. The methods employed for these allocations are discussed below.

A. Indirect Plant Costs

For simplification, items of indirect plant costs that can be related to capital investment are grouped under investment costs, as shown in Fig. 5-3. Annual charges for all of these items can be estimated as a given percentage of the fixed investment. The fixed investment is defined as that part of the capital employed to purchase land, equipment, furniture, tools, trucks, machinery, etc., as discussed in Chapter 4 page 81. The figures given in parentheses in Fig. 5-3 can be utilized as approximations where detailed data are unavailable. (The approximation of 10 per cent for depreciation assumes a service life of 10 years with no salvage value. A more precise estimation is 10 per cent annually for process equipment and 5 per cent for service facilities, actual figures being closer to 6 per cent over-all.) Taxes are for real estate and do not include a profits tax. Financial costs denote the interest charges on debts as well as clerical, printing, and legal expenses for financial matters, such as the issuing and transferring of stock and the borrowing of money. Although the charge here is variable, it can be estimated as essentially the interest charge on outstanding bonds. In a going plant, the accounting records provide the actual expenditures for these items. Such costs are illustrated later in the chapter.

The second group of items under indirect plant costs are the overhead costs. These are salaries for plant managers, superintendents, engineers, etc.; other overhead items include the cost of operating the office for clerical expense in accounting, in making out bills, in shipping reports, in filing records, etc. If top executives are not included, the over-all costs for overhead may be of the order of $1500 per man-month, or, alternatively, about twice the annual payroll costs for salaried workers. Rent is an overhead cost, and a company paying rent for land and buildings may deduct such costs as a legitimate expense. However, a company owning its own land and buildings outright is not permitted to deduct a cost equivalent to a rent expense for income tax purposes, since depreciation is allowed. The total of the indirect costs is called the burden that the products produced must carry as their share of the total manufacturing cost.

Where a plant produces many different products varying considerably in amount, that is, unit value, it becomes necessary to apportion the total indirect plant costs, or plant burden, to each product to obtain a more or less correct cost for each. The final profit made by the company is the same regardless of the accuracy of the allocation. However, the accuracy in a comparison of costs for different products varies with the accuracy of the allocation of indirect costs.

The methods for allocation of indirect costs will be mentioned briefly. (1) In the direct-labor cost method, the indirect costs apportioned to each product are the same fraction of the total indirect costs that the direct labor costs for all products. (2) In the direct man-hour method, the burden is allocated on the basis of the ratio of man-hours of direct labor for one product to the total man-hours of direct labor required for all products. Methods 1 and 2 are the same if the average man-hour labor wage rates are the same for different products. (3) In the machine-hour method, the basis used is the hours that a particular machine or piece of plant equipment operates; the basis may also be the relative utilities consumed. However, this method is more complicated where one or more machines are employed on a given product or where a given machine may produce more than one product. (4) Sometimes burden is apportioned on a basis which does not change with production rate, such as the relative capital investment for the particular items employed in making the product, or the relative square footage of space occupied by the equipment and the operations involved with the product.

In some cases, a portion of the plant burden is allocated on one basis and another portion on a second basis. However, the selection of methods employed for such allocations should not be so complicated that the cost involved in making the allocations exceeds the value of the resulting data. For example, the investment costs may be apportioned on a basis of relative investment or of relative area required and the overhead costs on a basis of relative man-hours of direct labor. Usually, however, the total indirect costs are allocated by either method 1 or 2, the latter being considered the most generally useful because the apportionment does not change with fluctuations in labor rates on individual jobs.

In many production-cost studies within a given area or center of production, one need not consider other fixed-cost items of managemental and selling expense, since they are essentially unaffected by modifications in production procedures. However, where a cost study involves either a comparison among different plants or the absolute values for financial returns on any given product, prorating these other expenses is logical. In some cases, any one plant or product may require specialized legal, research, or marketing considerations. Such costs should be charged only to the appropriate product if each product is to be analyzed for its contribution to the economic success of the company, which requires that all facts affecting the costs of individual products be available. However, a realistic view must be taken of such costs. If the value of such information is not worth the effort involved in obtaining it, management and selling expenses for the entire company operation are lumped together as an item of the total cost of sales for all products.

B. Management Expense

Figure 5-3 shows a second group of items under fixed costs, that is, management expense; these costs apply primarily to those companies owning several plants maintaining central executive offices where the major officers and the legal department are located. The research laboratories may be situated at some individual plant or at a separate location. The management costs are allocated to the individual plants in some manner selected by the accountants. If only one plant and location is involved, the management expense is similar to the plant overhead, but is an addition to it. If the plant makes a variety of products, the management expense is prorated in the same manner as indirect costs. The costs for market research and technical research usually must be apportioned to each plant, since the benefits of such expenditures accrue to all plants. However, research costs often are applied directly to the product involved.

The 5 per cent estimation of the management expense is subject to the same limitations noted for all approximations on cost of sales, since some variation in this item with type of plant is to be expected.

C. Selling Expense

The third group of costs that make up part of the fixed costs, as shown in Fig. 5-3, is the selling or distribution expense. This item has a varied interpretation. However, here selling expense refers essentially to those costs required to move the product from the production line to the consumer. The consumer may be some other operation of the producer (that is, "internal" or "captive" consumption), a wholesaler, a retailer, or the individual person. In each case, the cost of selling starting from zero may vary tremendously. In the diagram, these costs are considered to be at 10 to 20 per cent of the *cost of sales*, not of the ultimate sales price. However, this number should be scrutinized carefully for any given real world situation and modified according to the available information from past history or from experienced estimating. Furthermore, the basis for the calculation may be very critical as a simple problem will demonstrate.

Assume a manufacturer sells a product for $10 to a retailer who sells it to a customer for $11. The manufacturer makes a profit of $2, that is, his cost of sales is $10 − $2 = $8. Of this $8 about $1.20, or 15 per cent of the cost of sales, is required for all items of selling expense. Since the product finally sells for $11, the excess of $3 received by the retailer plus the manufacturer's $1.20 selling expense (that is, a total of $4.20) may also be considered as selling expense. This $4.20 is $4.20/$8.00 = 0.525, or 52.5 per cent of the cost of sales. It is, therefore, apparent that the calculation procedure used must be clearly specified when considering certain expenses as a fraction of cost of sales.

Note that the approximate figure given for selling expense is 15 per cent of the total cost of sales. Although this may be considered low for the distributing costs in many industries, it is believed to represent a sound general figure for the manufacturing industries, where the majority of products are not sold directly to the ultimate consumer. Many of these products are moved in large quantities, and the purchaser subsequently utilizes them in other processes or packages them in smaller quantities for final retail sale. When values for distribution costs are greatly higher than 15 per cent, this is usually because (a) an integrated company retails its product in small quantities to the ultimate consumer, with large advertising and customer service costs, (b) an accounting procedure is employed, which includes many items of overhead and management expense (Fig. 5-3) as part of the distribution costs, (c) there are relatively low, variable costs for certain items such as raw materials, etc., or (d) a low volume product is produced. Combinations of these causes can sometimes raise the distribution cost to a figure equal to the factory cost plus profit so that, for every dollar of final sales, 40 to 50 cents may represent the sales cost. These possible variations indicate the necessity for stating clearly the exact allocation of each item of cost in a cost analysis or a cost estimate.

One item of selling expense in Fig. 5-3 that perhaps needs clarification is technical service. This refers to work carried on by the company's personnel for the special benefit of customers; this work may include investigations of complaints and studies to help the customer obtain better results with the company's product.

If a company has only one plant but manufactures a variety of products, the selling expense must be prorated to each product in the same manner used for indirect costs. In some cases, the costs of selling any one product can be determined directly as follows: All of the items making up the selling expense for each plant are added to the management expense and indirect costs to obtain the total fixed costs. This total may then be apportioned to the various individual products.

PROFITS AND EARNINGS

The terms "profits", "earnings," "returns," and "income" are subject to variations in meaning as they are employed by diverse accountants and management personnel. However, in general, they are synonymous. Loosely, they all refer to the difference in the cash inflow of income and the cost of sales (the cost of doing business). Thus, if the cash inflow is the smaller of the two, there is a *negative profit*, or a *loss* for the operation. Some of these terms have more restrictive meanings or connotations, which will become apparent with further development of the subject.

The value of corporation profits in the United States for various years is shown in Table 5-1, where the data are in billions of dollars. For economic analysis studies the mere statement of profit dollars is insufficient. Since the use of a basis is necessary, profits are often expressed as some per cent of sales or as some per cent of investment. Then, operations from company to company in the same or different industries can be compared from an economic standpoint.

Since the United States government and some states impose a tax on profits, it is necessary to recognize a difference in profits *before taxes* and profits *after taxes*. In this text, the former is called *gross* profit[1] and the profit after taxes is called the *net* profit since in reality it is the only real dollars that are netted by the operation.

---

[1] This nomenclature may not agree with some accounting procedures. For example, in some merchantile establishments the costs of the goods is deducted from the cash inflow of sales to obtain the gross profit. The cost of doing business (such as overhead, rent, and all other expenses) is deducted from the gross profit to obtain the net profit before any profits tax is deducted.

TABLE 5-1. Corporate Profits, Taxes, and Dividends (Billions of Dollars)

	Gross Profits[a]	Net Profits[b]	Dividends	Undistributed Profits
1950	40.6	22.8	9.2	13.6
1952	36.7	17.2	9.0	8.2
1954	34.1	16.8	9.8	7.0
1955	44.9	23.0	11.2	11.8
1956	45.5	23.1	12.0	11.1
1957	43.2	22.3	12.6	9.7
1958	37.4	18.7	12.4	6.4
1959	47.7	24.5	13.7	10.8
1960	45.4	23.0	14.4	8.6
1961	45.6	23.3	15.0	8.3

[a] Profits before taxes.
[b] Profits after taxes.
Taken from "Statistical Abstract of the United States," 1963.

The current-profits tax rate varies with gradual increase in profit until an annual profit of about $50,000 and above is reached, in which case the government collects 52 per cent. Thus, this tax must be deducted (just as any cost) before the owners of an operation can consider the residual profit as their own. The calculation of the profits tax is complicated further by the fact that any monies paid out as interest or otherwise in connection with financing the venture are also deductible and not subject to tax. Such costs are called financial costs and will appear as a part of the net profit according to accounting procedures but actually will not be available to the owners. The relationship among the various profit and other financial terms will be clarified by a discussion of the income statement.

THE INCOME STATEMENT

The accountant's name for a periodic accounting of the cash inflows and outflows is the income statement. This is usually published in printed form once a year, but management has a periodic accounting more often, usually quarterly and sometimes monthly. The income statement compares the total income with total expense and shows the disposition of the difference. The income statement is usually made up at the same time as the balance sheet (Chapter 4, page 90), and the two are reported in an annual report prepared for the stockholders to summarize a company's operation. The balance sheet shows the distribution of the capital dollars.

A typical income statement is shown in Table 5-2, which uses the accounting terminology in this book. The terms will be amplified in the following discussion where the significance of certain items will be considered.

Most of the terms in Table 5-2 are self-explanatory, and the method of calculating them is made in a stepwise calculation starting at the top. The gross sales item is the actual dollar value of the products sold. This sum must be reduced by an amount equal to the sum of discounts allowed customers, the product losses in marketing, the excise taxes, etc., in order to obtain the actual dollars (net sales) the company receives. (NOTE: Excise taxes are those, such as for alcohol, oil, etc., levied on the manufacturer.) The difference between the net sales and the cost of sales is the gross profit (or loss). It is also called the net operating income, and if the company receives earnings from outside investment (nonoperating in-

TABLE 5-2. A Typical Income Statement (Millions of Dollars)

Gross sales	102.8	
Excise taxes, discounts, etc.	5.0	
Net sales	97.8	97.8
Cost of sales:		
Variable costs	45.0	
Fixed Costs[a]	40.5	
Total	85.5	85.5
Gross profit[b]		12.3
Profits tax[c]		6.1
Net profit (earnings)[d]		6.2
Distribution of profit:		
Financial costs[e]	0.5	
Dividends	2.0	
Earned surplus	3.7	
	6.2	
Net cash flow[f]		10.7

[a] Includes a depreciation cost of $5 million but does not include financial costs.

[b] Where a company obtains revenue from rents, royalties, dividends from investment of reserves, etc. (which are not the result of processing operations), the excess of revenue over the costs to produce it is added to the gross profit. As explained in the text, the term "gross profit" (shown here) is not an exact terminology.

[c] Gross profit does not take into account the influence of the source of the capital. To obtain the profits tax, however, the financial costs are deducted from gross profit to determine the taxable profit. The tax, therefore, is on 12.3 − 0.5, or $11.8 million, to give a net profit of $6.2 million at an assumed tax rate of 52 per cent. All the net profit is not available to the owners, since the $0.5 million for financial costs must be paid out first.

[d] In some cases the net profit is decreased by the amount of the financial costs to describe the earning. This decreased earning will be called the proprietary earning in this book.

[e] Actual interest paid out plus all expense in connection with borrowing money.

[f] The net cash flow is the sum of net earnings minus financial costs plus depreciation and depletion; it is also called the annual return.

come), they are added to obtain the net income. The net income represents an increase in assets resulting from successful operations, and is an important source of working capital used to pay the income tax and the interest on debts. Net income, after income taxes and interest costs have been deducted, may be retained within the company, should management decide that such net income is needed for expansion or for other reserve funds, in which case it becomes part of the surplus of Fig. 4-7 in Chapter 4. This is earned surplus. When the net profit in excess of financial cost is distributed to the stockholders or the owners, it is in the form of dividends. Most companies split the net profit about equally, one half going into dividends and the other being retained in the surplus account. The averages for all industries are indicated in Table 5-1.

In terms of simple equations, the income statement is computed as follows:

Gross profit equals sales minus costs (excluding financial costs)

$$Z = S - C_T \tag{5-2}$$

Taxable profit equals gross profit minus financial costs

$$Z_t = Z - C_f = S - C_T - C_f \tag{5-3}$$

Profits taxes equals tax rate times taxable profit

$$T = tZ_t \tag{5-4}$$

Net profit equals gross profit minus profits tax

$$Y = Z - tZ_t$$
$$Y = (S - C_T) - t(S - C_T - C_f) = (1 - t)(S - C_T) + tC_f \tag{5-5}$$

In this form, tC_f is a tax credit and applies to any tax-deductible item not included in C_T when calculating a "gross" profit. For example, if there are no financial costs and the net profit is computed where depreciation, D, is not included as a cost, then a special case of Eq. (5-5) is:

$$Y = (1 - t)(S - C) + tD \tag{5-6}$$

where C is all deductible costs except depreciation.

The net cash flow is the sum of net earnings (net profit), after financial costs have been paid, plus any other items not requiring an outlay of cash, such as depreciation and depletion. The net cash flow is being used more and more by accountants in income statements, and corresponds by analogy to the annual return as employed in this text.

Net cash flow represents the total real dollars available to the company for the operation, even though part of it is depreciation which is recovery or protection of original capital. The net cash flow, which was demonstrated in Table 5-2, has considerable usefulness in evaluating profitability, and is discussed in more detail in Chapter 11.

The income statement as illustrated in Table 5-2 is only one of many diverse forms used by accountants to show the flow of money into and out of the company treasury. For example, in some cases the financial costs as well as income tax are deducted from gross profit, and the residue is called the net profit. This method implies that the earnings are dependent upon the method of financing. In order to avoid confusion and the use of a cumbersome terminology of profits before tax and profits after tax, the term "gross profit" in this book will refer to the difference between net sales and cost of sales as illustrated in Table 5-2, and net profit will refer to the profit after taxes and include any financial costs which must then be deducted to obtain the *proprietary earnings*, which is what the proprietors (that is, the owners) actually receive.

INCOME STATEMENT RATIOS

Those who have an understanding of the accounting procedure normally are in a better position to understand the economic implications of the data given on an income statement when such data are used by the accountants in the form of ratios, or expressed as percentages, by which the operations of one company may be compared with those of others.

The general forms that these ratios take are given in Tables 5-3 and 5-4. In some cases, the calculations involve only income statement data; in other cases, data from both the income and the balance sheet are required. The calculated ratios are employed to evaluate both the business operations (profits and earnings.) The business operations are covered in Table 5-3; however, the profits and earnings ratios of Table 5-4 are discussed in the following sections. The terminology for the ratios described in Tables 5-3 and 5-4 is by no means standardized, although certain terms are generally employed by all accountants. For any specific company, the exact meaning of the ratios must be established for that company and for any other company when financial data for the two are to be compared.

Items 1 and 2 in Table 5-3 are measures of the gross profit expressed as basis of a unit dollar of sales. The sum of these two items is 100. Typical values for different industries in a period of good prosperity were as follows for item 2.

	Profit as Per Cent of Sales
Merchandising companies	7
Tobacco companies	10
Railroads	13
Steel industry	15
Chemical companies	19

Item 3 measures on the basis of total investment the annual volume of business being carried. However, instead of a total investment, the base could be total liabilities, which represents total capital in use, although some of it (current liabilities) does not belong to the company. The capital turnover varies with industries, being very high where obsolescence is highly probable and lower where a stable market exists. The relationship between product value and value added by manufacture affects the turnover, since large additions to value require greater facilities and more investments. For the same sales dollars, smaller turnovers result than for

TABLE 5-3. Schedule of Certain Ratios Used for Studying Income Statements and Balance Sheets[a] (Millions of Dollars)

1. Operating ratio $= \dfrac{\text{cost of sales}}{\text{net sales}} \times 100 = \dfrac{85.5}{97.8} \, 100 = 87.4\%$

2. Margin ratio $= \dfrac{\text{gross profit}}{\text{net sales}} \times 100 = \dfrac{12.3}{97.8} \, 100 = 12.6\%$

3. Capital turnover $= \dfrac{\text{net sales}}{\text{total investment}} = \dfrac{97.8}{66.0} = 1.48$

4. Interest turnover $= \dfrac{\text{net profit}}{\text{financial costs}^b} = \dfrac{6.2}{0.5} = 12.40$

5. Inventory turnover $= \dfrac{\text{net sales}}{\text{inventory (sales price)}} = \dfrac{97.8}{10} = 9.8$

6. Profit ratio $= \dfrac{\text{total investment}}{\text{net profit}} = \dfrac{66.0}{6.2} = 10.6$

7. Capital ratio $= \dfrac{\text{total investment}}{\text{net sales}} = \dfrac{66.0}{97.8} = 0.675$

8. Receivables ratio $= \dfrac{\text{accounts receivable}}{\text{net sales}} \times 100 = \dfrac{8.1}{97.8} \, 100 = 8.3\%$

[a] Based on data in Fig. 4-7 and Table 5-2.
[b] Financial costs here are considered to be the total costs for interest charges on debt for the period for which the income statement applies.

small additions to value. Similarly, working capital is a smaller proportion where greater process investments are required and lower turnovers occur.

The interest turnover computed in item 4 is a measure of the risk involved for the bondholders. Values for this ratio above 4 may be considered satisfactory for most industries. Item 5 indicates the number of times each unit of merchandise stored is sold each year, and item 6 computes the number of years required for the net profit to equal the investment. Instead of expressing the accounts in the form of turnover, one may express them as percentages, as in the case of item 8. Note that item 7 is the reciprocal of item 3. In the case of accounts receivable and inventories in items 5 and 8, the ratios computed from the income statement are the same as those that would be obtained by employing either the actual number of items or the volumes (weights) which are equivalent to the dollar values used.

ANALYSIS OF THE INCOME STATEMENT

In order to study the economic feasibility of an operation, certain ratios specifically related to earnings are employed, as in Table 5-4. The items in the table are based on profits after taxes, since those profits are the only real dollars available. If similar calculations are desired before tax deductions, the gross profit is used in place of net profit. The profits tax is based on gross profit minus the financial costs; the current rate is 52 per cent for profits above $50,000, but a different graduated rate is used for profits of less than $50,000. Items 1 to 6 are commonly employed by accountants. Items 7 to 10 may be utilized for special comparisons conducted on an economic basis.

Item 1 is useful for comparing processing companies of a like nature only, because where two companies are producing different materials, one may have a very large volume of sales with a small profit per unit (for example, gasoline or food products) where another may require a large profit on a few sales to earn a sufficient profit to stay in business (for example, heavy machinery). Item 2 is practical for comparing earnings of companies processing different materials, since the basis is the ownership capital that is invested. Net worth as a basis of comparison is satisfactory, since it tends to remain constant (as pointed out in Chapter 4) for a company that has been operating over a period of years, unless the company expands considerably.

Items 3 and 4 are financiers' methods of evaluating earning rates, and reflect the public's opinion concerning the value of a company. The world economic condition affects both these items because, first, net profits are

TABLE 5-4. Schedule of Ratios for Studying Profit, Rate of Return, and Yield[a]

1. Profit rate as % of sales $= \dfrac{\text{net profit}}{\text{net sales}} \times 100 = \dfrac{6.2}{97.8} \, 100 = 6.35\%$

2. Profit rate as % of net worth $= \dfrac{\text{net profit}}{\text{net worth}^{b}} \times 100 = \dfrac{6.2}{53.8} \, 100 = 11.5\%$

3. Stock earning rate $= \dfrac{\text{net profit per share of stock}}{\text{stock market price}} \times 100 = \dfrac{6.0}{36} \, 100$

   $= 16.7\%$ (assumed)

4. Stock yield rate $= \dfrac{\text{dividends per share of stock}}{\text{stock market price}} \times 100 = \dfrac{1.93}{36} \, 100$

   $= 5.2\%$ (assumed)

5. Total capital earning rate $= \dfrac{\text{net profit}}{\text{total liabilities}} \times 100 = \dfrac{6.2}{77.1} \, 100 = 8.0\%$

6. Capitalized earning rate $= \dfrac{\text{net profit}}{\text{total investment}^{c}} \times 100 = \dfrac{6.2}{66} \, 100 = 9.4\%$

7. Capitalized payout time $= \dfrac{\text{total investment}^{c}}{\text{proprietary earning}^{d} + \text{depreciation}}$

   $= \dfrac{66}{5.7 + 5.0} = 6.16 \text{ years}$

8. Risk earning $=$ net profit $-$ expected net profit[e] $= 6.2 - 5.3 = \$0.9$ million

9. Risk earning rate $= \dfrac{\text{risk earning}}{\text{total investment}} \times 100 = \dfrac{0.9}{66.0} \, 100 = 1.4\%$

   $=$ capitalized earning rate $-$ expected rate $= 9.4 - 8.0$

   $= 1.4\%$

10. Proprietary earning rate[f] $= \dfrac{\text{proprietary earning}}{\text{owners' capital}} \times 100 = \dfrac{5.7}{53.8} \, 100 = 10.6\%$

[a] Based on profits after deduction of income taxes and data in Fig. 4-7 and Table 5-2.
[b] Net worth computed as in Fig. 4-7.
[c] See Chapter 4, page 96. Process investment or invested capital may be used instead of total investment. The total investment here is the current value.
[d] Proprietary earning is net profit minus financial costs, or in this case
$$6.2 - 0.5 = \$5.7 \text{ million.}$$
[e] Ratio of net profit to total investment is considered to be expected for this type of venture; it is assumed here as 8 per cent. This ratio is multiplied by the total investment.
[f] Owners' capital is the total investment minus the funded debt.

decreased in a depression and, second, the stock market price is also depressed. The reciprocal of item 3 divided by 100 is sometimes called the price-earnings ratio.

Item 5 may be employed as a measure of earning efficiency where the basis is the total capital involved. Since the current liabilities item is capital not belonging to a company but being used by it, this ratio permits an estimation of the earning rate where the base allows for this nonowned capital being employed in the operations.[2]

The result in item 6 is a most important ratio because it not only expresses the earning rate of the venture but also because it has theoretical economy implications.

Whereas items 1 through 5 express the profit as some per cent of a base quantity (such as total sales, net worth, stock value, etc.) none of these bases are related to theoretical economy. However, when the base is total current investment as in item 6, the earning rate is then related to the value of money according to the principles of Chapter 2. This will be amplified by a brief discussion.

Chapter 2 demonstrated that the annual capital recovery, R, must be equivalent to an initial sum of P; that is, if an investment of I dollars was expended, an annual return of R would be necessary to recover in N years those dollars plus annual interest on the initial sum. This R is made up of two parts. One part recovers annually that portion exactly equivalent to the initial sum of dollars, P, and this is the sinking-fund depreciation of Chapter 3. Numerically, it is:

$$R_s = \frac{iP}{(1 + i)^n - 1} \qquad (5\text{-}7)$$

The second part of the capital recovery is the annual interest on P which amounts to iP; thus the total equivalent annual return, R, is[3]:

---

[2] In accounting, two other values of inventory are often employed which will be merely mentioned here. In the procedure known as LIFO (last in, first out) the costs or values of materials are considered at their latest or current values. Thus, materials made or procured at an earlier date and at a lower price are overvalued, which *understates* the profits. The inverse applies for materials of an earlier date that cost more because they are now being charged for at their latest values, which are below their cost.

In the second procedure of FIFO (first in, first out) all inventories are charged on the basis of their original costs, rather than current values. Thus, in an inflationary period, the past costs are undervalued and profits are *overstated*. In a deflationary period, the products or costs are overvalued relative to current operations, which results in profits at a low level in any financial analysis made.

[3] Note that Eq. (5-8), which is $R = R_s + iP$, can be reduced to:

$$R = \frac{iP}{(1 + i)^{n-1}} + iP = \frac{Pi(1 + i)^n}{(1 + i)^{n-1}} = P/a_{n/i} \qquad (2\text{-}2)$$

$$R = R_s + iP \qquad (5\text{-}8)$$

From the accounting data, the depreciation charge corresponds (by analogy) to the R_s term and the net profit (that is, the interest on the investment, $i_c I$) corresponds to iP. Thus, if the net profit, Y, is known for an investment, I, then the capitalized interest rate, i_c, can be solved for:

$$i_c = \frac{\text{net profit}}{\text{investment}} = Y/I \qquad (5\text{-}9)$$

This is the relationship expressed in item 6 of Table 5-4. It is also equal to the product of item 3 of Table 5-3 and item 1 of Table 5-4, or:

Capital turnover × profit as per cent of sales
= profit as per cent of total investment

$$\frac{\text{Net sales}}{\text{Total investment}} \times \frac{\text{net profit}}{\text{net sales}} = \frac{\text{net profit}}{\text{total investment}}$$

From the above relationships, illustrated in Fig. 5-4, it can be inferred that a small profit per sale with a large number of sales for a given investment is often more profitable than a large profit on few sales. This is the principle in the economic operation of meat and food processing indus-

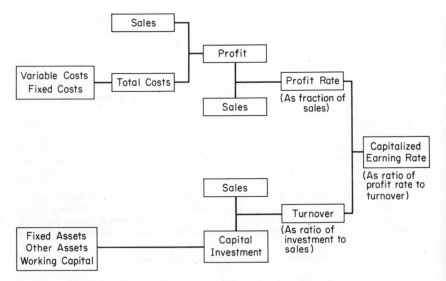

Fig. 5-4. Summary of balance sheet and income statement relations.

tries, etc. The capitalized earning rate is used in evaluation studies where the expected rate of return is given, where the annual net profit of a plant is known, and where it is desirable to estimate the capitalized value of the company.

Further consideration of Eq. (5-8) shows that the depreciation charged off as a cost is available to the company in the form of cash just as is the net profit. Accordingly the sum of the depreciation, D, and net profit, Y, represents the total annual returns actually available if there are no financial costs. However, since the accountant defines net profit as including the financial costs which are actually paid out, the real dollars available to the proprietors as an annual return are the proprietary earning, or:

$$\text{Proprietary earning} = \text{net profit} - \text{financial costs} = Y - C_f$$

Thus, the annual return in more general form is:

$$\text{Annual return} = (Y - C_f) + D = (1 - t)Z_t + D \qquad (5\text{-}10)$$

This annual return is utilized in item 7 to compute a capitalized payout time n_c.

$$n_c = I/[1 - t)Z_t + D] \qquad (5\text{-}11)$$

This calculation determines the number of years required to payoff the investment under current conditions[4]. The two ratios—earning rate and payout time—are preceded by the word "capitalized" to specifically identify them in this text, since many different earning rates and payout times are employed in economic evaluations. Throughout this book the term *annual return*[5] implies the sum of depreciation plus annual interest, and the term *earnings* applies only to the interest (or net profit) part of the annual return. In certain cases, the financial costs are deducted from the net profit, and the difference (proprietary earnings) is employed as the net profit or earnings for the ratios in Table 5-3 and 5-4. Thus, the earning rate for a company depends upon the method of financing that is used, since two identical companies with the same net profits and same investment would now show the same proprietary earning if one company employed borrowed money whereas the second did not (that is, the second would have no financial costs).

---

[4] In theoretical economy where the sum of depreciation plus interest earned (or profit) represents an annual return, R, and with no salvage considerations, then the ratio of

$$\frac{\text{Investment}}{\text{Earnings plus depreciation}} = P/R = a_{n/i}$$

[5] Annual return and net cash flow are synonymous terms, as discussed earlier.

In connection with item 7, it may be pointed out that annual depreciation charges tend to become a constant except for a rapidly growing operation. Therefore, the sum of the net profit (assuming that it is not abnormal) and the depreciation charge less the financial costs might be considered as the real dollars annual payment to be expected in perpetuity. Item 7 is the ratio of the investment to this annual payment. This ratio of investment to assumed perpetual annual return is arbitrarily called a capitalized payout time, and the increase in value of the original investment is not considered. Numerous other payout-time calculation methods are available, but none of them have been adopted universally. This subject is amplified in Chapter 11. If the ratio for process investment alone is desired, the total investment in item 7 may be replaced by the process investment. For example, the capitalized payout time for the process investment of $30 million in Table 5-4 is 30/10.7 = 2.8 years.

Item 8 is a method of computing the risk earnings. Risk earnings may be defined as the excess earnings, over an accepted rate which the investor should receive for taking the risk in investing his money. The basis is the net profit expressed as a percentage of total investment. This may be an excess profit above a minimum acceptable base net profit. The latter is arbitrarily selected according to the type of industry involved. It is the annual net profit in dollars minus some fraction of the investment, where that fraction may vary usually from about 0.03 to 0.35. This excess profit is the risk earning in item 8, which is related to the capitalized earning rate through item 9.

Item 10 is a method of computing the real dollars available annually to the owners where the interest paid on debts is deducted from the net profit. When the proprietary earning is employed in item 10, a capitalized earning rate based on the owners' capital may be computed. If the company has no debts and there are no financial costs, then item 10 equals item 6. Both items 6 and 10 are analogous to item 2, but the total investment is equal to the owners' capital only if there is no funded debt.

In the case of most of the calculations in Tables 5-3 and 5-4, the income has been defined as net income after taxes to simplify the presentation. In some cases the gross profit before taxes is used for calculations, but the inescapable fact remains that it is only the profit after taxes that provides the real dollars that go to the investors.

The importance of the information shown by accountants on the balance sheet and income statements cannot be overestimated. Although a company is showing satisfactory gross profits, the effect of financial costs and profit taxes on the real dollars available for dividends and surplus may be such as to prove the venture unsound from an investment viewpoint.

If the company shows only a small profit year after year, meeting changes in market trends and expanding its operations will be difficult, for investors are reluctant to invest their capital unless a potentially high return is indicated. If the company operates at a loss year after year, it gradually uses up the invested capital and, when the people (creditors) to whom it owes money will no longer extend credit to the company, the latter is likely to be declared bankrupt. As a result, the assets (property and capital) of the company may be taken over by a new management (through legal action) which is agreed upon by the creditors of the company. An alternative is to liquidate the assets by selling them for what they will bring in the market, and the money received thereby is divided proportionately among the creditors. Such bankruptcy proceedings often require an evaluation of the company's operations to determine what constitutes a fair market price for the company's assets.

INTEREST AS A COST

In the preceding section it was pointed out that the net profit represents the interest being earned by an investment in a venture in accord with economic principles. Accordingly, the owners of the venture could charge the operation with an interest charge as a cost and thereby lower the profit returned. However, the interest charged off would have to be paid to the owners so that they would net the same total dollars. In principle, this is correct. Nevertheless, in practice such a procedure is not allowable, since such interest charges are not deductible for profits tax computation unless the interest for borrowed capital *is actually paid out* for notes, mortgages, and bonds. Where interest is paid out in this way, the owners do not get the interest dollars; therefore, the real earning rates are lower, based on proprietary earning. This is demonstrated by the following analysis of hypothetical financial statements in a real problem.

Problem

Compare the earning rates of a potential operation involving an investment of $200,000 having a gross profit of $55,000 when (a) no money is borrowed for investment and (b) $50,000 of the investment capital is raised by selling bonds requiring annual financial costs equal to 4 per cent of the amount borrowed. Profits tax is 52 per cent.

Solution

A comparison of the financial data using the previous nomenclature and calculation procedures is shown in Table 5-5.

TABLE 5-5. Comparison of Financial Analysis with and Without Borrowed Capital

Financial Data	(A) No Borrowed Capital, Dollars	(B) Some Borrowed Capital, Dollars
Total investment, I	200,000	200,000
Funded debt	none	50,000
Annual gross profit, $Z = S - C_T$	55,000	55,000
Annual financial costs, C_f	none	2,000
Annual financial costs, C_f	none	2,000
Taxable profit, $Z - C_f$	55,000	53,000
Profits tax, $0.52(Z - C_f)$	28,600	27,560
Annual net profit, Y	26,400	27,440
Annual proprietary earning, $Y - C_f$	26,400	25,400

Capitalized Earning Rates	Per Cent	
Based on net profit, Y/I	13.2	13.7
Based on proprietary earning, $(Y - C_f)/I$	13.2	12.7

From these results it is apparent that the company pays a smaller profits tax when it borrows some capital but the earnings are also reduced. It is interesting to note that if only the owner's investment of $150,000 is considered under plan (B) the capitalized earning rate based on proprietary earning is $25,440/150,000 = 17.0$ per cent compared to the 13.2 per cent under plan (A). If all of the investment capital has been borrowed, the proprietary earning rate would be infinity, since the denominator for owners investment would be zero.

From this analysis, it is obvious that one cannot discuss interest as a cost in simple terms. Scovell[6] has developed the subject in detail in an interesting manner and defends interest as an item of cost that should be included in the cost of sales. The essence of the argument hinges upon whether the word "profit" should include interest at the going rate as a cost for the use of capital. If this cost is deducted, profit is the excess earning rate for the total capital (both owners' and borrowed) invested in the company over the economic going rate for the capital. However, for profit tax purposes, the federal government views profit as the difference between the sales dollars (income) and the total of direct and fixed costs, neglecting interest on the owners' investment. Profit ordinarily is defined as the total of the economic value of capital (the going rate) plus the excess over the going rate; thus interest on the total capital is not deducted

---

[6] C. H. Scovell, "Interest as a Cost," The Ronald Press Company, New York, 1942.

as a cost. The distinction points up the difference in viewpoint of the economist and the accountant.

It seems logical to consider one economic viewpoint, that is, money is always worth the going rate. Thus, the profit for taking a risk should be the excess above the going rate, but standard accounting procedure treats the earning rate as the sum of the two. Interest deductions on borrowed capital are called financial costs. However, they are not classified as part of the cost of producing a unit of product, since the cost of making and selling a unit of product should be independent of how the company is financed. As shown in Table 5-2, the financial costs are deducted later from the net profit to determine the actual dollars the company receives for its operations. The influence of real interest paid out should be recognized in economic analysis. For instance, a more efficiently operated plant employing borrowed capital may show less proprietary return (but more net profit) than a less efficient plant of the same size and type which did not use borrowed funds (consult the problem above).

Considering the use of borrowed money provides one interesting aspect of the effective cost. For example, in the preceding problem where *some* capital was borrowed the financial charges were $2000 on a $50,000 debt, that is, essentially a 4 per cent interest rate. However, as shown by Eq. (5-4), a tax credit of tC_f, or 0.52 (2000) = $1040, would be allowed so that the real cost of the operation was $2000 − 1040 = $960. This amount is at an effective rate because the profits tax of 960/50,000 = 0.0192; hence, the interest rate is 1.92 per cent rather than the apparent 4 per cent. Thus, the tax credit is (0.52)(0.04), or 2.08 per cent of the amount borrowed; the effective rate is (1 − 0.52)(0.04), or 1.92 per cent. These relations for tax credits apply to any allowable deduction for profits tax. The effective real cost is (1 − 0.52) times the amount involved.

COST ANALYSIS

In addition to providing an accounting of costs for the income statement, management is concerned with the study of costs in a continuing attempt to lower them and thereby increase profits. Such cost studies take a variety of forms; however, the general attack is to observe the specific costs and their magnitude and how these factors apply to each step in the operations. Efforts are made to ensure that each product is carrying its fair share of the costs. Typical examples of such studies are as follows.

Consider the distribution of costs in Table 5-6.

A study of the process shows that by certain changes the material costs can be reduced by 12 cents. This represents a reduction of 10 per cent in

TABLE 5-6. Costs for Producing One Unit of Product X

	Dollars	Per cent of Total Cost
VARIABLE COSTS:		
Material	1.20	12.0
Labor	2.70	27.0
Services	0.27	2.7
Spoilage	0.73	7.3
All other	0.10	1.0
Total variable	5.00	50.0
FIXED COSTS:		
Indirect costs	3.50	35.0
Management expense	0.50	5.0
Selling expense	1.00	10.0
Total fixed	5.00	50.0

material costs and an over-all cost reduction of 1.2 per cent. On the other hand, a saving of 44 per cent in service costs would be required before the same over-all cost reduction could be attained. It probably would be considerably more difficult to reduce service costs by 44 per cent than to reduce material costs by 10 per cent. Thus, the relative magnitude of one type of cost with respect to others may be studied so that efforts can be made to trim costs where the greatest possibility for reduction exists.

In the studying of costs and the establishing of values for them, one must take care that they reflect true values; that is, the various company operations must be sufficiently stabilized when the costs are recorded for representative data to be obtained. When this is done, *standard costs* can be established for various segregated operations. In observing the economics of plant operations, one can compare *actual costs* with these *standard costs*. Thus, any variance between the actual and the standard values draws attention to the phase of the operation from which the variation stems. The operating personnel are required to explain the abnormality. Often, some peculiar circumstances occurring during the period reported will account for the variance. However, if the variance cannot be explained readily and continues to appear, then a cost study must be made to determine the cause. In some cases, the established standard costs are in error or have been altered some because of changes in operation that were not supposed to affect costs. With standard cost systems, both detrimental developments and beneficial procedures can be evaluated readily through the variances they produce. This control of costs is one important type of cost study.

A second type of cost study that is of considerable value applies where several products are being made. The over-all operations may be profit-able, and a study will permit management to determine which products are making the least profit (or perhaps being sold at a loss). For example, a company shows sales of $1 million in manufacturing two products *Y* and *Z* having a total cost of $800,000. Cost studies show the following for a total annual profit of $200,000:

	Product *Y*, Dollars	Product *Z*, Dollars
SALES INCOME	600,000	400,000
COST OF SALES:		
Variable costs	260,000	320,000
Fixed costs	120,000	100,000
Total	380,000	420,000
PROFIT (OR LOSS)	220,000	(20,000)

The loss shown for product *Z* indicates that a detailed analysis of the operations should prove fruitful in increasing the company's profit. Merely to stop making product *Z* is not a solution to the problem since if the fixed costs of $100,000 continued anyway, the total profit would be reduced to yield only $120,000 for product *Y*. It should be emphasized that the use of cost data presumes that such data are based on a reasonable and fair allocation of all costs that cannot be assigned directly to any one product. However, if the cost accounting system is not reasonably accurate, the efforts in improving plant efficiency could be directed to the wrong process.

There is a third important use of cost data; these data are of value in establishing prices for the products manufactured. For many process-industry commodities, the selling price is fixed by the market supply and demand. For a newly developed product, however, it is necessary to establish a selling price based on the cost of sales and a reasonable profit. In either case, cost accounting procedures provide the means for making reasonable estimates of the profit to be expected. Chapter 11 applies the principles of cost accounting to expected profit.

SUMMARY

Management and production personnel are greatly concerned with the cash flow, or life blood, of a company. A proper cash flow must ensure that the cash outflows (costs) are less than the cash inflows (income). If this condition is not maintained, it is impossible to make a profit, which is the economic interest on the investment.

There are a variety of models used to study cash flows, but the income statement is the most important. The major components of the income statement and its relations to the balance sheet are summarized in Fig. 5-4.

Analyses of various items on the income statement and balance sheet provide guides to the strength of an economic venture. These analyses take a variety of forms, most of which relate costs and profits to the investment on a percentage basis. By having income statement ratios compared with theoretical economy relations, management can be advised when and how action should be taken to improve operations.

This chapter has emphasized that the amount of profit, the basis employed, a consideration of profits tax, the influence of borrowed capital, and other factors must all be considered in studying cost and earning for the cash flow model. To communicate adequately the accounting of the past and the results of estimates for the future, all data and calculations must be clearly defined.

PROBLEMS

5-1. In the food industry, the variable costs for canning whole corn may be as follows: raw materials 49 per cent, packaging 24 per cent, direct labor 17 per cent, and others 10 per cent. (a) If the fixed costs are $100,000 and are equivalent to one-half the direct costs at full production, what is the profit on operations if sales are $400,000? (b) What would be the increase in dollars of profit if packaging costs are reduced 25 per cent?

5-2. A company has a total investment of $2 million, of which $1.5 million is fixed investment. Two products A and B are produced with A requiring 12 man-hours of direct labor per unit of product, while B requires 6 man-hours. Total annual variable costs are $270,000 per year. Assuming that the overhead costs equal the investment costs, calculate the production costs per unit for items A and B, omitting management and selling expenses when production is 400,000 pounds per year for each.

5-3. In heat treating certain special steel gears, a furnace costing $20,000 and having a service life of 10 years is required. One man per shift operates the furnace 24 hours per day, 4 days per week. Twelve gears are treated at one time on an 8-hour cycle for charging, heating, cooling, and discharging. Fuel consumption per cycle amounts to 20 million Btu. Fuel costs are $0.20 per million Btu. All costs other than the above, including the indirect costs chargeable to this operation, are $2400 per year. What are these annual total costs, and what are the heat-treating costs per gear?

5-4. Prepare a condensed income statement for a company doing an $8 million annual business with no excise taxes. The company has a depreciation reserve of $1.2 million and a total investment of $4 million, of which 60 per cent is fixed; it has a debt of $1 million. Assume a profits tax of 52 per cent and the approximate figures in Table 5-1 for estimating cost of sales. Dividends to the extent of 50 per cent of the net profits (after financial costs at 4 per cent of the debt as the going rate for money) will be paid to the stockholders. Calculate any ratios of Table 5-3 that are applicable.

5-5. A chemical company having a net worth of $12 million and total investment of $18.3 million showed a gross profit of $4.3 million on net sales of $19 million. Financial costs were $0.8 million, and the fixed assets at cost were $15 million; the fixed assets were depreciated at 8 per cent per year. If the going rate for money is 4.5 per cent, compute all items of Table 5-4 that are applicable.

5-6. A company having a net worth of $2 million shows a net profit of $180,000. Financial costs are at 4 per cent of the total debt, and there is a depreciation charge of $85,000. The total investment is $3 million. What is the capitalized payout time and the proprietary earning rate.

5-7. The following data are available for a company making a synthetic textile fabric.

	Dollars
Total sales	5,000,000
Cost of sales	4,500,000
Dividends paid	300,000
Current assets	2,000,000
Fixed assets (net)	3,000,000
Depreciation	400,000
Current liabilities	800,000
Funded debt	100,000
Stock	4,000,000
Surplus	100,000

Prepare a set of tabulated ratios for studying the financial condition of the company. There are 100,000 shares of stock, and financial cost is 4 per cent of the funded debt.

5-8. Compare the earning rates for a company with $500,000 investment and no debt with one of the same investment and a $200,000 debt. Financial costs are at 4.5 per cent. The annual sales for each are $400,000, and all variable costs are $150,000 per year. Fixed costs are equal to about 60 per cent of the variable costs, and the profits tax for Federal and State totals 58 per cent.

5-9. A cost study of a plant has developed the following cost-per-unit-of-sales information.

Conversion costs	$7.00
Prime costs	$4.00
Services	$0.80
Total variable costs	$8.00
⁻Total fixed costs	about 54.4% of variable costs

(a) Using the relations of Fig. 5-3 where needed for cost estimations, determine which items you would study first for possible reduction in total cost of sales. All costs are given per day of production. (b) If daily production would be doubled, would the cost of sales per unit be halved? Why?

PRICING: THEORY AND PRACTICE 6

According to the previous conceptual development, profits are the earning, or interest, on capital. However, in order to determine profits, revenues must be established. *Revenue* is the cash inflow to the company and depends upon the price of the product or service rendered.

The price of an economic good or service depends upon a number of interrelated factors. Some of these factors apply because they fall within economic-theory applications and others because real world situations do not fit economic theory. The objective of this chapter is to present the elements of pricing policy with respect to supply and demand economic theory and to show its relation to the real world by simple models.

In the real world the price commanded for an economic want varies with its supply, Q_S, and its demand, Q_D. This results in two viewpoints that reflect two relations between price and availability. Thus, the customer desires to pay a low price, P_D. However, a low price can apply only if a large volume of sales for an item makes possible an inexpensive production of it in large quantities. The *customer demand* or demand quantity is high with low prices and vice versa, as shown by the line DD in Fig. 6-1.

However, the supplier desires to receive a high price, p_S, for his product and will produce a large supply only if the expected price is high. The *producer supply* or supply quantity, Q_S, is high with high prices and vice versa as shown by the line SS in Fig. 6-2.

Thus, the two plots are in conflict because at high prices the demand is low (the customer is not willing to buy large quantities), yet the supplier is willing to make large supplies available if it is expected that a high price will be received. However, to sell a large quantity the price must be reduced because of the demand relation between price and quantity. The net result is that these diverse viewpoints are reconciled at some equilibrium price and quantity. This reconciliation is explained by the cobweb model which will be elaborated upon in a later section.

NOMENCLATURE

a = a constant
b = a constant, subscripts refer to demand, D, or supply, S
C = costs
C_D = direct costs
C_f = financial costs
C_m = materials cost
C_T = total costs
d = the trade discount expressed as a decimal
ϵ = an elasticity ratio
E = demand elasticity, the negative percentage rate of change of demand relative to percentage rate of change of price
F = fixed cost
I = investment, dollars
i = return on investment, expressed as a decimal
i_s = return on sales, expressed as a decimal
i'_s = return on gross sales before trade discount, expressed as a decimal
k = a constant
m = a constant
$\bar{p}$ = an equilibrium price, dollars per unit
p = a price, dollars per unit. Subscripts refer to time points, or to demand, D, or to supply, S
p' = a list price
$\bar{Q}$ = an equilibrium quantity
Q = a quantity. Subscripts refer to time points, or to demand, D, or to supply, S
R = capital ratio
S = net sales dollars
S' = gross sales dollars before trade discounts
T = capital turnover
t = tax rate as a decimal fraction, or time period
V = variable cost, dollars per unit
Y = net profit, dollars
Z = gross profit, dollars

ECONOMIC THEORY AND THE DEMAND CURVE

The plot shown in Fig. 6-1 is an idealized plot, since it rarely is a smooth curve in the real world. Also, the plot applies only to the aggregate sum of all demands at a given price, since any one individual customer's willingness to purchase at a price may vary from the idealized line shown. Furthermore, it applies only to one specific demand schedule of prices and quantities. If for some reason there is an increase of over-all demand at any one price, the entire demand curve, DD, is shifted to the

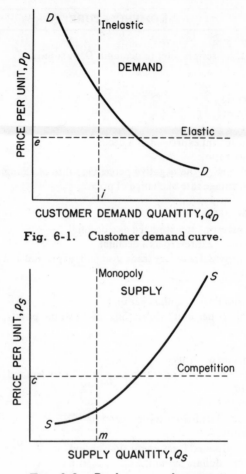

Fig. 6-1. Customer demand curve.

Fig. 6-2. Producer supply curve.

right or left, corresponding to a new demand schedule. Thus, the curve refers to an instantaneous situation at a specific time and is not concerned with changes in demand from year to year.

The total sales are the product of $p_D Q_D$, which has a definite value at each demand level according to the position of the line on the plot and the coordinates on the line.

Referring to a given demand curve such as Fig. 6-1, one finds that the total revenue is a measure of elasticity of demand. If the line is a rectangular hyperbola where total sales dollars (the product of $p_D Q_D$) is a constant (see curve B of Fig. 6-3), the demand is said to have unit elasticity. This means that the price-demand schedule is such that the total revenues are always a constant number. However, if the demand does not vary with changes in price such as indicated by the vertical line (slope is infinite) at point i in Fig. 6-1, the demand is said to have no *elasticity* (that is, the demand is inelastic). A classical example of a commodity that

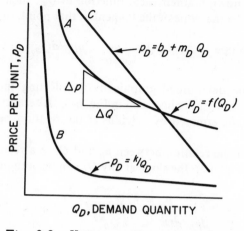

Fig. 6-3. Various types of demand curves.

most approximates this category is common table salt. No matter what the price may be consumers will need and buy just so much salt. Certain other basic foodstuffs and personal demands such as cigarettes and tobacco have low elasticities because the customer demand stays constant even though the prevailing price may change over a limited range.

Opposite from no elasticity is a *completely elastic* demand described by the horizontal line (slope is zero) at point e in Fig. 6-1. Where very small changes in price cause a wide change in demand the demand is said to be highly elastic. An example schedule is shown in Table 6-1.

TABLE 6-1. Commodity Prices at Various Demands for an Elastic Demand

Price per Unit, p_D	Demand, Q_D Units	Sales Value, $p_D Q_D$
22	40	880
16	130	2080
14	270	3780
13	400	5200

DEMAND CURVE MODELS AND ELASTICITY

Three typical demand curves are illustrated in Fig. 6-3 to which real world relations might be compared. The slopes at any point on these curves are related to reciprocal instantaneous elasticity. Thus, on curve A if p_D is some unknown mathematical function of Q_D, then the derivative at any point, p, Q, is the slope of the tangent at that point, or:

$$\text{slope} = \Delta p_D / \Delta Q_D \underset{\text{limit } \Delta Q \to 0}{=} dp_D / dQ_D \qquad (6\text{-}1)$$

This slope can be determined graphically by plotting the tangent and measuring Δp_D and ΔQ_D as illustrated in Fig. 6-3. Alternatively, the slope dp_D / dQ_D may be determined by calculus if the relation between p_D and Q_D is known.

For example, if the relation between p_D and Q_D is a rectangular hyperbola (curve B of Fig. 6-3), the slope at any Q_D can be obtained.

$$p_D = k / Q_D$$
$$dp_D / dQ_D = -k Q_D^{-2} \qquad (6\text{-}2a)$$

At low values of Q_D the slope is steep, denoting low demand elasticity, whereas at large values of Q_D a small slope occurs and high elasticity is indicated. For this special case, the total revenues are constant at any Q_D, since $p_D Q_D = k$.

A second special case is the straight line C of Fig. 6-3 which indicates a linear relation between Q_D and p_D or:

$$p_D = b_D + m_D Q_D$$
$$dp_D / dQ_D = m_D \qquad (6\text{-}2b)$$

Thus, the slope is constant *but the elasticity is not constant* because the total revenues, S_T, are not. The total sales dollars are:

$$S_T = p_D Q_D = b_D Q_D + m_D Q_D^2 \qquad (6\text{-}2c)$$

which will go through a maximum as Q_D increases[1]. The slope of the plots

---

[1] This is readily demonstrated by setting the derivative of Eq. (6-2c) equal to zero and solving for the optimum Q_D for maximum sales.

$$dS_T / dQ_D = b_D + 2 m_D Q_D = 0$$
$$Q_{D_{\text{opt}}} = b_D / (2 m_D)$$

in Fig. 6-3 is not an absolute measure of elasticity, since the latter is concerned with *relative* changes in total revenues as the price changes. For the real world situation, the demand curve usually passes from an inelastic condition to an elastic one as the price changes from a high value to lower ones over a very wide range. However, in general the problem is concerned with describing elasticity over a narrow range of price changes for which either an elastic or inelastic condition may be considered as applying for that range. As a result, a comparison of total revenues at two price schedules may be used for estimating elasticity according to the following:

$$\text{Elasticity ratio, } \epsilon = \frac{p_D Q_D, \text{ high price}}{p_D Q_D, \text{ lower price}} \qquad (6\text{-}3)$$

When ϵ equals one, unit elasticity applies as in curve B of Fig. 6-3. If ϵ exceeds unity, the situation is relatively inelastic. If ϵ is smaller than one, the situation is relatively elastic, since more revenue is generated at the lower price.

Elasticity of demand is also quantitatively defined as the negative ratio of the percentage change in quantity relative to the percentage change in price, or

$$E = -(\Delta Q/Q)/(\Delta p/p) \underset{\Delta p \to 0}{=} -(dQ/Q)/(dp/p) = -(dQ/dp)(p/Q) \quad (6\text{-}4)$$

Thus, from Fig. 6-1 in moving from high prices to low prices on the demand curve, one finds that the reciprocal slope, dQ/dp, increases, but the ratio of p/Q decreases so that the absolute value of E depends upon both the shape and position of the demand curve.

A rearrangement of Eq. (6-4) where E is a constant yields:

$$dp/p = -(1/E)dQ/Q$$

and integrating results in:

$$\log p = -(1/E)\log Q + C \qquad (6\text{-}5)$$

which would be a straight line with a slope of $1/E$ when plotted on logarithmic coordinates.

SUMMARY OF DEMAND ELASTICITY

The properties of customer demand curves vary considerably with the shape of the curve as demonstrated in Table 6-2. These properties may be employed to evaluate the elasticity of demand, which for the real world problems may vary from inelastic to elastic depending upon the range of demands covered, although usually only small changes from a prevailing condition are considered.

TABLE 6-2. Properties of Different Demand Curves (See Fig. 6-3)

Curve	Relation	Slope, dp_D/dQ_D	Sales Total, $S_T = p_D Q_D$
A	$p_D = f(Q_D)$	varies as a function of Q_D	varies as a function of Q_D
B	$p_D = k/Q_D$	$-kQ_D^{-2}$	k
C	$p_D = b_D + m_D Q_D$	m_D	$b_D Q_D + m_D Q_D^2$

The essential points on elasticity and demand may be summarized briefly. *Elasticity* is a measure of relative revenues obtained as prices change. It varies from low values, or *inelasticity,* to a *highly elastic* condition.

Highly Inelastic: A change in the price per unit has little effect on demand quantity which usually occurs in regions of steep slope on the customer demand curve. The total revenue decreases with decrease in price when inelasticity applies. A test to apply for inelasticity is that a ratio of more than unity should result from:

$$\frac{p_D Q_D \text{ at high price}}{p_D Q_D \text{ at lower prices}}$$

Highly Elastic: Small changes in price per unit have considerable effect on demand quantity in regions of small slope on the demand curve. The total revenue increases with decrease in price when the condition of elasticity prevails. The ratio of total revenues at high prices to those at lower prices should be less than unity for high elasticity.

In connection with the demand curve, it is useful to consider that movements along the curve would refer to an instantaneous situation with regard to a given price if exactly that demand quantity were available. The price-demand schedule may vary with time, since there are changes from year to year. Such general changes cause a shift in the absolute position of the entire demand curve, either to the right or left. Thus, the demand may be considered as a conceptual term to describe a potential at any given price. It will be shown later that when the supply potential and demand potential are equal an equilibrium price is established.

ECONOMIC THEORY AND THE SUPPLY CURVE

As pointed out in the introduction of this chapter, the producers supply curve of Fig. 6-2 differs from the consumers demand curve because of a difference in viewpoint. The supply curve describes the quantity that a producer is *willing to supply* at a given price; the higher the price the greater will be the amount supplied to the market according to a specific price-supply schedule. If a new schedule applies (for example, 1965 instead of 1960), the whole curve will be shifted to the right or left.

In general, the same remarks on elasticity and mathematical relations apply for the supply curve as to the demand curve. One difference is that changes in price with quantity are positive for supply; whereas they are negative for the demand schedule.

Two other economic theory concepts regarding price may be introduced at this point. One is the concept of monopoly where a producer who controls all supply quantity may charge any price selected. At a given Q_S on Fig. 6-2, this is represented by a vertical line at point m. The second concept is the opposite of monopoly and is described as pure competition. Under this condition, the quantity made available by any one producer has no effect on the price and is indicated by the horizontal line at point c of Fig. 6-2. This, often stated as the "perfect" competitor, operates under conditions of the completely elastic demand curve (at point e of Fig. 6-1).

DEMAND AND SUPPLY AT EQUILIBRIUM

In the introduction to this chapter, it was pointed out that the two viewpoints (of the producer and the consumer) must be reconciled at some price and quantity. This equilibrium condition is determined where the supply and demand curves cross as shown in Fig. 6-4 at point E to give an equilibrium price, $\bar{p}$, at an equilibrium quantity, $\bar{Q}$. Thus, at a price, p_t, at a given time, t, there is a potential demand quantity, Q_{D_t}, from customers. Corresponding to this is a producer supply potential of Q_{S_t}. The difference in potential is indicated by the difference between the supply and demand curves as shown by a horizontal line at any price, p. However, the separate demand and supply curves apply at different time periods because of the lag in time required to actually produce a supply. Thus, a one time period lag refers to a supply curve that is one step behind a demand curve. This is clarified by referring to Fig. 6-4 and noting the symbols for supply and demand quantities. Thus, Q_{S_0}, any initial selected quantity to be supplied at time period zero, implies that it will be sold in the next period; that is, Q_{D_1}, the customer demand in time period one, will equal

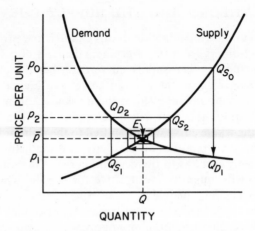

Fig. 6-4. Equilibrium of supply and demand with one time period lag.

Q_{S_0}, the supply quantity at time period zero. These relations are indicated in Fig. 6-4 by the vertical lines between any quantities $Q_{D_{t+1}}$ and Q_{S_t}.

The relation that determines the selected supply quantity is shown by the price lines which are horizontal and connect Q_D and Q_S for corresponding time periods. Note that these *corresponding* time periods differ in absolute time by a one time period lag. Thus, a study of supply quantities for three steps of Q_{S_0}, Q_{S_1}, and Q_{S_2} corresponds to demand quantities of Q_{D_1}, Q_{D_2}, and Q_{D_3}, respectively. Where Q_{S_t} is less than Q_{D_t} the price incentive for the supplier was decreased which caused the decreased production. Conversely, an increase in price results in an incentive for more supply.

Thus, if a price of p_0 is expected, then from Fig. 6-4 the producer will supply Q_{S_0} initially which becomes the demand quantity of Q_{D_1} for the first period. At this available quantity, however, the indicated price of p_1 is of sufficient interest to produce only a supply of Q_{S_1} for the next period. This limited supply results in an increase in price to p_2, etc., until a state of equilibrium is reached at point E of the chart. The model in Fig. 6-4 is known as the "cobweb" model because of its resemblance to a cobweb. The cobweb model may be developed for operations where a time lag of more than one period is involved, such as with wood forest production and certain agricultural products.

The foregoing may be restated as follows: the cobweb model is a chart that relates p_t, a price at time t, to a supply quantity, Q_{S_t}, that will be produced and a customer demand, $Q_{D_{t+1}}$, which causes a price change allowing for the time lag between production and consumption. Equilibrium is attained at a quantity for which the supplier's expected price is equal to the demand price for the same quantity. Where the difference between the equilibrium price and the price at any time can be shown graphically or analytically, the model may be analyzed mathematically. This will be illustrated for the elementary case in the next section.

COBWEB MODEL RELATIONS

For purposes of presentation, the cobweb model outlined in Fig. 6-4 will be simplified by taking the supply and demand curves as straight lines in the following presentation:

Problem

A commodity has supply and demand price curves defined by the following relations:

$$p_D = 30 - 6Q_D, \text{dollars per unit}$$
$$p_S = 10Q_S + 17, \text{dollars per unit}$$

where p = price per unit
Q = thousands of units per year
D = demand
and S = supply

These equations are plotted in Fig. 6-5, and the mathematical relationships involved can be demonstrated by the following problems. (a) Calculate the equilibrium price and quantity. (b) Compute the third-period price and quantity both graphically and algebraically if the first-period demand is 2000 units. Is the situation damped or explosive?

Solution

Part a. To obtain the equilibrium values the given equations may be solved simultaneously, neglecting the subscripts since the equations merely represent two relations of p and Q which have a common point called the equilibrium point. Thus,

$$p = 30 - 6Q$$
$$\underline{p = 17 + 10Q}$$
$$16Q = 13$$
$$\bar{Q} = 0.81 \text{ thousand units}$$

and $\qquad \bar{p} = 30 - 4.86 = \$25.1 \text{ per ton.}$

Both these results check the graphical solution of Fig. 6-5.

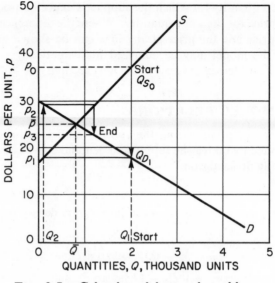

Fig. 6-5. Cobweb model example problem.

Part b. A graphical solution to determine the third-period price and quantity is obtained by a stepwise (iterative) procedure. The chart is entered at a first-period demand, Q_1, of 2000. This quantity was supplied by the manufacturer on the basis of a previous period price of $p_0 = \$37$. At Q_1, however, the demand price p_1 is only \$18 so the manufacturer reduces the supply for the next period to Q_2; this results in a higher demand price p_2 of \$29, etc., until for the third period the price drops to \$22.5 per unit. It will be noted that the price at period t is alternately above and below the equilibrium value, $\bar{p}$, and the difference, $p_t - \bar{p}$, becomes progressively smaller if the algebraic sign is neglected. Thus, a converging or *damped* condition exists for the alternate values of p_t. This is always the case when the slope of the supply curve is greater numerically than that for the demand curve. This implies that for this situation the demand curve should indicate the higher elasticity. If the reverse is true (that is, the slope of the demand curve is the higher), then a divergent or *explosive* condition exists. In this situation, the alternating values of p_t get further and further away from $\bar{p}$ and never do reach equilibrium. (Problem number 6-3 at the end of the chapter is an example of this case.)

When the slopes of the demand and supply curves are equal numerically, the value of p_t merely oscillates at a constant difference above and below the equilibrium value.

The results for the graphical solution to the example problem may be confirmed algebraically as will be shown.

ALGEBRAIC ANALYSIS OF COBWEB MODEL

For linear relations where:

$$p_D = b_D + m_D Q_D, \text{ demand curve} \tag{6-6}$$

$$p_S = m_S Q_S + b_S, \text{ supply curve} \tag{6-7}$$

the equilibrium $\bar{p}$ and $\bar{Q}$ are given by:

$$p_D = p_S$$

and
$$b_D + m_D \bar{Q} = m_S \bar{Q} + b_S$$

or
$$\bar{Q} = \frac{b_D - b_S}{m_S - m_D} \tag{6-8}$$

and
$$p = \frac{m_S b_D - m_D b_S}{m_S - m_D} \tag{6-9}$$

Equations (6-8) and (6-9) are generally true and may be solved directly. Thus, in the previous example problem:

$$b_D = 30, b_S = 17, m_D = -6, m_S = 10$$

and
$$\bar{Q} = \frac{30 - 17}{10 + 6} = 0.81 \text{ thousand units}$$

$$\bar{p} = \frac{(10)(30) + (6)(17)}{10 + 6} = \$25.1 \text{ per unit}$$

The general algebraic solution for values of p at any period, t, is relatively straightforward. From Eqs. (6-6) and (6-7) an equation may be written for demand at period, t, and supply at the previous period, $t - 1$, since they are identical:

$$Q = (p_D - b_D)/m_D = (p_S - b_S)/m_S \tag{6-10}$$

and p_D is p_t and p_S is p_{t-1} (see Fig. 6-5), and at equilibrium

$$\bar{Q} = (p - b_D)/m_D = (\bar{p} - b_S)/m_S \tag{6-11}$$

If Eq. (6-11) is subtracted from Eq. (6-10) after substituting p_t and p_{t-1} in Eq. (6-10), there results:

$$Q - \bar{Q} = \frac{p_t - b_D - \bar{p} + b_D}{m_D} = \frac{p_{t-1} - b_S - \bar{p} + b_S}{m_S}$$

thus
$$p_t - \bar{p} = (m_D/m_S)(p_{t-1} - \bar{p}) \tag{6-12}$$

If the data are available at $t = 0$, then by iteration[2], for any t

$$p_t = (m_D/m_S)^t(p_0 - \bar{p}) + \bar{p} \qquad (6\text{-}13)$$

If this equation is applied to the previous example problem, for p_3 the result is:

$$p_3 = (-6/10)^3(37 - 25.1) + 25.1$$
$$p_3 = (-0.6)^3(11.9) + 25.1 = \$22.5$$

This answer agrees with the graphical solution for the price at the 3rd period.

PRICE SETTING THEORY

A price setting theory relating the real world accounting to elasticity, E, defined as the negative of the percentage change in quantity divided by the percentage change in price, results in an equation for optimum price to give the best profit, if certain assumptions are made.[3] This equation for optimum price is as follows in terms of variable cost, V, and elasticity:

$$p_{\text{opt}} = VE/(E - 1), \text{ optimum price per unit} \qquad (6\text{-}14)$$

Thus, when the relationship between price and demand quantity can be expressed analytically as a linear or exponential function, then the optimum price can be ascertained from Eq. (6-14).

------

[2] Proof of the iteration is as follows:

$$\begin{aligned}
\text{at } t = 1 \qquad & p_1 - \bar{p} = (m_D/m_S)(p_0 - \bar{p}) \\
\text{at } t = 2 \qquad & p_2 - \bar{p} = (m_D m_S)(p_1 - \bar{p}) \\
& \qquad\quad = (m_D/m_S)[(m_D/m_S)(p_0 - \bar{p})] \\
& \qquad\quad = (m_D/m_S)^2(p_0 - \bar{p}) \\
\text{at } t = t \qquad & p_t - \bar{p} = (m_D/m_S)^t(p_0 - \bar{p})
\end{aligned}$$

[3] The equation results from setting the derivative of the profit with respect to sales price equal to zero and solving for the optimum price, p_{opt}. (This mathematics procedure is discussed in detail in Chapter 9, so only the steps will be shown in this footnote.)

$$Z = pQ - (QV + F)$$

$$\frac{dZ}{dp} = Q + p\,dQ/dp - Q\,dV/dp - V\,dQ/dp - dF/dp = 0$$

Based on the assumptions stated following Eq. (6-14), dV/dp and dF/dp are zero, which will then reduce to Eq. (6-14), since F is fixed cost.

The assumptions for which Eq. (6-14) applies are as follows:

(1) The conditions of analysis are considered to apply to a time period for which the variable costs per unit are constant.
(2) The market potential Q does change with time; that is, the price demand curve is not shifted to right or to the left.

The significance of Eq. (6-14) is that the p_{opt} is independent of fixed costs and provides an estimate of the magnitude of the fixed costs that may be permitted without incurring a loss.

For the case where the elasticity is unity over the entire curve, then Eq. (6-14) gives a value of p_{opt} that is infinite. For this case other market and business factors must be considered to determine the model to be employed for optimum price.

PRICES, COSTS, AND THE REAL WORLD

The previous material in this chapter dealt with elementary economic theory relating to prices as affecting supply and demand. The remainder of this chapter considers certain real world situations with respect to prices.

It should be emphasized from the start that the objective of the producer is to make a maximum profit. Accordingly, the margin (excess of sales dollars over costs) should also be a maximum. However, except for a monopoly operation, an excess price results in a relatively small total quantity sold and small revenues. In addition, a high price promotes competition which, in turn, results in smaller revenues when the total demand quantity is divided among numerous suppliers.

Since profit is a relative term for the difference between sales dollars and costs, there are many factors other than the market considerations that enter in fixing the real world price of a commodity. Thus, any reduction in costs permits a producer to reduce prices also. A consideration of the value and sale of by-products as a cost reducer may permit a price reduction. In times of poor sales, a reduction in inventories (which affect costs because of interest and warehouse charges) by lowering prices to increase demand may be advantageous. Any procedure that causes cost changes to affect availability will also affect the price accordingly. Increased costs result in increased prices at which suppliers produce more to cause a subsequent fall in prices until an equilibrium is reached via the cobweb model.

When prices increase at constant demand, there is a shift of the total demand curve schedule, and, therefore, movements on the supply and demand curve must be interpreted correctly. A demand curve such as Fig. 6-1 refers essentially to a constant schedule, that is, to relations between price and quantity under a given set of conditions. External factors (such as competition, speculation, advertising, etc., that change and are different from the original conditions of a consumption or demand schedule) cause a shift of the entire curve. Movements up and down the curve are interpreted as potential prices at potential quantities under the schedule for an existing set of conditions.

The curves in Figs. 6-1 and 6-2 are representative of numerous real life commodities such as wheat, automobiles, sugar, etc. However, numerous other examples of real world situations fit the economic-theory models on prices. Certain of these having a generalized usefulness are presented to indicate their practical value.

THE EXCLUSION CHART

A schematic illustration of a real product demand relation is shown in Fig. 6-6 where the unit prices and annual production of a group of chemicals *of a similar nature and/or utilization* are plotted. A "shotgun" scattering of these data results, but a demand line *DD* may be drawn which approximates the upper limits of the area covered by the data. This chart has certain practical applications.

First, the chart shows the range and distribution of prices for production quantities of a general class of products that may be considered as a single commodity.

Secondly, assuming the various products represent a single commodity, one finds that a practical price-demand curve is simulated by line *DD*. Any conclusions made from such a plot of course should recognize the

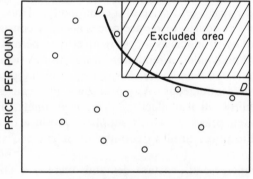

PRODUCTION, POUNDS PER YEAR

Fig. 6-6. Demand plot for pesticides.

underlying assumptions and theoretical limitations as to the accuracy of the curve. Nevertheless, the line may serve a very useful purpose in a semiquantitative sense.

Another practical value of a plot such as Fig. 6-6 is that it serves to emphasize the hazard of introducing a new pesticide at a price range and quantity above the DD line, since no other pesticides are found in this excluded area. Thus, it is quite doubtful that a new proposed plant capacity and sales price in this area would be successful, although this is not an absolute statement. The shaded area provides a warning, and strong justification would be required to enter it with a new product. Similar charts could be drawn for other classes of commodities, using data from U. S. Government sources.

The boundaries of the excluded area of Fig. 6-6 are drawn arbitrarily, but usually as a rectangle which excludes all plotted points. Occasionally, a single point or two may fall in the excluded area where they can be justified on the basis of a brand new product, a special variety of the general class of commodity, or for other reasons. Although judgement and experience are required in establishing the excluded area, the chart is quite useful.

THE FOUR-TENTHS FACTOR

A study of price-quantity curves by some authors has suggested that a valid relation for different items in a general classification (such as discussed above for the exclusion chart) is:

$$p = kQ^m, \text{ dollars per unit} \qquad (6\text{-}15)$$

where Q is annual production, units per year, and k and m are constants. On log-log coordinates this equation will give a straight line with a slope m, since:

$$\log p = \log k + m \log Q \qquad (6\text{-}16)$$

The student may recognize that for a rectangular hyperbola (curve B of Fig. 6-3) $m = -1$. Equation (6-15) may be thought of as fitting many real world data situations, where m is considered to be some type of a modulus to measure economic elasticity in a quantitative manner.

As has been pointed out previously, certain investment costs vary as a power of the annual capacity, or:

$$(I_2/I_1) = (Q_2/Q_1)^a \qquad (6\text{-}17)$$

which states that the relative investments, I_2 and I_1, for two plants vary as the ratio of their productive capacities raised to the a power. This relationship has been found reasonably valid in the chemical industry, with a equal to 0.6. It can be expected to apply to data for a variety of products that are of a similar general nature, such as electrical motors, other electrical components, metallic products, similar machinery factors, etc. However, the exponent a may vary from the 0.6 figure and show a characteristic value for each type of industry, such as 0.7 or 0.8. Generally, it would be expected to be less than one.

It was also demonstrated in Chapter 4 by Eq. (4-11) that total revenue (sales dollars) can often be related to the capital investment, I, or:

$$pQ = TI \qquad (6\text{-}18)$$

where T is the capital turnover ratio equal to the ratio of the revenues to the investment.

Thus, for similar products or in a given industry the ratio of the investments involved for two plants with two different total sales may be written as:

$$\frac{I_2}{I_1} = \frac{p_2 Q_2 / T_2}{p_1 Q_1 / T_1} \qquad (6\text{-}19)$$

and where $T_2 = T_1$ (which would be expected for similar type operations), Eq. (6-19) reduces to:

$$I_2/I_1 = p_2 Q_2 / (p_1 Q_1) \qquad (6\text{-}20)$$

Combining Eqs. (6-17) and (6-20) yields:

$$p_2/p_1 = (Q_1/Q_2)(Q_2/Q_1)^a = (Q_2/Q_1)^{a-1} \qquad (6\text{-}21)$$

This equation states that the price varies as the $(a - 1)$ power of the annual production. As given by Eq. (6-15) the price ratio as a function of demand could be written as:

$$p_2/p_1 = (Q_2/Q_1)^m \qquad (6\text{-}22)$$

where m is the slope of a price demand schedule with log-log coordinates. This exponent is related to the $(a - 1)$ in Eq. (6-21) as follows:

$$a - 1 = m \qquad (6\text{-}23)$$

Concerning the real world results, these relations have been validated for the chemical industries actual data, since Schuman[4] showed for a plot based on Eq. (6-16) that m had a value of -0.35 and -0.55 for prices and production in 1940 and 1948. (The variation in the slope indicated a shift in the demand schedule for the two years.) The average of these two values is -0.45 which if substituted in Eq. (6-23) gives a value of 0.55 for the exponent a in Eq. (6-17). It will be recalled that an acceptable value of a was 0.6 as found for many plants from investment data. Thus, the theory and real world situations appear to be in good agreement.

The preceding discussion of price basis deals primarily with economic theory relations for demand and with the real world situation in the most elementary form. Many other facets of this problem such as the competitive situation, product quality and durability, company leadership, etc., are not discussed here. However, certain other pricing procedures are presented briefly for the purpose of general observation of real world procedures.

PRICES AS A RETURN ON INVESTMENT

Price can be set on the basis of obtaining a maximum rate of return on investment. This requires that (a) the costs of operation be determined from accounting data and (b) the acceptable rate of return be established.

In procedures of this type the costs of production should be based on the long run average to allow for subnormal and abnormal operations which will occur from time to time. These considerations can usually be estimated from a production cost chart at some percentage of plant capacity (for example, at 75 to 80 per cent.)

The acceptable rate of return on investment varies with all the economic factors applicable to a given situation, product, company, or industry. From the over-all cumulative influence of these factors some number will emerge distinctive for each. Historical values are the best guide to the value of these earning rates. They average from 5 to 6 per cent for the manufacture of durable goods and up to 40 or 50 per cent for a new pharmaceutical product. A general average may be 12 to 15 per cent; this range can serve as a criterion in setting prices.

In order to visualize the mathematical calculations, the following relations are useful where t is the income tax rate:

Return on investment, i = net profit/investment = Y/I
Return on sales, i_s = net profit/sales dollars = Y/S

---

[4] S. C. Schuman, "New Look at Economics of Pricing," *Chem. Eng.*, **62** (3):180 (1955).

Gross profit, Z = sales dollars minus costs = $(S - C_T)$
Net profit, Y = gross profit minus taxes = $(S - C_T)(1 - t)$
Capital ratio, R = investment/sales dollars, I/S
Capital turnover, T = sales dollars/investment = S/I

$$\text{Return on investment} = \frac{\text{sales dollars}}{\text{investment}} \times \frac{\text{net profit}}{\text{sales dollars}}$$

$$i = (S/I)(S - C_T)(1 - t)/S = (S - C_T)(1 - t)/I \qquad (6\text{-}24)$$

Since $S - C_T$ is the mark-up, the above equation may be solved for the mark-up in terms of total dollars for a given amount of sales, S, at cost, C_T, excluding financial costs, in various forms:[5]

$$S - C_T = iI/(1 - t) = Y/(1 - t) \qquad (6\text{-}25)$$

and the fractional mark-up or margin based on total costs would be:

$$\frac{S - C_T}{C_T} = \frac{iI}{C_T(1 - t)} \qquad (6\text{-}26)$$

Alternatively, the average mark-up in dollars per unit where Q is the units sold is:

$$(S - C_T)/Q = i(I/Q)/(1 - t) \qquad (6\text{-}27)$$

Thus, where the costs, C_T, are known and the acceptable rate of return, i, can be established, the mark-up and final price can be determined for the desired management policy. Note that mark-up is synonymous with gross profit, Z. Care must be exercised that the wording of mark-up is clearly defined. For example, if the fractional net profit based on sales is to be an i fraction of the capital ratio, R, it would mean that:

$$\frac{(1 - t)(S - C_T)}{S} = i(I/S) \qquad (6\text{-}28)$$

However, if the wording is such that the *fractional net profit* on net sales is a certain amount, i_s, then the fractional mark-up based on net sales is

---

[5] The form of Eq. (6-25), where financial costs are considered, is more complicated, since from Chapter 5 with C_f not part of C_T

$$S - C_T = (Y - tC_f)/(1 - t) = iI/(1 - t) \qquad (6\text{-}25a)$$

defined differently:

$$\frac{(1 - t)(S - C_T)}{S} = i_s \tag{6-29}$$

Note further that if calculations are made on the basis of a list price total sales, S', which is after an additional retailer mark-up above the sales dollars received by the manufacturer, then still another relation must be considered. Where the *fractional profit on gross sales,* or S', is denoted by i'_s and the trade discount is d based on the list price, the relations are:

$$\frac{[S'(1 - d) - C_T](1 - t)}{S'} = i'_s \tag{6-30}$$

Thus, the mark-up for the manufacturer as a fraction of the net sales is obtained from $S = S'(1 - d)$ as follows:

$$\frac{S - C_T}{S} = \frac{i'_s}{(1 - t)(1 - d)} \tag{6-31}$$

Any of these equations may be solved for any one of the variables when all the others are known or specified. The values for prices and costs may be for total values of per unit of output. It should be observed further that the mark-up $(S - C_T)$ can be based on any component or combinations of components desired by management, but diligence should be applied that no costs are omitted. Furthermore, the basis should be a logical one, utilizing modern costing principles so that erroneous price levels are avoided which may impair the number of sales (if set too high) or which may produce losses in the long run (if priced too low).

MULTIPLE PRODUCT PRICING

Where a company operates to produce two or more products, two problems arise: how they should be priced to return the maximum in the long run and how to meet competition. There are numerous procedures that may be followed to allocate costs and establish prices, all of which can be defended on a basis of economic theory or straight accounting. Without an involved discussion of marketing incentives, by-product accounting, or economic-theory reasoning, a brief presentation is given now of the principal methods that can be employed.

Uniform Margin Basis

In one method ("cost plus") all products are priced so that they have the same percentage mark-up over their total costs for all products. An alternative to this method is to use a basis of incremental costs (that is, costs that vary with the production level) instead of total costs. Thus, for each respective product, the percentage mark-up of their incremental costs is uniform. One other modification is a uniform percentage mark-up based on the "value added" in production. This value-added cost is sometimes called the conversion cost and is the total cost minus the original materials costs. Thus, for a company involved in marketing a service, the conversion cost is a high percentage of total cost, and this is also true of the mark-up. Conversely, where a valuable material is processed, the original material cost may be a large percentage of the total cost with a relatively small mark-up in pricing the final product.

By-Product Basis

In this method the main product may carry all the production charges and the second (or by-product) may be considered all profit and priced accordingly. Usually, however, the by-product has some necessary direct charges for operations to make it marketable, and the price must allow for these. This results in the more realistic approach of allocating a portion of operating charges to the by-product, with a policy of some uniform margin mark-up procedure for pricing.

Competitive Market Basis

In the real world market situation, the prices for any product must be competitive, with differentials allowed for quality, quantity, location, channels, and timing. Accordingly, some ceiling on price exists above which the firm would be unlikely to make sales. The floor for any price system is a price equal to the incremental cost or marginal cost at which for a lower price the cost of production is greater than the return. This minimum price varies with different levels of production.

PRICING PROCEDURE CALCULATIONS

In order to demonstrate certain pricing procedures a typical example is presented.

Problem

In manufacturing products M and N in amounts of 20,000 and 16,000 units annually a company analysis shows the following data for a profits tax rate of 52 per cent.

Product	M	N
Apportioned Investment (dollars), I	200,000.00	140,000.00
Production (annual units), Q_A	20,000.00	16,000.00
Cost (dollars):		
Materials, C_m	60,000.00	20,000.00
Direct, C_D	120,000.00	60,000.00
Others, Fixed	40,000.00	80,000.00
Total, C_T	220,000.00	160,000.00
Unit Costs, C_T/Q_A	11.00	10.00

Compute the prices for each product where management has selected a pricing policy of a uniform mark-up for (a) a 20 per cent return on investment, (b) a 30 per cent mark-up based on total cost, (c) a 20 per cent profit on net sales price, (d) a 15 per cent profit based on list sales when the trade discount is 35 per cent, (e) a 40 per cent mark-up based on incremental costs, and (f) a 50 per cent mark-up based on value added by manufacture.

Solution

Part a. A 20 per cent return of investment, I, requires calculation of the required sales dollars, S, from Eq. (6-25). This equation is rearranged:

$$S = iI/ (1 - t) + C_T$$

And:

FOR PRODUCT *M:* $S = 0.20(200,000)/(1 - 0.52) + 220,000 = \$303,500$

Thus, for 20,000 annual units the unit price would be $303,500/20,000 = \$15.18$ each. Then:

FOR PRODUCT *N:* $S = 0.20(140,000)/(0.48) + 160,000 = \$218,400$

Unit price would be $218,400/16,000 = \$13.60$ each.

Part b. A 30 per cent mark-up on total costs would mean that the difference between sales dollars and total costs must equal 30 per cent of the total costs, or:

$$(S - C_T) = 0.3C_T$$
$$S = 1.3C_T, \text{ annual dollars}$$

Then, the unit price would be:

FOR PRODUCT *M:* $p = \dfrac{S}{20,000} = \dfrac{1.3 \times 220,000}{20,000} = \$14.30$ each

FOR PRODUCT N:
$$p = \frac{S}{16,000} = \frac{1.3 \times 160,000}{16,000} = \$13.00 \text{ each}$$

Part c. A 20 per cent profit on sales would mean that the net profit as a fraction of the sales price must be 0.20, that is, $Y/S = i_s = 0.20$, or from Eq. (6-29):

$$(1 - t)(S - C_T)/S = i_s \tag{6-32}$$

$$S = \frac{(1 - t)C_T}{(1 - t) - i_s}$$

The unit price is:

FOR PRODUCT M:
$$p = \frac{S}{20,000} = \frac{(1 - 0.52)\,220,000}{20,000\,(0.48 - 0.20)} = \$16.80$$

FOR PRODUCT N:
$$p = \frac{S}{16,000} = \frac{(1 - 0.52)\,160,000}{16,000\,(0.48 - 0.20)} = \$17.15$$

Part d. A 15 per cent profit on list sales after allowing for the trade discount of 35 per cent requires the use of Eqs. (6-30) and (6-31) to give two unit prices. One is p, the price the manufacturer receives, and the other is p', the list price for the ultimate consumer. The relations for S and S' are:

$$S = \frac{C_T}{1 - \dfrac{i_s'}{(1 - t)(1 - d)}} \tag{6-33}$$

$$S' = S/(1 - d) \tag{6-34}$$

Thus:

FOR PRODUCT M the unit price for the manufacturer, p, would be $S/20,000$:

$$p = \frac{220,000/20,000}{1 - \dfrac{0.15}{(1 - 0.52)(1 - 0.35)}} = \frac{11}{0.556} = \$19.70$$

The unit list price, p', would be $p/(1 - d)$:

$$p' = \$19.70/(1 - 0.35) = \$30.40$$

FOR PRODUCT N, the unit price for the manufacturer, p, would be $S/16,000$:

$$p = \frac{160,000/16,000}{1 - \dfrac{0.15}{(1 - 0.52)(1 - 0.35)}} = \frac{10}{0.556} = \$17.98$$

The unit list price, p', would be $p/(1 - d)$:

$$p = 17.98/(1 - 0.35) = \$27.74$$

Part e. A 40 per cent mark-up on incremental costs means that

$$S - C_T = 0.4(C_m + C_D)$$

Hence:

FOR PRODUCT M, the sales dollars are:

$$S = 1.4(C_m + C_D) + F \qquad (6\text{-}35)$$

The unit price is:

$$p = \frac{S}{20,000} = \frac{1.4(60,000 + 120,000) + 40,000}{20,000} = \$14.60 \text{ each}$$

FOR PRODUCT N, the result is:

$$p = \frac{S}{16,000} = \frac{1.4(20,000 + 60,000) + 80,000}{16,000} = \$12.00 \text{ each}$$

Part f. For a 50 per cent mark-up based on value added, the basic relation is:

$$S - C_T = 0.5(C_T - C_m)$$

Thus:

FOR PRODUCT M, the sales dollars are:

$$S = 1.5C_T - 0.5C_m \qquad (6\text{-}36)$$

The unit price is:

$$p = \frac{S}{20,000} = \frac{1.5(220,000) - 0.5(60,000)}{20,000} = \$15.00 \text{ each}$$

FOR PRODUCT N, the unit price is:

$$p = \frac{S}{16,000} = \frac{1.5(160,000) - 0.5(20,000)}{16,000} = \$14.38 \text{ each}$$

A summary of the unit prices is:

	Product M, Dollars	Product N, Dollars
a. Based on investment	15.18	13.60
b. Based on total costs	14.30	13.00
c. Based on net sales	16.80	17.15
d. Based on list price	19.70	17.98
e. Based on incremental costs	14.60	12.00
f. Based on value added	15.00	14.38

It should be obvious that the relative values as well as their absolute unit prices will vary considerably, depending upon the base used and the variations in the magnitude of the base itself. These variations obviously will also affect profits and the return on the investment.

Since there are numerous variations on the pricing procedures illustrated, any policy requires clear statement to avoid misunderstanding of exactly how a final price was established. In general, margins based on selected parts of the total cost are defended from the standpoint of economic theory. However, they are not usually acceptable. Thus, the results in part a and part b are perhaps the most realistic.

SUMMARY

The material in this chapter indicates (1) that the approach to the equilibrium point relating price and quantity for supply and demand can be explained by the cobweb model. (2) Actual real world situations follow the economic theory in a general way. This provides a theoretical basis for certain practical methods of evaluating pricing procedures. (3) Pricing models of the type proposed in this chapter should be useful tools in (a) justifying prices, (b) evaluating potential profits of a new venture and (c) estimating effects of future competition on production quantities, prices, and profit. (4) The bases selected for determining mark-up can give a wide variety of final unit prices.

PROBLEMS

6-1. Given the following data for wheat, plot the supply and demand curves and discuss the elasticity at both ends of the curves and in the middle region.

Price, $/bu	5	4	3	2	1
Demand, million bu/mo	8	9	11	14	19
Supply, million bu/mo	20	17	13	8	1

6-2. In economic theory the demand curve is a plot of unit price as the ordinate versus quantity as abscissa. (a) If elasticity is defined as the ratio of percentage change in quantity to percentage change in price show how its value can be obtained from the slope of a plot of log price versus log quantity. (b) If the arithmetic plot is a recentagular hyperbola, what is the numerical value of the elasticity at any price?

6-3. A commodity has demand and supply price-quantity curves related as follows.

$$P_D = 50 - 10Q_D, \text{dollars per unit}$$

$$P_S = 6Q_S + 10, \text{dollars per unit}$$

Q_D and Q_S are in thousands of units per year: (a) Compute the equilibrium values. (b) If at the end of the first period, the demand quantity is 2000 units, what are the prices at period 0 and period 3? (c) Is the situation damped or explosive?

6-4. Construct a generalized demand plot for the commodity group classed as rubber processing chemicals as given by the annual report of the U. S. Tariff Commission, "Production and Sales of Organic Chemicals."

6-5. Analysis of certain economic data for a certain group of materials has shown the price demand schedule to follow the relation:

$$p = 48/Q^{0.3}, \text{dollars per pound}$$

Where Q is pounds per year: (a) Draw the demand curve. (b) What is the price at a demand $Q = 200,000$ pounds? (c) *Compute* the price at twice the production rate of (b). (d) If a plant for $Q = 200,000$ pounds per year cost $80,000, what would the investment be for part (c)?

6-6. (a) Determine the proper prices for two products A and B, basing price on a return on investment of 15 per cent. (b) At the same price as in (a) what is the markup based on (1) total costs (2) incremental costs, and (3) value added in production. Available data for one year's operation, with an income tax rate of 52 per cent and 10,000 annual units of each product, are as follows:

Product	A, Dollars	B, Dollars
Proportional Investment, P	90,000	120,000
Costs:		
Materials, C_m	20,000	40,000
Direct, C_D	30,000	15,000
Others, fixed	50,000	15,000
Total, C_T	100,000	70,000

ECONOMIC
PRODUCTION
CHARTS

7

In management, the economic production chart is a particularly useful model. Basically, the economic production chart is a *physical* model capable of incorporating a large amount of cost and price data into a single graph. Thus, the chart provides management with a summary of how policies of increasing or decreasing production affect profits. There are numerous variations of the economic production chart (sometimes known as the break-even chart), and many of them can be analyzed mathematically in addition to being presented graphically.

All of these charts employ data and calculations developed in preceding chapters. Hence, one sees graphically the interrelationship of fixed costs, variable costs, profits, and of sales dollars received as revenues. Having these relationships illustrated graphically enables management to make relevant decisions more confidently, since any suggested changes in operation can be evaluated directly by a glance at the chart. Such charts are powerful and useful tools in management economy studies.

A variety of these charts and their analyses are developed in this chapter. Moreover, other presentations in later chapters employ the basic principles of this chapter.

THE BASIC ECONOMIC PRODUCTION CHART

The relations among sales dollars, costs, and profits developed by the accountant can be expressed mathematically if certain useful assumptions are made. For the purpose of this section, these assumptions are (1) that the variable costs are substantially directly proportional to production rate in the range of 0 to 100 per cent capacity, (2) that the fixed charges are constant regardless of the annual production, (3) that there are no financial costs, (4) that there is no income other than that from operations, and (5) that all units produced are sold at a constant price per unit. Thus,

NOMENCLATURE

(AC) = average unit cost equal to $V + F/n$, dollars per unit

C_T = total cost dollars per unit time

E = elasticity defined by Eq. (6-4)

F = fixed cost, dollars per unit time

i = earning rate, fraction of investment

i_m = marginal profit rate dY/dI

(MC) = marginal costs $d(nV + F)/dn$, dollars per unit

(MR) = marginal revenues or marginal sales, $d(nS)/dn$, dollars per unit

n = number of production units, or fraction of productive capacity; same as Q

n' = a special value of n relating to operations above normal

n'' = a special value of n relating to number of units dumped

I = investment, dollars

ΔI = incremental investment

p = price per unit; see S

Q = quantity or units; see n

S = average sales price, dollars per unit, or total sales dollars per time unit

S' = a special case of S relating to a dumping procedure

t = tax rate, fractional

V = average variable cost, dollars per unit, or total variable costs, dollars per time unit

V' = a special case of V at operations above normal capacity

Y = net profit after taxes, dollars

ΔY = incremental profit, dollars

Z = gross profit before taxes, dollars

if the total costs, C_T, are the sum of the variable costs, nV, and the fixed costs, F, on an annual basis, then

$$C_T = nV + F, \text{ dollars per year} \qquad (7\text{-}1)$$

where n is the number of units produced per year. Where the sales price is S dollars per unit, the gross profit is:

$$Z = nS - (nV + F) = n(S - V) - F, \text{ annual gross profit} \qquad (7\text{-}2)$$

where Z = gross profit, dollars

n = number of units produced per year

S = net sales price, dollars per unit

V = variable cost, dollars per unit

F = annual fixed cost, dollars

(NOTE: If S and V are total sales dollars and total variable-cost dollars, respectively, at 100 per cent capacity, then n is fractional capacity.) If the profits tax, t, is included, then, since it was assumed there were no financial costs, the net profit can be computed from

$$Y = Z(1 - t), \text{ annual net profit, dollars} \qquad (7\text{-}3)$$

Where t is the tax rate expressed as a fraction, if t equals 0.52, Y equals $0.48Z$ for values of Z above \$50,000. The student is urged to study Eqs. (7-2) and (7-3) carefully, since they are the basis for many economic studies in industry.

The relations expressed by Eqs. (7-1), (7-2), and (7-3) are illustrated in Fig. 7-1. The fixed costs, F, are shown as a horizontal line indicating a constant value up to 100 per cent capacity. The variable costs, nV, are shown as a straight line increasing at a constant rate per unit. This rate is

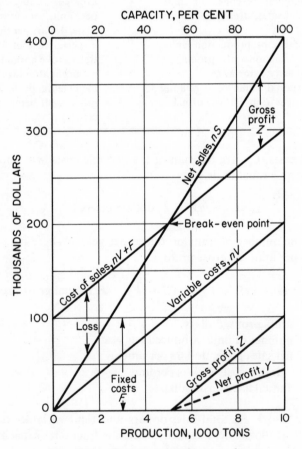

Fig. 7-1. Economic production chart.

V dollars per unit and is the slope of the variable-cost line. The total-cost line is shown as the sum of $nV + F$ and is linear with capacity, and the sales-dollars line is drawn with a slope of S, which is the rate of increase of total dollars. The difference between the net sales line and the cost-of-sales line is the profit or loss for the operation.

The relations as plotted give a graphical representation of costs and profits, and, furthermore, the capacity can be determined at which the gross income just equals the cost of sales when all items produced are sold. This is called the break-even point, and such plots are a useful tool in analyzing a company's operations. Above the break-even point, a profit results. Mathematically, from Eqs. (7-2) when $Z = 0$, the break-even point is:

$$n = \frac{F}{S - V} = \frac{F}{S(1 - V/S)}, \text{ annual units, or fractional capacity} \quad (7\text{-}4)$$

(NOTE: The value of n is in fractional capacity if S and V are in total dollars at 100 per cent capacity.) The break-even point is the same regardless of whether or not profits tax is included. The only difference is that the dollars available to the company are less when the profits tax applies.

The gross profit from Eq. (7-2) is a function of the production rate and may be plotted as shown. If the profits tax is t (expressed as a fraction), the net profit may also be plotted readily, since Eq. (7-3) is a straight line that starts at the break-even point and has a slope of $(S - V)(1 - t)$. If n is fractional capacity, S and V must be total dollars at 100 per cent capacity. In the present discussion, no correction is made for the variation of income tax rate at low profit levels. The tax rate is considered constant starting at zero profit. This assumption, used in the interest of simplification, makes only a slight difference in the shape of the profit curve below $50,000.

These relationships are shown in Fig. 7-1. The quantitative effect of various selling prices on the gross and net profits can be visualized readily by the use of such economic production charts, since a separate net-sales line for each different unit-sales price may be superimposed on the chart.

These relations, as illustrated graphically in Fig. 7-1, are for a typical operation where the annual fixed costs are $100,000, which is the intercept F of the cost-of-sales plot. The variable costs for the product are $20 per ton, and the design capacity of the plant is 10,000 tons per year. Thus, the slope of the variable-cost line is 20, and extends from the origin to $200,000 at 100 per cent capacity. The total cost can be plotted readily as shown in Fig. 7-1. The value of the production, nS, at $40 per ton for the selling price can also be plotted, since at 100 per cent design capacity S

is the slope of a line extending from the origin to $400,000. The break-even point is shown graphically as occurring when 5000 tons are made and sold (50 per cent of capacity), which agrees with the result obtained from Eq. (7-4).

When the market for a product requires a plant to operate below the break-even point (that is, at a loss), it does not mean the plant should be shut down, because the fixed costs must be paid in any event. Therefore, shutdown point occurs when the annual loss equals or exceeds the value of F. Where the variable costs are assumed to be zero at zero production, the shutdown point occurs also at zero production. However, if the variable-cost line shows an initial rapid increase at low production, the shutdown point occurs at some production rate greater than zero, that is, where the loss equals F. In other words, a loss equal to F plus the initial high variable cost occurs at low capacities and thus a loss equal to F should be the allowable limit for operations, since a still greater loss will be sustained where the variable-cost line increases rapidly at the origin and then becomes essentially linear.

In connection with variable costs, it should be noted that where the net-sales and variable-costs lines are linear starting at the origin, there is a constant relation between V and S because the ratio of V/S is constant. A similar constant ratio exists for any component item of the total variable cost (such as direct material, direct labor, etc.) that varies linearly with production when each item of variable cost is a fixed fraction of the total variable cost.

The plots discussed here may be utilized for analysis during any given period of time and do not have to be made on an annual basis. Furthermore, a graph may be developed for one product when a plant also produces other products. It is merely necessary to prorate the pertinent costs as discussed in Chapter 5 (page 110). However, where multiple production units are used, each at rated capacity, then the variable costs per unit produced will be essentially constant, regardless of whether 1, 2, or 10 units of equipment are in operation, and the idealized charts in the present chapter will apply.

Operations for different plants may be placed on the same chart, if it is desirable to compare their operations. For example, two companies making different products showing the same profit at 100 per cent capacity may have entirely different net sales, fixed costs, and variable costs. The study of this data can be facilitated by the use of economic production charts.

The present discussion deals with linear (straight line) relations on the economic production chart. The charts may be utilized to study operations for maximum profit (or minimum loss) under the limiting assumptions made at the start of this section. More elaborate analyses have been proposed where the fixed costs and variable costs are each broken down into more than one component. Then these components are graphed separately (some with finite values at zero production but increasing with production) on the economic production chart, the total costs being the sum of all components. In addition, working-capital variations have been considered by some authors. Although these refinements result in very complicated charts, they may be useful for highly detailed analyses.

For many real world conditions, the cost relations are not linear but instead vary in a more complicated manner with production. The analysis of such complicated relations is taken up later in this chapter. A differential or incremental analysis approach is employed to give these charts a wider and more generalized application.

ECONOMIC PRODUCTION CHART FOR ABOVE NORMAL OPERATIONS

Usually, a plant operates at its maximum efficiency or at minimum cost when producing at 100 per cent of its normal capacity. However, this does not mean the plant makes the maximum total profit at this capacity, since the total profit is cumulative, and operations above 100 per cent of capacity can result in increased profits, even though the profit per unit of production is less. These points are illustrated in Fig. 7-2, where the cost of sales up to 100 per cent of capacity is the same as for Fig. 7-1 (that is, $300,000 or a gross profit of $10 per ton to give a profit of $100,000). Above 100 per cent of capacity, however, the average variable costs are assumed to be linear (but somewhat higher because of less-efficient operation) and to be equal to $35 per ton. This means that although the profit at a production of 12,000 tons per year is $110,000, the profit per ton is only $9.17 per ton as compared to $10 for a 100 per cent of capacity operation. Thus, the total profit continues to increase at operations above 100 per cent of normal capacity as long as the costs of production do not exceed the sales price per unit, whereas the profit per unit decreases. The variation in unit costs above 100 per cent of production depends upon whether the new variable costs exceed the old unit cost at 100 per cent of capacity. The effects on gross profit and net profit are shown in Fig. 7-2, and the effect of variations in unit sales price can be evaluated by superimposing a set of similar lines for each different sales price.

If the two parts of the plots are considered in two steps, all of the above relations can be expressed by equations of the same form as Eqs. (7-2) and (7-3), the proper terms being included for operations above 100 per cent of

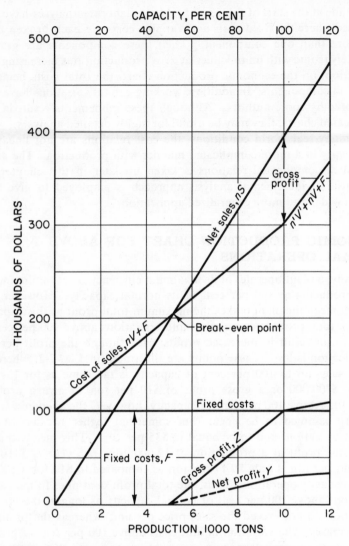

Fig. 7-2. Economic production chart above 100 per cent design capacity.

capacity. For example, gross profit may be written:

$$Z = S(n + n') - (nV + nV' + F)$$

$$= n(S - V) + n'(S - V') - F \qquad (7\text{-}5)$$

where n' and V' are the number of production units and the variable costs above 100 per cent of normal capacity, respectively, and the other terms are the same as in Eqs. (7-2) and (7-3). Net profit, Y, is $Z(1 - t)$ as in Eq. (7-3).

If the slope V' exceeds the unit sales price, the cost-of-sales line approaches the net-sales line, indicating less profit. Eventually, the two lines intersect, and operations above the capacity at that intersection result in a loss. This is an illustration of the law of diminishing returns, which occurs not only because of increased variable costs per unit, but also because of possible increased fixed costs. The total profits diminish with production rate whenever the slope of the cost-of-sales line (this slope is called the marginal cost) is greater than the slope of the net sales line (this slope is called the marginal revenue). This point (at which the slope changes) is sometimes called a marginal point, which will indicate the start of decreasing profits.

In a later section, the economic production chart model is used to consider curved lines where the profit goes through a maximum (the critical production point). Finally, the situation is considered where profits may be decreased to zero, thereby reaching a second break-even point called a "profit-limit" point.

ECONOMIC PRODUCTION CHART FOR DUMPING

The principles discussed in connection with Fig. 7-2 apply to the practice of "dumping." This occurs when a manufacturer sells a portion of his production at one sales price, S, but, because the demand is not sufficient to take all of the production, he sells the remaining production at a lower price, S'', which may actually be below average unit cost. Thus, he obtains greater total profit by running his plant at near capacity because he obtains lower unit costs than would result from producing a lesser capacity and making a higher profit per unit. If n is the number of units sold at price, S, and n'' is the number sold at price, S'', the gross profit is:

$$Z = nS + n''S'' - (nV + n''V + F)$$

$$= n(S - V) + n''(S'' - V) - F \qquad (7\text{-}6)$$

The net profit Y is $Z(1 - t)$ as before.

A plot is shown in Fig. 7-3 for the case where the data employed are the same as for Fig. 7-1, up to 70 per cent of capacity, but, above that, n'' units are sold at a price of $25 per ton. As shown, the gross profit is only $40,000 when operating at 70 per cent of capacity for a unit profit of $5.72 per ton. However, by operating at 100 per cent of capacity and dumping

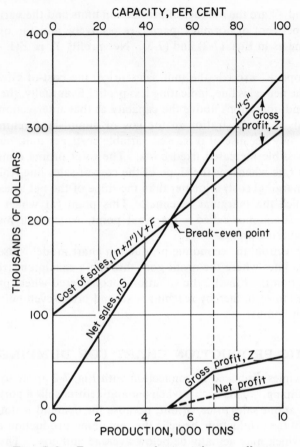

Fig. 7-3. Economic production chart to illustrate dumping.

the additional 30 per cent of capacity, the total profit is $55,000. It will be noted that the shapes of the profits curves are similar in Figs. 7-2 and 7-3.

A SIMPLE BREAK-EVEN POINT EXAMPLE

Problem

The annual fixed costs for a plant are $100,000, and the variable costs are $140,000 at 70 per cent of capacity with net sales of $280,000. What is the break-even point in units of production if the selling price per unit is $40?

Solution

With an understanding of Fig. 7-1, all problems of this type can be solved from the basic equation (7-2). Thus, from Eq. (7-2) where n equals $280,000/40 = 7000$

units, the slope $(S - V)$ of the gross profit line in dollars per unit may be computed as:

$$Z = 280,000 - (140,000 + 100,000) = 7000(S - V) - 100,000$$

$$(S - V) = \frac{140,000}{7000} = \$20 \text{ per unit}$$

From Eq. (7-4), the break-even point is:

$$n = \frac{\$100,000}{\$20} = 5000 \text{ units}$$

The per cent of capacity is $100 \times 5000/(7000/0.7) = 100 \times 0.5$, or 50 per cent of capacity. An alternate solution is to let n equal the fractional capacity in Eq. (7-2); then the slope $(S - V)$ in dollars may be computed as:

$$Z = 280,000 - (140,000 + 100,000) = 0.7(S - V) - 100,000$$

$$(S - V) = \frac{140,000}{0.7} = \$200,000$$

and the break-even point is, from Eq. (7-4), $n = 100,000/200,000 = 0.5$, or 50 per cent of capacity. The break-even units of production are:

$$\frac{0.5 \times 7000}{0.7} = 5000 \text{ units}$$

AN EXAMPLE WITH PROFITS TAX

Problem

A company has fixed costs of $100,000 with variable costs equal to 50 per cent of net sales. The company is planning to increase its present capacity of $400,000 net sales by 30 per cent with a 20 per cent increase in fixed costs. The profits tax rate is 52 per cent. (a) What new sales dollars are required to obtain the same gross profit as for the present plant operation? (b) What would be the net profit if the plant were operated at full capacity? (c) What would be the net profit for the enlarged plant if net sales remained the same as at present? (d) What would be the net profit for the enlarged plant if net sales were $200,000?

Solution

It will be noted from Fig. 7-1 that nV/nS is a constant ratio. For this problem, nV/nS is 0.5, nS being the net sales dollars. Thus, from Eq. (7-2)

Part a:

$$Z = n(S - 0.5S) - 100,000$$
$$Z = nS(1 - 0.5) - 100,000$$

but since nS = \$400,000, the gross profit is:

$$Z = \$400,000 \times 0.5 - 100,000 = \$100,000$$

The new value of nS required to show this profit is:

$$100,000 = nS \times 0.5 - 1.2 \times 100,000$$

$$nS = \frac{220,000}{0.5} = \$440,000$$

Part b: $Y = Z(1 - 0.52) = (1.3 \times 400,000 \times 0.5 - 1.2 \times 100,000)0.48$

$$Y = 0.48 \times 140,000 = \$67,200$$

Part c: $Y = (400,000 \times 0.5 - 1.2 \times 100,000)(0.48) = \$38,400$

Part d: $Z = 200,000 \times 0.5 - 120,000 = -\$20,000$

Thus, a loss, not a profit, would result if sales were only \$200,000, since the break-even point in sales dollars from Eq. (7-4) is:

$$nS = \frac{120,000}{1 - 0.5} = \$240,000$$

AVERAGE UNIT COSTS

The cost of a unit of production such as one ton, 1000 square yards, one barrel, or one item, etc., often in the real world varies with production. There are two unit costs with which management is concerned: average cost per unit (AC) and marginal cost per unit (MC).

The average unit costs are the total costs divided by the total production and for the basic economic production chart the average unit costs at any production would be:

$$(AC) = (nV + F)/n = V + F/n, \text{dollars per unit} \qquad (7\text{-}7)$$

where V = the *average* variable cost in dollars per unit
       F = the total fixed costs
and n = the production units

This relation is a generally valid one, regardless of whether V is a constant or varies with the production n, and is plotted according to the principles of Fig. 7-1. For the general case, even F may be permitted to vary with n, although it is usually considered a constant. The case for V and F each being constant was considered in the previous sections where the economic production charts showed straight lines. The average unit-total-cost lines will normally be curved (regardless of whether V is constant or not) because of the F/n term.

For the case of operations above normal as in Fig. 7-2, the average costs are:

$$(AC) = \frac{n'V' + nV + F}{n' + n}, \text{ dollars per unit} \qquad (7\text{-}8)$$

where the terms are the same as in Eq. (7-5). A plot of the equation was shown in Fig. 7-2. The Eq. (7-8) is a general equation that applies at less than 100 per cent capacity where n' is zero.

For the situation illustrating dumping as in Fig. 7-3, the average unit cost is:

$$(AC) = V + F/(n + n''), \text{ dollars per unit} \qquad (7\text{-}9)$$

The marginal cost per unit (MC) differs from the average unit cost (AC). The former is defined as the cost of manufacturing one *additional unit* of product after the first one is made. Thus, if the first unit must absorb all the fixed costs, the unit costs will be a very large number; however, the average costs for two units will cost considerably less because all of the fixed costs per unit are halved. Where the cost relations are straight lines in economic production charts, the marginal costs (MC) are equal to V and are constant. However, since the average unit costs (AC) are obtained by prorating the fixed costs over all of the units produced, they decrease rapidly as demonstrated in Fig. 7-4.

Where a change in the costs occurs such as shown in Fig. 7-2 so that single straight lines no longer apply, both the average unit costs and marginal costs are affected. As shown in Fig. 7-5, the average unit cost passes through a minimum at 100 per cent capacity.

The marginal costs are constant at $20 up to 100 per cent of capacity but then jump to $35 per ton.

Where the models for revenues and costs are straight lines (as discussed in connection with Figs. 7-1 to 7-5), the economic interpretation for practical decisions is relatively simple. However, in many instances, straight lines are not applicable, and the real world economic models employed must be of a complex type that requires sophisticated analysis. Such nonlinear models are explained in the next section.

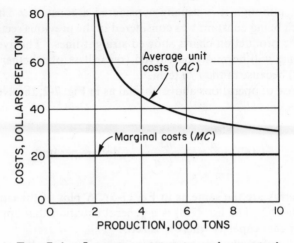

Fig. 7-4. Average unit costs and marginal costs for data in Fig. 7-1.

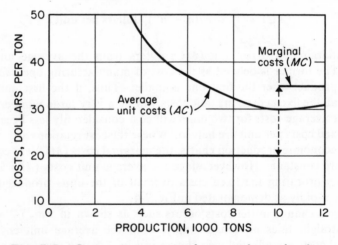

Fig. 7-5. Average unit cost and marginal cost for data in Fig. 7-2.

NONLINEAR ECONOMIC PRODUCTION CHARTS

The discussion up to this point has been based on the assumption that plots of data on an economic production chart are straight lines. In the real world, this is not generally true, since available data frequently plot as a curved line. However, very often lines on an economic production chart can be approximated by simple mathematical relations. Such mathematical models are useful in an economic analysis. With simple mathematical approximations, requiring only an elementary knowledge of calculus, analyses are possible that are not much more difficult than those employed on economic production charts with straight lines. The analytical solution of these equations appear complicated because of the algebraic symbols that must be used, but in reality the procedures are quite simple where the reader has some basic training in the steps to follow.

Figure 7-6 illustrates a very general case where the average variable cost, V, changes with production; the effect of this variation on the cost

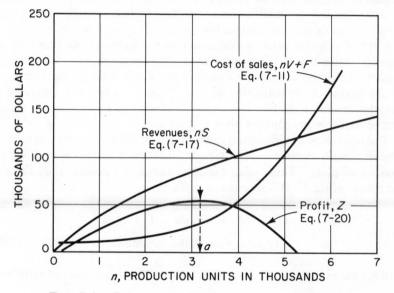

Fig. 7-6. Economic production chart for total values.

of sales is also shown. Consider, for example, that the variable costs of metal stampings for a child's toy varies as follows with the number made, n:

$$V = n^2 \times 10^{-6} - n/500 + 3, \text{ dollars per unit} \qquad (7\text{-}10)$$

and also consider that the company apportions $10,000 of all of its annual fixed costs to this particular toy. The total costs, C_T, for n units would be $nV + F$, or:

$$C_T = n^3 \times 10^{-6} - n^2/500 + 3n + 10,000, \text{ dollars} \qquad (7\text{-}11)$$

The average unit costs from Eq. (7-7) would be:

$$(AC) = n^2 \times 10^{-6} - n/500 + 3 + 10,000/n \qquad (7\text{-}12)$$

This equation is plotted in Fig. 7-7, and as is shown for this case, it goes through a minimum at 2140 units as noted at point b.

The application of differential calculus to these nonlinear economic relations is a good example of how such mathematical techniques are of value. In differential calculus, changes are considered to occur in very small or infinitesimal steps. Thus, the slopes of the lines on economic production charts may be thought of as small increases in the ordinate (y axis) divided by small increases in the abscissa (x axis). Therefore, where the plots are not straight lines with constant slopes as in Fig. 7-6, the use of calculus permits evaluation of the slope at any point on the curves. After this is accomplished by differentiation, the differentiated equations may be employed to analyze the original model by computers at any degree of sophistication desired. There are two main differential calculus formulas used in this chapter, which are given below[1]. Others can be found in any calculus book.

The utilization of differential calculus can be shown very well as follows: The minimum average costs of Eq. (7-12) are determined without the making of plots. Thus, since the minimum (AC) occurs when the first derivative of Eq. (7-7) is set equal to zero, or:

$$d(AC)/dn = dV/dn - F/n^2 = 0$$

$$dV/dn = F/n^2 \text{ at the minimum} \qquad (7\text{-}13)$$

and for the example given from Eq. (7-10) and the fixed costs of $10,000:

$$2n \times 10^{-6} - 1/500 = 10,000/n^2$$

Multiplying by n^2 gives:

---

[1] The simple differentiation of a single variable y with respect to x where a and b are constants is given as follows:

$$y = ax^n + b$$
$$dy/dx = nax^{n-1}$$

When the dependent variable y is a function of two variables, the form of the differentiated equation is as follows:

$$y = uw$$
$$dy/dx = u\,dw/dx + w\,du/dx$$

$$2n^3 \times 10^{-6} - n^2/500 = 10,000$$

By trial and error solution, $n = 2140$ units

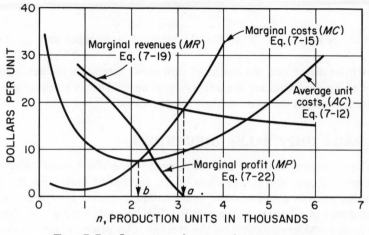

Fig. 7-7. Average and marginal cost relations.

MARGINAL OR INCREMENTAL COSTS

The slope of the total cost line as plotted on an economic production chart is $d(nV + F)/dn$ and is the *marginal* (or *incremental*) cost with the same dimensions as the average unit cost. In mathematical terms the marginal cost (MC) is:

$$(MC) = d(nV + F)/dn, \text{ dollars per unit} \qquad (7-14)$$

It is the rate of change of total costs with production.

From the special case of economic production charts which are straight lines (that is, V is a constant regardless of production), the marginal cost (the slope of the total cost curve) is a constant equal to V, since the fixed cost is also constant (with a zero slope)[2].

However, in the real world the total costs tend to increase with production because of the overtime premiums, extra maintenance, lower efficiency, etc., that accompany an increase in production. Under these

---

[2] This is demonstrated as follows. If V and F are constant, then by calculus:

$$(MC) = d(nV + F)dn = n \, dV/dn + V \, dn/dn + dF/dn$$

$$(MC) = 0 + V + 0 = V, \text{ dollars per unit}$$

conditions, the slope of the total-cost curve ordinarily increases. Thus, the marginal costs (MC) may go through a minimum and then increase with production, as illustrated in Fig. 7-7 for costs based on Eq. (7-11), which when differentiated gives the following marginal cost equation to be plotted.

$$(MC) = 3n^2 \times 10^{-6} - 2n/500 + 3, \text{dollars per unit} \qquad (7\text{-}15)$$

It will be noted that the marginal-cost curve intersects the average-cost curve at the minimum for the average unit cost. This is a requirement of these curves from the basic mathematics involved[3].

AVERAGE UNIT SALES (REVENUES)

In the simple case of the economic production chart model where the sales price, S, is constant regardless of the number of units sold, the average sales price is also constant, that is, equal to S. However, as was pointed out in Chapter 6, the relationship between supply and demand may have considerable effect on the average sales price because of competition, customer demand, and the economic state. Accordingly, the average sales price will usually decrease with the supply available or the production achieved. This sales price, S, can often be expressed in the form:

$$S = 300/n^{0.3}, \text{dollars per unit} \qquad (7\text{-}16)$$

Thus, the total sales revenue at any production would be nS, or:

$$nS = 300n^{0.7}, \text{dollars} \qquad (7\text{-}17)$$

---

[3] As shown in Eq. (7-13) the minimum on the average unit cost curve occurs numerically when the derivative of the average unit costs or $d(V + F/n)dn$ is zero; that is, $dV/dn - F/n^2 = 0$, and $dV/dn = F/n^2$. For constant F, the marginal cost is always:

$$(MC) = V + n \, dV/dn \qquad (7\text{-}14a)$$

or at the optimum production for a minimum average unit cost where $dV/dn = F/n^2$,

$$(MC) = V + n(F/n^2) = V + F/n$$

Thus, $(MC) = (AC)$ at the exact production at which the average total unit cost is a minimum.

By similar reasoning, it could also be demonstrated that if the average variable cost curve, V, by itself passes through a minimum (at $dV/dn = 0$) the marginal cost from Eq. (7-14a) would be:

$$(MC) = V + n(0) = V$$

or, the marginal cost would also intersect the average variable cost curve at the identical production at which the variable unit costs only went through a minimum.

The expression for nS can go through a maximum if the average sales price decreases more rapidly than the effect of additional sales on increased revenues. Equation (7-17) is plotted in Fig. 7-6 along with Eq. (7-11) and the difference, which is gross profit, is also plotted in the same figure. The profit is shown to go through a maximum at point a.

MARGINAL REVENUES

Corresponding to marginal costs is another important property of the economic production chart model. This is the marginal revenue, or marginal sales. Marginal revenue is the slope of the total revenue line $d(nS)/dn$ and may be defined as the *additional revenue or sales dollars resulting from selling one more unit of sales at a specified level of production.* (Here production units and sales units are considered to be synonymous, since all of the units produced are sold.)

As has been noted previously, S is the slope of the total revenue line on an economic production chart. If the total-sales (revenue) line is a linear relation as in Fig. 7-1, the marginal revenue (MR) is a constant S (dollars per unit) and equals the average sales price. However, where S varies with production, the slope of the sales-price curve, or the first derivative $d(nS)/dn$, is the marginal revenue. Thus, the value of (MR) for Eq. (7-17) would be:

$$(MR) = d(nS)/dn \qquad (7\text{-}18a)$$

$$(MR) = 210n^{-0.3}, \text{dollars per unit} \qquad (7\text{-}19)$$

This equation is plotted in Fig. 7-7, and intersects the marginal-cost curve (MC) at a production corresponding to point a. It will be noted that this is the same value for production shown in Fig. 7-6, at which the gross profit is a maximum. This production is called the critical production rate and will be discussed further.

It is interesting to note a relationship between revenues and elasticity as discussed in Chapter 6. If Eq. (7-18a) is rewritten in parts by calculus:

$$(MR) = d(nS)/dn = S(dn/dn) + n(dS/dn)$$

or $$(MR) = S + ndS/dn \qquad (7\text{-}18b)$$

Since demand elasticity as defined by Eq. (6-4) may be written for price p (instead of S) and quantity Q (instead of n), then:

$$E = -(dQ/dp)/(p/Q)$$

and Eq. (7-18b) may be written:

$$(MR) = p + ndp/dQ \tag{7-18c}$$

From Eq. (6-4),

$$-dp/dQ = p/(QE)$$

Therefore, since dS/dn is equivalent to dp/dQ then:

$$(MR) = S + S/E = S(1 - 1/E) \tag{7-18d}$$

This equation states that marginal revenues are equal to the average sales value minus the average sales value divided by the demand elasticity.

THE CRITICAL PRODUCTION RATE

The objective of the economic production chart model is to describe operating conditions so that management can study several major situations. These are:

(1) *the break-even point,* or minimum production above which operations are profitable;
(2) *The effect of operations above the break-even point* in determining whether increasing profits are made or whether at some production (the critical rate) a maximum profit is obtained; and
(3) if a maximum profit is obtained, *the possible effect of production above the critical rate* actually resulting in losses because a second break-even point of zero profit is reached (This is called the profit limit point).

In addition, the mathematical analysis can provide information on unit prices, costs, and profits per unit at various levels of production. Although these data are useful, the most important purpose of the model is to determine the critical production rate. This rate is determined at the point of maximum difference between the total-sales curve and the cost-of-sales curve, since the difference is profit. Mathematically, the profit is Z (assuming zero profits tax):

$$Z = nS - C_T = nS - (nV + F), \text{dollars} \tag{7-20}$$

For Z to go through a maximum as demonstrated in Fig. 7-6, the first derivative should equal zero or:

$$dZ/dn = d(nS)/dn - d(nV + F)/dn \tag{7-21}$$

If this is equated to zero, then:

$$d(nS)/dn = d(nV + F)/dn$$

and from the previous Eqs. (7-18a) and (7-14):

$$(MR) = (MC)$$

The expression dZ/dn in Eq. (7-21) is the marginal profit (MP) and represents the slope of the profit line in Fig. 7-6. The marginal profit decreases with increasing profit and at the production for maximum profit passes through zero as shown at point a in Fig. 7-7. From this point on, the marginal profit is negative, indicating a decreasing profit with additional production. Since the marginal profit is the difference between the marginal revenues and the marginal costs, it must be zero at the critical production rate. Thus, the critical production rate may be determined directly from differentiation of the profit equation when all variables are expressed in terms of production units n.

For the problem being discussed the marginal profit may be written analytically as follows:

$$dZ/dn = (MR) - (MC)$$

$$(MP) = 210n^{-0.3} - 3n^2 \times 10^{-6} + 2n/500 - 3 \qquad (7-22)$$

When this result is equated to zero as directed by Eq. (7-21) and solved by trial and error, it is found that 3150 units is a correct solution.

$$\frac{210}{(3150)^{0.3}} - 3(3.150)^2 + \frac{2(3150)}{500} - 3 = 0$$

$$\frac{210}{11.4} - 30 + 14.6 - 3 = 0$$

$$0 = 0$$

Summarizing, one finds that the *critical production rate* occurs when all of the following are true (since they each state the same thing):

(1) The slope of the total revenue curve equals the slope of the cost-of-sales curve.
(2) The marginal revenue equals the marginal cost.
(3) The additional sales return in dollars for one more unit exactly equals the additional costs to make one more unit.
(4) The marginal profit is equal to zero.

The numerical and graphical proof of these statements is demonstrated in Fig. 7-7 at 3150 units (point a) where the (MR) and (MC) curves intersect and (MP) is zero. Figure 7-6 shows that the maximum profit occurs at the same production rate of 3150 units.

The critical production does not necessarily occur at the rate corresponding to the minimum average unit cost, as shown at point b on Fig. 7-7. Similarly, the critical production rate does not necessarily occur at the point of maximum profit per unit. The average profit per unit is Z/n, defined as the difference between the unit sales price and the average unit costs (AC), or:

$$Z/n = S - (V + F/n), \text{ dollars per unit} \qquad (7\text{-}23)$$

The maximum value for Eq. (7-23) can occur at a different value of n than that for the maximum profit given by Eq. (7-20). Considering the cost data given for illustration purposes in Figs. 7-6 and 7-7 starting with Eq. (7-23) will give the optimum production for maximum profit *per unit:*

$$Z/n = 300/n^{0.3} - (n^2 \times 10^{-6} - n/500 + 3 + 10,000/n)$$

$$d(Z/n)/dn = -90/n^{1.3} - 2n \times 10^{-6} + 0.002 + 10,000/n^2 = 0$$

or by trial and error:

$$n_{opt} = 790 \text{ units}[4]$$

This optimum is different than that for the maximum profit noted at point a on Figs. 7-6 and 7-7.

The relations discussed are assembled in Table 7-1 for ready reference. This table includes some additional items that relate to profitability evaluation; these are covered in Chapters 11 and 12. Actual cost and revenue curves may show a wide variety of shapes and are not limited to those treated here, since special situations can produce curves of very peculiar appearance.

The foregoing analysis implies that production is varied within the capacity limits of a given size plant having a fixed investment but operating at various production levels, n. Accordingly, the capitalized earning rate, or ratio of Y/I, varies in a manner similar to net profit and maximizes at the same production as does the net profit. Where investment is

---

[4] Note that the same result is obtained by equating the derivative dS/dn to the derivative of the average unit cost, $d(V + F/n)$, since this is analogous to the analysis for total profit. In other words, the maximum profit per unit occurs when the slope of the average-price curve equals the slope of the average-unit-cost curve.

not linear with production rate, as is usually true, the earning rate and net profit will not maximize at the same investment levels. This is illustrated and discussed in Chapter 11. The same general relations shown in Table 7-1 are utilized when investment and production rate are nonlinear, except that the differentiation is made with respect to investment instead of production units, all of the cost and revenue relations being expressed in terms of investment, I.

EARNING RATES

The general subject of profitability evaluation is deferred until Chapter 11 as has been mentioned. However, in order to complete the summary of economic equations in Table 7-1, two earning rates will be defined.

The first is the capitalized earning rate, i_c, listed on item 6 in Table 5-4 and defined as the ratio of the annual profit, Y, to the investment, I, or:

$$i_c = \frac{100\,Y}{I} = 100(1 - t)\,\frac{[nS - (nV + F)]}{I} \qquad (7\text{-}24)$$

All terms have been defined previously.

The second earning criterion that will merely be mentioned here is the marginal profit rate, i_m, defined as the rate of change of the profit with investment, or:

$$i_m = 100\,dY/dI \qquad (7\text{-}25)$$

This rate will be discussed in more detail in Chapter 11; it is presented here so that the reader can compare it with other rates, once its implications are understood.

EARNING RATES AND THE SALES PRICE

Chapter 6 demonstrated that the demand for a product has a great influence on the sales price. Accordingly, the revenues returned to a plant operation may vary considerably with the size of the operation. The economic production chart analysis may be employed to benefit in such situations, particularly where a new product is contemplated. By estimating the costs for different size plants and then setting the earning rate required, one can propose a price-quantity curve as a model for management decisions. From these relations, management can decide what plant size allows the best economic scale of operations. An illustration follows.

TABLE 7-1. Summary of Relations for Economic Production Charts[a]

Term	Symbol and Equation	Units	Equation Number
Average net sales or revenues	S	dollars per unit	
Average variable costs	V	dollars per unit	
Average fixed costs	F/n	dollars per unit	
Average unit costs	$(AC) = V + F/n$	dollars per unit	7-7
Average gross profit per unit	$Z/n = S - (V + F/n)$	dollars per unit	7-23
Average net profit per unit	$Y/n = (1 - t)Z/n$	dollars per unit	
Marginal costs	$(MC) = d(nV + F)/dn$	dollars per unit	7-14
Marginal revenues (sales)	$(MR) = d(nS)/dn$	dollars per unit	7-18a
Marginal profit	$(MP) = dZ/dn$ or dY/dn	dollars per unit	7-21
Total sales	nS	dollars	
Total variable costs	nV	dollars	
Total fixed costs	F	dollars	
Total costs	$nV + F$	dollars	7-1
Total gross profit	$Z = nS - (nV + F)$	dollars	7-2
Total net profit	$Y = (1 - t)Z$	dollars	7-3
Capitalized earning rate[b]	$i_c = 100\,Y/I$	per cent	7-24
Marginal profit rate	$i_m = 100\,dY/dI$	per cent	7-25
Break-even point when:	$nS = nV + F$ or $S = V + F/n$		
Minimum unit variable cost when:	$dV/dn = 0$		

TABLE 7-1. (Continued)

Term	Symbol and Equation	Units	Equation Number
Minimum unit cost when:	$d(V + F/n)/dn = 0$		
Maximum unit gross profit when:	$d(Z/n)/dn = 0$		
Maximum unit net profit when·	$d(Y/n)/dn = 0$		
Minimum total cost when:	$d(nV + F)/dn = 0$		
Maximum gross profit when:	$dZ/dn = 0$		
Maximum net profit when:	$dY/dn = 0$		
Maximum earning rate when:	$di/dn = 0$		
Profit limit point when:	$nS = nV + F$ or $S = V + F/n$		

[a] Gross profit is defined as the difference between revenue and all costs for producing and selling the product. Net profit is the profit after taxes, t being the tax rate expressed as a fraction. The investment on which earning rate is computed is I. The number of units produced and sold is n.

[b] Capitalized earning rate may be based on Z or Y.

Consider Fig. 7-8 where an estimate has been made of the cost of sales at various plant sizes with 100 per cent production. Over the range being considered, the average unit costs (AC) are shown to decrease with increased capacities. This is because the larger plants operate more efficiently in the range considered, and their investments increase as some fractional power of capacity, etc. Since the earning rate is defined by Eq. (7-24), this equation may be arranged in the following form for S, the average costs, and t, the tax rate, to give [c.f., Eq. (6-26)]:

$$S = \frac{iI}{(1 - t)n} + \frac{(nV + F)}{n}, \text{ dollars per unit} \qquad (7\text{-}26)$$

where the terms are those that have been employed before. Equation (7-26) may be recognized to constitute the following statement. The average sales price for a *given earning rate* is the margin or gross profit plus the average unit costs, all based on one unit of production. Thus, another form of Eq. (7-26) is:

$$S = Z/n + (AC) \qquad (7\text{-}27)$$

Values of S for different earning rates may then be plotted as in Fig. 7-8 to provide a graphical model to show management the relationship between costs, capacity, earning rates, and prices. In this form, the economic production chart is a useful tool for supplying a large amount of information in a small space.

The chart in Fig. 7-8 demonstrates typical demand-price curves at two earning rates. As shown by the relation in Eq. (7-26), the difference in values for sales price and cost per unit is the net profit per unit. This difference will be proportional to the earning rates selected at a given production.

Using a combination of a model such as Fig. 7-8 for earning rates and Fig. 7-7 for total dollars profit, as well as any other charts, management can decide what investment, plant size, and sales price are needed to meet minimum economic requirements.

Where desired, earning-rate curves may be superimposed on any of the

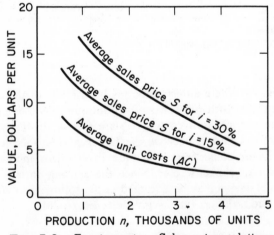

Fig. 7-8. Earning rate. Sales price relations for various plant sizes at 100 per cent production.

economic production charts as a function of sales price. Thus, in Fig. 7-1, 7-2, and 7-3 (if the sales price is constant) earning rates are straight lines directly proportional, by the relation $(1 - t)$, to the net-profit lines. For a given plant with a given investment, I, the earning-rate curves will have the same shape as the profit lines. However, where the investment changes at a given level of production (or sales, assuming all production is sold), the earning-rate curves show a different shape from that for profit. Furthermore, these shapes are all subject to the variation, if any, in unit price.

A ONE-ACTIVITY-MULTIPLE-PROCESS PROGRAM

A variation of the application of incremental analysis concerns two or more different processes (machines, plants, etc.) being used for making the same product, where costs for the different processes may vary. The problem is to select the level at which each machine (or plant) should operate to give the maximum profit. Here again the solution to the problem involves the principle of equal incremental or marginal costs. The best policy is to operate all processes at conditions where each shows the same incremental costs for one additional unit, since the same revenue is obtained regardless of which process is employed as a source of the item. The mathematical analysis for minimum total costs demonstrates that optimization occurs at operating levels where the first derivatives of all of the cost equations for the individual processes are equal.

Consider three production facilities, A, B, and C making a single product in a total amount, $n = n_A + n_B + n_C$, where n_A represents the units made on facility A, n_B represents the units made on B, and n_C those manufactured on facility C. The fixed and variable costs on these different facilities vary because operations for different processes may have different labor costs, raw-materials costs, service costs, and investment costs. However, from a knowledge of the cost relations on each facility, management can assign the optimum producing rate for each facility, thus obtaining the desired production at a minimum cost. In addition, the analysis will show how production on different facilities should be varied up or down as product demand varies from time to time.

As a general rule, operations at the same unit incremental cost for all production facilities (assuming that the sum of the output of all equals the desired total production) will give the optimum result. This analysis may be performed by trial and error from graphs, or by mathematical analysis, where the total costs for each facility may be expressed in terms of the output. At any given total production, the sum of the total costs for all facilities must be a minimum. The total cost for any producing facility is the fixed cost plus the product of the average unit cost times the output. The total cost, C_T, for three facilities, A, B, and C, would be:

$$C_T = n_A V_A + F_A + n_B V_B + F_B + n_C V_C + F_C \qquad (7\text{-}28)$$

If V and F can be expressed as functions of the outputs, if the total desired constant annual production is n (with $n = n_A + n_B + n_C$), and if each fixed cost is constant, then Eq. (7-28) can be written for a real case. (This is explained in the next section.)

At first glance, the reader may decide that marginal costs decrease with increased production, which is true at low levels of operation. However, a number of real world factors cause marginal costs to rise with production. In general, a plant operation if successful is producing at or near some best capacity. Accordingly, losses in technical efficiency, extra out-of-pocket costs, increased service charges, and other economic losses actually increase the total costs at an increasing rate (illustrated in Fig. 7-9) as the plant output is pushed beyond its normal or design capacity. Under these conditions, the marginal costs are useful in the selection of optimum plant operations.

There are, of course, situations where decreasing marginal costs apply, but this generally occurs where the production rate is considerably below the design capacity of the plant. Frequently, the marginal costs may pass

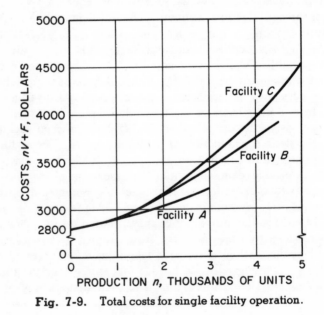

Fig. 7-9.   Total costs for single facility operation.

through a mimimum. However, at normal operating levels the marginal costs tend to increase with production.

The analysis presented here will apply in the real world, and, in fact, the analysis *concept* applies to all situations. Nevertheless, for simplicity, the analysis per se and the examples of it are usually applied only to increasing marginal costs. In the real world, this means that where all economic factors are considered, the operations are essentially at a constant cost for the initial units; then, for other production units costs slowly rise.

AN EXAMPLE OF A MULTIPLE PROCESS PROGRAM

Based on Eq. (7-28) total costs in three plants, A, B, and C, producing n_A, n_B, and n_C annual units, respectively, are:

$$C_T = (0.06n_A^{1.1} + 1200) + (0.02n_B^{1.3} + 1000) + (0.002n_C^{1.6} + 600), \text{ dollars} \tag{7-29}$$

where the values for the constant terms may represent individual fixed costs. The total cost equation with combined fixed costs would be:

$$C_T = 0.06n_A^{1.1} + 0.02n_B^{1.3} + 0.002n_C^{1.6} + 2800 \tag{7-30}$$

The minimum value of this total cost occurs when this equation is differentiated partially with respect to each variable, and the resulting equations are each set equal to zero and solved simultaneously for the optimum value of each of the three variables or some other criterion.

The combined cost equation, Eq. (7-30), is plotted in Fig. 7-9; it has been assumed that production will occur in one plant alone up to a limit of capacity of 3000, 4500, and 5000 units for facilities A, B, and C, respectively.

Relationships between costs and production are curved lines, that is, they are nonlinear. Thus, problems of this type cannot be considered in linear programming, which is discussed in Chapter 8.

All curves show essentially the same costs for production up to 1000 annual units, and any of the three facilities may be employed for this amount. However, for production up to 3000 units, the total costs are least in facility A; hence, A would be used for the operation provided the cost for making the last 2000 is less for A than it is for making the 2000 in any proportion on lines B and C. The additional cost of additional production in any facility is the sum of the costs for each additional unit of the additional production. The additional cost, therefore, is related to the marginal cost (MC) previously discussed, and rigorously it is the integral $\int (MC)\, dn$. Thus, it is obvious that for a given amount of additional production, the facility with the lowest marginal costs (MC) should be utilized. However, a complication arises when the marginal costs are not constant *because if they rise with production* it becomes economically feasible to allocate the production among the facilities. If this is done in such a manner that the marginal costs are equal for each increment of production, the minimum cost will be obtained.

For the curves in Fig. 7-9 the marginal costs are the slopes of the lines, or the first derivatives:

FACILITY A: $\quad dC_T/dn_A = 0.066 n_A^{0.1}$, dollars per unit
FACILITY B: $\quad dC_T/dn_B = 0.026 n_B^{0.3}$, dollars per unit
FACILITY C: $\quad dC_T/dn_C = 0.0032 n_C^{0.6}$, dollars per unit

These are plotted in Figs. 7-10 and 7-11 and demonstrate some small variations at productions of 200 annual units or less, which are not apparent in Fig. 7-9.

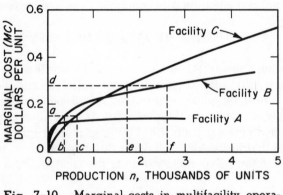

Fig. 7-10. Marginal costs in multifacility operation.

For example, these relations show that the marginal cost of the first unit on each facility for $n = 1$, is:

$$A = \$0.066, B = \$0.026, \text{ and } C = \$0.0032$$

Thus, the production levels on C at which the marginal costs equal the marginal cost for the first ones on B and A, respectively, are:

$$n_C = \left(\frac{0.026}{0.0032}\right)^{1/0.6} = 8.1^{1.67} = 33 \text{ units}$$

$$n_C = \left(\frac{0.066}{0.0032}\right)^{1/0.6} = 20.3^{1.67} = 156 \text{ units}$$

This is demonstrated on Fig. 7-11 by points b and c, respectively.

Similarly, the number of units that can be made on facility B at the same marginal cost as the first one on A is:

$$n_B = \left(\frac{0.066}{0.026}\right)^{1/0.3} = 2.5^{3.33} = 22 \text{ units}$$

as demonstrated at point a on Fig. 7-11.

Thus, the first units should always be made on facility C, since its marginal costs are least, as is further demonstrated in Fig. 7-11. After the first 33 units, however, the marginal costs on Facility C rise to equal those on Facility B, and some units should be made on Facility B. When the marginal cost of \$0.066 per marginal unit is reached, production on facility A can be made at same cost as on B and C. Thus, the least amounts that would be made for *all* machines, running at a marginal cost of \$0.066

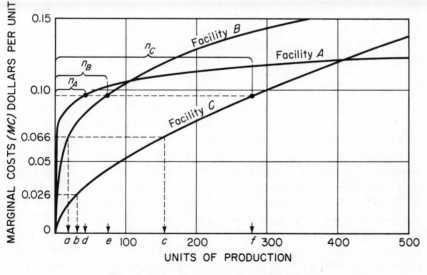

Fig. 7-11. Expanded plot of Fig. 7-10.

per unit and with minimum total cost, are: on A, one unit; on B, 22 units; and on C, 156 units; this gives a total of 179 units. By a similar procedure using trial and error until the total units equal the desired amount, a production order of a given total quantity is distributed among the three facilities.

Problem

A company wishes to manufacture 400 total items; what is the production schedule?

Solution

A straight edge is placed by trial and error on Fig. 7-11 in such a manner that $n_A + n_B + n_C = 400$ as shown, equimarginal costs being \$0.096 to give the following schedule for production and costs at points d, e, and f.

	Production	Variable Costs, Dollars
FACILITY A:	$n_A = 43$	3.78
FACILITY B:	$n_B = 76$	5.60
FACILITY C:	$n_C = \underline{281}$	$\underline{17.10}$
Total	400	26.48

Note that if all 400 units were made on any one machine, the variable costs would be:

		Dollars
FACILITY A:	$0.06(400)^{1.1} =$	44.40
FACILITY B:	$0.02(400)^{1.3} =$	44.00
FACILITY C:	$0.002(400)^{1.6} =$	30.00

From Fig. 7-10 it is observed that above 400 units the marginal costs for operation of facility A are rising slowly but are always lower than for the other facilities. If the variations in the initial period are neglected, all productions up to 3000 annual units probably are made in the actual situation with facility A, which is at its maximum capacity at 3000 units. (The first 1000 costs about the same on any facility, see Fig. 7-9; the incremental cost of the next 2000 on facility A is not much greater than production on facilities B or C in any proportion totaling 2000. Strictly, the marginal costs are lower on B up to 100 units and on C for initial productions up to about 400 units, as shown in Fig. 7-11. However, the small gain is offset by other complications caused by the necessity of running two or three facilities.)

Where more than 3000 annual units are to be made, the excess should be produced on facilities B and C at equal marginal costs. For example, if it is desired to produce 4000 annual units, what facilities should be used? The answers for production and variable costs are as follows:

Facility	A	B	C	Total
Production, n	3000	375	625	4000
Costs, $	395	44	60	499

As shown in Fig. 7-10, the extra required units above 3000 is allocated to Facilities B and C at constant marginal costs but with the restriction that the sum of the units made must equal 1000. From Fig. 7-10 by trial it is established that a horizontal line at a for common marginal costs of about $0.15 per unit will give a product at b of 375 units on facility B and a product at c of 625 units on facility C. The costs are computed from the individual component parts of Eq. (7-30).

Consider another example where 7300 units are to be made. Facility A is employed for 3000 units, leaving 4300 to be made in the other two facilities. From Fig. 7-10 by trial it is found that a horizontal line drawn at d for constant marginal costs will locate points e and f for 2600 units on Facility B and 1700 units on facility C. The tabulated results for costs are as follows:

Facility	A	B	C	Total
Production, n	3000	2600	1700	7300
Cost, $	395	530	300	1225

The total variable cost is $1225. This can now be compared with the suggestion that facility B only should be used for the excess 4300 units, since they might be made more cheaply than in facility C alone. Computing the value of $0.02n_B^{1.3}$ for 4300 units gives the additional cost as $1050, compared to the sum of $530 + $300 = $830 in the tabulation. Thus, the marginal cost model gives the lower result as it should from theory. Any combination of production units totalling 4300 that is different from the one shown above will give a higher total cost.

PROOF OF OPTIMIZATION FOR THE MULTIPLE PROCESS OPERATION

If the total cost relation, Eq. (7-30), is differentiated partially with respect to each variable, then at constant n_C:

$$\left(\frac{\partial C_T}{\partial n_A}\right)_{n_C} = 0.066n_A^{0.1} + 0.026n_B^{0.3}\left(\frac{\partial n_B}{\partial n_A}\right)$$

Since at constant $n = n_A + n_B + n_C$, the differential form becomes at constant n_C:

$$\left(\frac{\partial n}{\partial n_A}\right)_{n_C} = 1 + \left(\frac{\partial n_B}{\partial n_A}\right)_{n_C} + 0 = 0$$

or $\qquad\qquad\qquad \partial n_B / \partial n_A = -1$

Thus $\qquad \left(\frac{\partial C_T}{\partial n_A}\right)_{n_C} = 0.066n_A^{0.1} + 0.026n_B^{0.3}(-1)$

Equating this result to zero for the minimum cost shows:

$$0.066n_A^{0.1} = 0.026n_B^{0.3}$$

An analogous result is obtained if the original total cost equation, Eq. (7-30), is differentiated at constant n_B and constant n_A:

At constant n_B: $\qquad 0.066n_A^{0.1} = 0.0032n_C^{0.6}$

At constant n_A: $\qquad 0.026n_B^{0.3} = 0.0032n_C^{0.6}$

or for the condition of minimum annual cost, C_T:

$$0.066n_A^{0.1} = 0.026n_B^{0.3} = 0.0032n_C^{0.6}$$

It will be noted that each of the terms in this equality is the first derivative of the total cost expression for each plant or the instantaneous slope of a plot of annual costs versus production units, which is the marginal cost for each plant.

The relation states that the minimum total cost in producing a constant output of a given product n by three processes having different cost relations occurs at those productions where the slopes of the total-cost lines are equal. In other words, the decision for the optimum production with each process is operation at those production levels where the marginal or incremental costs are all equal.

For the relationships between the slopes for each facility, the ratio of production on two can be found in terms of any selected one (say n_A) and a solution obtained for the desired total production: Thus:

$$n_B = \left(\frac{0.066n_A^{0.1}}{0.026}\right)^{1/0.3} = 22n_A^{0.33}$$

$$n_C = \left(\frac{0.066n_A^{0.1}}{0.0032}\right)^{1/0.6} = 155n_A^{0.167}$$

Therefore, the total production in terms of one variable at minimum cost is:

$$n = n_A + 22n_A^{0.33} + 155n_A^{0.167}$$

This relation may be solved for n_A by trial and error for a stipulated value of n and the values for n_B and n_C determined from their relationship to n_A. For example, a previous illustration (page 188) for minimum cost operation required 400 total items. Setting n in the previous problem equal to 400 and trying $n_A = 43$ results in:

$$n = 43 + 22(43)^{0.33} + 155(43)^{0.167}$$

$$n = 43 + 76 + 281 = 400$$

Thus, the proper value of n_A is 43 and the second and third terms on the right side of the equation give n_B and n_C, which confirms the earlier analysis.

COMMENT ON MULTIPLE PROCESS OPERATION

If the unit marginal costs are constant at any output and if fixed charges (including standby expense of all plants) must be borne by those that are operating, then the total cost equation has no analytical minimum. The lowest production costs are obtained in the plant with the lowest unit increment costs. This situation applies where certain plants other than the one with lowest costs must be kept operating at some minimum output to maintain labor forces, to prevent freeze-up of lines, or for other reasons. These costs would become part of the over-all total costs for any other plant or plants producing the majority of the output.

When the unit-increment-cost curves are constant or linear with output (this occurs only if the respective total costs vary as the first and second power of the output) and when fixed charges on all plants must be borne by the plant operating, then the plant showing the lowest unit incremental costs also shows the minimum total costs up to an output of equal unit incremental costs. Above this output, any additional production should be made at equal unit incremental costs.

Moreover, when all plants are kept running, the optimization is made at equal unit increment costs for any production; this holds for all conditions other than linear or constant unit incremental costs with output. It is assumed in the foregoing analysis that the marginal-cost curves have the same algebraic sign; otherwise, trial and error solutions for the mimimum total costs are necessary. Thus, graphical or analytical methods may be employed to plot the marginal costs versus production for each plant. The sum of the areas of such plots for each plant's production is the total costs for all plants for their respective productions. This sum of areas is a minimum at the optimum production for each plant.

Production can be balanced among plants by varying outputs; however, in all such efforts the outputs must be within the capacity limits of the respective plants. In starting up operations, the production is varied first in the plant showing the lowest unit marginal costs (that is, the smallest slope of the total-cost curve) until an output is reached where the plant with the next lower unit incremental costs is the same as the first. Excess production is then at equal unit incremental costs. When the unit-incremental-cost curves have the same slopes at any equal incremental costs, the excess production required is distributed equally in all plants. When multiple facilities are being used, management may employ economic production charts to envision more readily the economic effects of changing the capacities up and down in individual machines.

Where only one plant is to be built for a given production and thus only one fixed charge is involved, the economic analysis is made solely on the basis of total costs, as discussed in earlier sections of this chapter. On the other hand, the incremental costs can control the total costs and can affect the selection of plant facilities where consideration is being given to operating over a range of capacities. This consideration is particularly important if fixed costs vary sharply for the alternatives.

The preceding discussion, which applies to the conditions required for minimum costs, is applicable also to the achieving of maximum profit, if the sales price per unit of output is constant. However, the effect of any price change with output must be included in the analysis of the model when maximum profits are considered.

The utilization of economic production charts and marginal cost analysis is not limited to analyses made to achieve maximum profit. For example, operations may be at a loss because of uncontrollable factors, and it may be desirable to ascertain at what conditions the losses will be at a minimum. Where the average-unit-cost curve goes through a minimum and the unit sales price drops to a value equal to the minimum cost, the operations should be carried out at the minimum cost. However, where the sales price drops below the minimum average unit cost, the least loss may result (assuming it is desired to keep the plant running) at some plant capacity below that for the minimum average unit cost because the minimum total dollar loss may be less at the lower capacity, even though the loss per unit is greater.

THE PRODUCT-MIX PROBLEM

The analysis that has been presented up to this point implies that production involves one type of item or activity. A more elaborate problem occurs when there are two or more activities, that is, a mixture of products, which compete for production time on equipment. This is known as the product-mix problem. Thus, where a textile-cloth weaving machine can be utilized for product A or B, but the time to produce A differs from that for B and the sales value of A and B differs, then there may be some combination of production for A and B that will give the maximum profit. Furthermore, products A and B may compete for more than the weaving-process time. For example, the printing or dyeing of the cloth may be involved with different times for each product being required in the dyeing process. The analysis of such problems requires programming for the best combinations of two products A and B (called *activities*) to be arranged for two operations—weaving and printing. When there are multiple activities and processes and the relations among the activities, processes, and profits vary in a linear or first order manner, the analysis is called linear programming. The solution of such problems has been developed to a very high degree by operations-research techniques for situations that otherwise would be impossible (practically) for solution. An elementary approach is given in a later chapter, but a comprehensive treatment is beyond the scope of this text. Where there are two activities and only one process involved, an analysis can be graphed on the regular economic production chart as is demonstrated in the example below. In this type of problem if all the relations among plant capacities, costs, and prices are linear with production, no optimum exists because the maximum profit occurs when the plant production is utilized for only one product. However, if any one relation is nonlinear, a maximum profit may be possible at some optimum operating policy. This optimum can occur from nonlinearity in either prices or costs for any of the products being made.

A TWO-ACTIVITY-ONE-PROCESS EXAMPLE

Problem

In a plant making either item A at a rate of 1000 pounds per hour or item B at 2000 pounds per hour, a maximum profit is desired. Both items A and B can be made in the plant to use the available production time of 2400 hours per year.

Annual fixed costs are $80,000, and other cost and sales data are as follows:

Activity	Variable Costs, Dollars per Pound	Selling Price, Dollars per Pound
Item A[a]	$6n \times 10^{-8} + 0.2$	0.50
Item B	0.15	0.20

[a] n is annual production; note that this is a nonlinear relation with all other costs or revenues directly proportional to production.

Solution

A *stepwise solution* will be shown; the results are given in Table 7-2 for different productions of item A, the production of item B taking up the remaining time. The stepwise solution will be followed by an *analytical procedure*.

Stepwise Procedure. A tabulation of values for annual costs, annual rates, and annual profits is made up as shown in Table 7-2, with the contribution to costs for each item as shown. The first step is to select the production of one of the items as a basis for production in order to draw a chart as in Fig. 7-12, where item A is the basis. Then, every hour available from not making item A is employed to manufacture item B; the costs and revenues are computed accordingly.

As an example of calculations, consider the 25 per cent A column. For a total of 2400 hours, 25 per cent is 600 hours for A and 1800 hours for B. Thus, the annual production for A is $(1000)(600) = 6 \times 10^5$ pounds and for B is $(2000)(1800) = 36 \times 10^5$ pounds. The mathematical relation is:

$$n_B = 2000(2400 - n_A/1000)$$

TABLE 7-2. Stepwise Calculations for the Two-Activity–One-Process Example

	Fractional Production Time Used For Item A				
	0	25	50	75	100
Annual production:					
Pounds of A	0	600,000	1,200,000	1,800,000	2,400,000
Pounds of B	4,800,000	3,600,000	2,400,000	1,200,000	0
Variable costs, $:					
For A	0	141,600	326,400	554,400	835,600
For B	720,000	540,000	360,000	180,000	0
Fixed costs, $	80,000	80,000	80,000	80,000	80,000
Total costs, $	800,000	761,600	766,400	814,400	915,600
Revenues, $:					
For A	0	300,000	600,000	900,000	1,200,000
For B	960,000	720,000	480,000	240,000	0
Total revenues, $	960,000	1,020,000	1,080,000	1,140,000	1,200,000
Profit, $	160,000	258,400	323,600	325,600	284,400

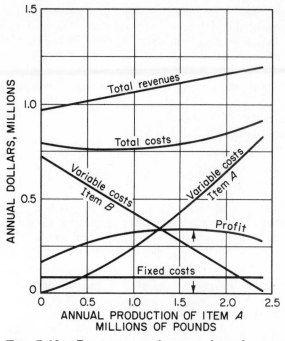

Fig. 7-12. Economic production chart for two activity example.

where n_B and n_A are annual production in pounds of items B and A, respectively. This relation is based on 2400 hours per year, the time for producing item B being time not used for producing A and vice versa.

The annual variable costs for n pounds per year is $6n^2/10^8 + 0.2n$, which is a nonlinear relation shown in Fig. 7-12. Continuing the calculations, one finds that the annual cost at 25 per cent production of A or n equal to 0.6×10^6 pounds per year gives:

$$6(0.6 \times 10^6)^2/10^8 + 0.2n = 21.6 \times 10^4 + 1.2 \times 10^5 = \$1.416 \times 10^5$$

The costs for item B are constant at $0.15 per pound, that is, for 3.6×10^6 pounds the annual costs for this item are $3.6 \times 10^6 \times 0.15 = \$540,000$. The total costs are the sum of the variable costs for each item plus the fixed costs, or a total of $761,000; this is plotted in Fig. 7-12.

The total revenues are the sum of annual income for both items A and B, as shown in Table 7-2, and the profit is the difference between the total costs and total revenues, or $258,400, as given in the table.

When all the results are plotted, the maximum profit is indicated at about 1.67 million pounds per year of item A, and this fixes production of item B at 1.47 million pounds.

Analytical Procedure. As indicated in a previous section, this optimum occurs when the slopes of the total-costs line equals that for total revenues, which permits an analytical solution if the equations for these two relations are known. Thus, the total annual costs are:

$$C_T = 6n_A^2/10^8 + 0.2n_A + 0.15n_B + 80,000, \text{ dollars per year} \qquad (7\text{-}31)$$

where n_A = annual production of item A
and n_B = annual production of item B
The total revenues are:

$$S_T = 0.50n_A + 0.20n_B, \text{ dollars per year} \qquad (7\text{-}32)$$

However, from the conditions of the problem it is possible to express n_B in terms of n_A. A total of 2400 hours is available for production, and the hours used for item A would be $n_A/1000$, since it is made at 1000 units per hour. Thus, the time available for item B is $(2400 - n_A/1000)$ hours; since it is made at 2000 units per hour,

$$n_B = 2000\,(2400 - n_A/1000), \text{ annual production}$$

Clearing this and substituting in the above equations gives:

$$C_T = 6n_A^2/10^8 + 0.2n_A + 0.15\,(4.8 \times 10^6 - 2n_A) + 80,000$$
$$= 6n_A^2/10^8 - 0.1n_A + 800,000 \qquad (7\text{-}33)$$
$$S_T = 0.50n_A + 0.20\,(4.8 \times 10^6 - 2n_A)$$
$$= 0.10n_A + 960,000 \qquad (7\text{-}34)$$

Thus, since the derivatives of these equations must be equal at the maximum profit from the principle of incremental or marginal values, they can be written as:

$$dC_T/dn_A = 12n_A/10^8 - 0.10$$
$$dS_T/dn_A = 0.10$$

Equating these last two equations results in:

$$12n_A/10^8 - 0.10 = 0.10$$
$$12n_A/10^8 = 0.20$$
$$n_A = 1,667,000, \text{ pounds per year}$$

An alternative analysis is to solve for maximum profit Z from Eqs. (7-33) and (7-34):

$$Z = S_T - C_T = 160{,}000 + 0.2n_A - 6n_A^2/10^8, \text{ dollars per year} \quad (7\text{-}35)$$

If the derivative of this equation is set equal to zero, there results:

$$dZ/dn_A = 0.2 - 12n_A/10^8 = 0$$

$$n_{A_{opt}} = \frac{0.2 \times 10^8}{12} = 1{,}667{,}000, \text{ pounds per year}$$

This confirms the result by the previous method to give the optimum annual production of item A for maximum annual profit.

NOTE: In this problem, the maximum profit does not occur at the minimum total cost. This variation is the result of total revenues varying with the product mix.

COMMENTS ON SELECTION OF A CRITERION FOR ECONOMIC ANALYSIS

The economic production chart is a model which illustrates how the criterion of annual profit varies with different costs, prices, and *production rate*. The model is described by Fig. 7-6. These models apply regardless of whether a graphical or an analytical solution for the exact optimum production is employed to give a maximum annual profit. As such, they are more useful procedures in general than optimization for minimum unit cost or maximum unit profit, as has been demonstrated. There are certain special situations, however, where these unit values may be used as criteria.

In summary, the criterion equation for the model *with costs and prices varying with production* is:

$$Z = nS - (nV + F), \text{ profit dollars per year} \quad (7\text{-}2)$$

At some optimum production rate, this relationship may maximize to give management the answer for the best operation.

Sometimes the average-sales-dollars curve varies as a rectangular hyperbola (that is the price-demand curve has unit elasticity); in such special cases, the nS term is a constant. Under these conditions, the minimum value of $(nV + F)$ for annual costs generates the maximum profit. Here, minimum annual cost can be employed as a criterion.

If a unit profit equation is written as:

$$Z/n = S - (V + F/n), \text{ profit dollars per unit} \quad (7\text{-}23)$$

it will be observed that this expression is of no special value as a criterion equation. If Eq. (7-23) is used to determine the optimum production level for maximum unit profit, the total annual profit from Eq. (7-2) may be considerably less than that obtainable by optimization of Eq. (7-2). This was demonstrated in Figs. 7-6 and 7-7 at points a and b on the charts. However, as will be developed in Chapter 9, there are examples where minimum unit cost and maximum unit profits are useful. These occur when the sales price is independent of the production rate and the costs vary with some *operational technique* or *equipment design variable* but not with *production rate*. Thus, the maximum profit for these specified conditions is made at the minimum unit cost, which also is at a maximum unit profit.

The over-all analysis for the best operation level depends, of course, on many interrelated factors. Since the analyses in this chapter do not allow for the time value of money, they constitute simplified approaches to economic feasibility. However, a more sophisticated analysis, involving the time value of money and investment level, is given in Chapter 11 on profitability.

SUMMARY

For simplicity, management in many instances may regard the costs of operation to be in either of two categories: those which vary linearly with the volume of production and those which are constant regardless of the total production. The sum of these two groups of costs is the cost of sales. Deducting the cost of sales from the income received by the company after selling the product gives the gross profit. By a consideration of profits taxes and other financial costs, management can determine the actual profit dollars received by the company. The computations made by the accountant provide a means of evaluating the economic condition of the company operations.

The relations among profits, sales, cost of sales, plant production, and selling price may be illustrated graphically by means of economic production charts for the purpose of studying and evaluating a plant's operations. By assuming the variable costs to be linear with the production rate, one may derive simple mathematical expressions to ascertain the effect of capacity upon costs and profits. Where the various cost functions are nonlinear but can be expressed in terms of production units, a differential mathematical treatment of economic production charts provides considerable information for an evaluation of plant production to determine the most economical operation.

If an empirical mathematical analysis of production charts is being utilized, two warnings should be taken into account. First, where a point of zero slope on a curve is found and employed as a criterion, the result should be tested by trial and error substitution on either side of the optimum point to ensure that the stationary point represents the maximum (or minimum) sought. Alternatively, a test for a maximum may be made by taking the second derivative of the equation. If the second derivative is negative for all values of the controllable variable, then a maximum is indicated, since the second derivative is positive for a minimum.

The second warning concerns solution of empirical equations that are not linear, or first order. In such cases, one or more answers may be possible, and some of them may be absurd. Accordingly, before an answer is accepted as final, its reasonableness should be considered; in short, is it logical from the standpoint of the conditions of the particular problem?

PROBLEMS

7-1. (a) What is the annual profit for a plant making 2000 tons per year of a product selling for $0.80 per pound. There are variable costs of $2 million at 100 per cent capacity and fixed costs of $700,000. (b) What is the fixed cost per pound at the break-even point? (c) What is the cost of sales per pound at 100 per cent of capacity?

7-2. A plant operating at 40 per cent capacity has annual fixed costs of $60,000. The variable costs are $60,000, net sales are $110,000. (a) What capacity is required to show a profit? (b) If profits tax rate is 52 per cent, what is the net profit at 100 per cent of capacity?

7-3. A plant operation shows fixed costs of $200,000. Plant capacity is 10,000 electrical appliances per year. The variable costs are $40 per unit and the product sells for $90. (a) Construct the economic production chart. (b) Compare the profit when the plant is operating only at 90 per cent of capacity with the plant operation at 100 per cent capacity where 90 per cent of capacity is sold at $90 per unit and the remaining 10 per cent of production is dumped at $70 per unit.

7-4. A plant operation with a total investment of $1 million for a production of 10,000 units per year has variable costs of $25 per unit with annual fixed costs of $120,000. The sales price per unit S varies with production as $S = 80 - 0.004n$, where n is the annual units of production. Profits tax is 52 per cent. Construct an economic production chart that shows how the profit after taxes and earning rate vary with plant capacity.

7-5. A company with annual fixed cost F of $20,000 shows the following analysis of operations for n in pounds per year, S in average sales dollars per pound, and V in average variable costs per pound.

$$\text{Annual sales dollars, } nS = 200n - 0.005n^2$$

$$\text{Annual cost of sales, } nV + F = 1.83n^3 \times 10^{-7} - 0.002n^2 + 20,000$$

Compute the production in pounds per year for the following: (a) critical production (that is, production at maximum profit), (b) Minimum unit cost of sales, (c) Maximum profit in dollars, (d) Maximum unit profit, and (e) break-even point.

7-6. Three different plants, A, B, and C are producing one million pounds of a plastic product annually. The production n, the variable costs, and the other constant costs in each plant are related by the following equations, where n is given in thousands of pounds.

Total Costs, Dollars per Year

PLANT A:	$120\,n_A^{1.2} + 60,000$
PLANT B:	$65\,n_B^{1.4} + 30,000$
PLANT C:	$14\,n_C^{1.3} + 100,000$

(a) Determine the optimum production at each plant.

(b) Determine the total annual costs for the million pounds at the optimum operation.

(c) Arbitrarily elect to produce one third million pounds in each plant; determine the total annual costs to compare with those in (b).

(d) Arbitrarily select any two of the three plants and produce one half million pounds in each plant; compare the total annual costs with those in (b) and (c).

7-7. A plant produces two products X (at 1200 units per hour) and Y (at 1800 units per hour) with any combination of amounts of each being made, up to the available production time of 3000 hours per year. Annual fixed costs are $100,000 per year. Other costs are as follows:

	Variable Costs, Dollars per Unit	Selling Price, Dollars per Unit
PRODUCT X	0.18	0.25
PRODUCT Y[a]	$5n \times 10^{-8} + 0.25$	0.48

[a] n is annual production units of Y.

What is the optimum distribution of production?

7-8. You own a plant operating three production lines A, B, and C, which have different monthly costs and capacities according to the following table, where N_A, N_B, and N_C, are production of the same product made on each line. (Note that certain costs are fixed regardless of production.)

Plant	A	B	C
Monthly costs, $	$0.05N_A^{1.1} + 1200$	$0.014N_B + 1000$	$0.003N_C + 600$
Maximum output, lbs/month	3000	4000	5000

(a) An analysis of this problem is called an "operations-research" type. Explain why in one short paragraph.

(b) Plot the total cost equation and explain from this graph (and any others necessary) how you would cut back production on each line; start with a maximum capacity of 12,000 units per month, then go down to a 6600 rate, and finally to a 3000 rate. State which line would be operating for the three rates, the capacities of each line, and the basis for your answers.

LINEAR PROGRAMMING

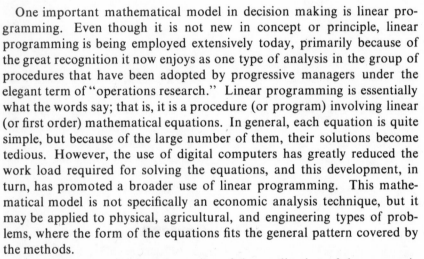

One important mathematical model in decision making is linear programming. Even though it is not new in concept or principle, linear programming is being employed extensively today, primarily because of the great recognition it now enjoys as one type of analysis in the group of procedures that have been adopted by progressive managers under the elegant term of "operations research." Linear programming is essentially what the words say; that is, it is a procedure (or program) involving linear (or first order) mathematical equations. In general, each equation is quite simple, but because of the large number of them, their solutions become tedious. However, the use of digital computers has greatly reduced the work load required for solving the equations, and this development, in turn, has promoted a broader use of linear programming. This mathematical model is not specifically an economic analysis technique, but it may be applied to physical, agricultural, and engineering types of problems, where the form of the equations fits the general pattern covered by the methods.

Linear programming is an extension of the application of the economic production charts explained in Chapter 7. In that explanation, the economic analysis dealt with contexts where a limited number of variables were under managemental control at one time. For example, profits were analyzed over a range of production for *one specific* operation, that is, one revenue situation. From this simple case, the analysis went to variation in the profits for different sales prices and for different sizes of plants; at the most, two different products were considered in a specific program. In Chapter 7, the relations handled were limited to linear or nonlinear ones in terms of an individual situation. On the other hand, the present chapter treats true linear programming, that is, analyses dealing with *combinations* of operations where the known relations are linear.

The basic principle in linear programming involves a set of two or more linear equations where more than one possible answer exists for each variable, that is, management has a choice of situations, operations, or procedures. The problem is to find the best or optimum values for the variables under conditions of constraint. This optimum is ordinarily for a minimum cost or maximum profit in management studies, but linear programming is not necessarily limited to an economic application where optimum monetary considerations are sought. Only the simplest situations are considered here to demonstrate the method and principles; the reader is referred to the bibliography for other references and a more extensive treatment. The application is best explained by an illustration.

A TWO-PROCESS–LINEAR-PROGRAM EXAMPLE

A company producing electronic devices utilizes a certain machine, I, to produce two different items, A and B. Then, these items are processed on a second machine II to render them resistant to corrosion. The capacity factor for machine I is 97.4 per cent (that is, it is in production 97.4 per cent of the time), or $0.974 \times 24 \times 60 = 1400$ minutes per 24 hour day. Machine I, requires 10 minutes of machine time for item A and 8 minutes for item B. Similarly, the capacity of machine II is 1200 minutes per day; item A requires 5 minutes of finishing machine time, and item B uses 15 minutes. The company obtains a profit of $4 for each item A and $2 for each of item B. What is the optimum schedule for production for maximum profit? (That is, what are the best daily production rates for items A and B?)

In the general language of operations, research items, A and B, are *activities* controllable by management and machines I and II are *processes* or *resources* which because of their restricting capacities provide *constraints* on the mathematical relations. The profits on each item provide a means of writing a decision equation which may be maximized in order to tell management what quantities of A and B should be produced to yield maximum profit. This decision equation is often called a *criterion* equation, or the *functional* equation, or the *objective* equation.

The first step in solving the problem involves writing an equation for each of the constraints. These are called the constraint equations, or restricting equations. Since the total quantities of both products is limited by the capacity of the machines, the equations are inequalities, that is:

MACHINE I: $\qquad 10x_A + 8x_B \leq 1400$ minutes $\qquad\qquad$ (8-1)

MACHINE II: $\qquad 5x_A + 15x_B \leq 1200$ minutes $\qquad\qquad$ (8-2)

where x_A and x_B are the units of item A and item B produced each day.

Equation (8-1) states that the total time to produce x_A quantity of item A (at 10 minutes per unit) plus the total time to produce x_B quantity of item B (at 8 minutes per unit) may not exceed the total time per day for which Machine I is available (1400 minutes). Likewise, the total time required of Machine II may not exceed the available time of 1200 minutes.

The two preceding equations may be made into equalities by inserting fictitious products or activities; these activities take up the slack when items A and B do not utilize all of the capacity. They are called slack variables, that is, they use up nonoperating or slack time, and are designated as x_1 and x_2.

$$10x_A + 8x_B + x_1 = 1400 \text{ minutes} \qquad (8\text{-}1a)$$

$$5x_A + 15x_B + x_2 = 1200 \text{ minutes} \qquad (8\text{-}2a)$$

The criterion equation may now be written for daily profit where x_1 and x_2 contribute no real profit, since they are fictitious.

$$Z' = 4x_A + 2x_B + (0)x_1 + (0)x_2, \text{ dollars per day} \qquad (8\text{-}3)$$

The solution to this problem, then, is to find the values of x_A and x_B for which Z' is a maximum. This cannot be carried out by calculus because certain possible solutions may occur at the boundary conditions as will be demonstrated below. Also there are three equations and five variables so that more than one set of values for the variables may satisfy the equations.

The equations as given constitute a set having certain mathematical properties that can be illustrated in a two-dimensional diagram such as Fig. 8-1. Their solution then follows a step by step (*iterative*) procedure to give the values for the maximum profit. As more products (activities) and processes are involved, the solution becomes more complex and recourse to more formal mathematical procedures for the iterative steps is required. These complex solutions are explained briefly later, but here the remainder of this discussion omits consideration of slack variables. Their inclusion is not necessary to an elementary understanding of linear programming, and they are included here merely to demonstrate how the inequality may be made into an equality when this is necessary. When the problem is set up without slack variables [that is, Eqs. (8-1) and (8-2) are made equalities], it is possible to analyze the feasible solution area graphically as shown in Fig. 8-1; in other words, special limiting conditions are assumed, namely, that all machine time is utilized for production on each machine. This assumption is made here only for purposes of explanation;

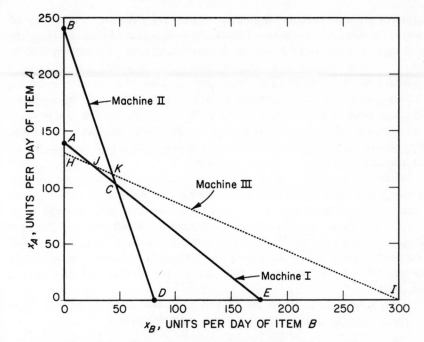

Fig. 8-1. Constraint equations for linear programming.

later it will be pointed out that optimum operations do not make use of all machine time when slack time is permitted.

Returning to the equations of the example problem and plotting them in Fig. 8-1 uncovers a number of principles. For instance, when all of the capacity of machine I is to be employed to manufacture both items for all of its available time (no slack), Eq. (8-1) and Eq. (8-2) may be considered equalities. When rearranged, these equations result in:

MACHINE I: $$x_A = 140 - 0.8x_B \qquad (8\text{-}1b)$$

MACHINE II: $$x_A = 240 - 3x_B \qquad (8\text{-}2b)$$

These are linear (first order) equations, which may be plotted as in Fig. 8-1; they bound an area $OACD$, which represents all possible or *feasible combinations* of daily production of items A and B *on the two machines*, with or without slack on either or both machines. The area $OACE$ represents possible production on machine I, but since no production of item B in excess of 80 (point D) can be made on machine II (even when zero items per day of item A are made), the area to the right of line CD is outside the limits of possible answers. Similarly, this holds for the area above line AC where the capacity of machine I limits the production of item A. Thus, the line segments AC and CD are important boundaries. Further, it will be shown that the points A, C, and D are important points (called extreme points) and represent possible final answers when the criterion equation for maximum profit is superimposed on Fig. 8-1. (The reader should note that any lines or areas that lie outside the limits of possible answers, as defined by constraining equations, are not within the feasible area; that is, any points outside the area $OACD$ are not realizable under the constraints or limitations of the problem as formulated.)

Where the conditions of no slack permitted on either machine are proposed, this further limits the possible operations. In fact, there is only one combination of productions (at point C) where both machines are to be used to full capacity. This point is obtained by simultaneous solution of Eqs. (8-1b) and (8-2b). All other possible production rates on line segment AC correspond to slack time on machine II, whereas, the operation of machine I is not brought to full capacity for production rates on line segment CD.

If Eq. (8-3) is transposed into a new form for any conditions of production where x_1 and x_2 contribute nothing to the profit:

$$x_A = Z'/4 - (2/4)x_B$$

or $\qquad\qquad x_A = 0.25Z' - 0.5x_B \qquad\qquad\qquad (8\text{-}3a)$

there results the *isoprofit* line for any value of daily profit Z' in dollars per day. Thus, at any one constant value of Z' all values of daily production of x_A and x_B that produce the same daily profit will be on a straight line in Fig. 8-2 as given by Eq. (8-3a). For example, assume x_A equal to 140 (at point A); then x_B is zero, and $Z' = 140/0.25 = \$560$. This profit could also occur when x_A is zero and $x_B = 140/0.50 = 280$ units (point F on the diagram), if it were not for the restriction of the solution area to $OACD$. The same profit could be achieved by any combination of A and B represented by the dashed line AF, if these quantities could actually be produced. Similarly, at point D where $x_A = 0$ and $x_B = 80$, the profit $Z' = (80 \times 0.5)/0.25 = \$160$; this same profit is made for all combinations of production on the isoprofit line DG where G is the production of x_A when $x_B = 0$. Note that all the isoprofit lines are parallel with a slope of 0.5; this is the ratio of the coefficients of the two activities x_B and x_A in the profit equation. Note also that since all profit lines are parallel, the greatest profit is achieved where a feasible solution intersects an isoprofit line at the highest value of the isoprofit line. For this problem, the intersection occurs at point A where no item B units are

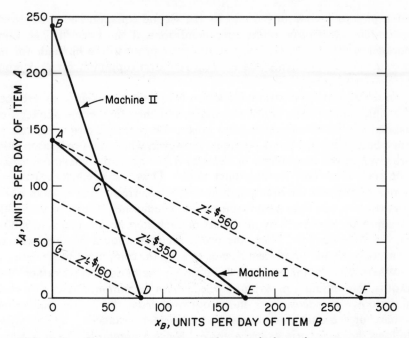

Fig. 8-2. Constraint equations and isoprofit lines for a two activity linear program.

made. However, if the slope of the isoprofit lines had been greater than that for line BD, then the greatest profit would be obtained when no item A units were produced, as would be designated by the intersection of the maximum isoprofit line at point D. (These conclusions result from the fact that the highest profit lines occur at the greatest distances outward from the origin.) In addition, if the slope of the profit lines had been less than that for BD but greater than that for AC, then the optimum program for maximum profit would occur at point C, where 104 units of A and 45 units of B would be manufactured each day. Thus, the extreme points A, C, and D for feasible solutions become significant.

Should the slope of the isprofit lines be the same as either the line segments AC or CD, which bound the feasible area of Fig. 8-2, then any and all combinations of productivity that lie on the particular line segment will yield the same profit; that is, there are a large number of possible numerical answers for the mathematical model of the real problem.

From the foregoing it is apparent that in a linear programming operation, choices must be available to management, different economic results being applicable for each selection. Thus, in Fig. 8-1 if two constraint equations superimposed on a line such as AE, there would be no additional restriction of the area by the second equation and it would be redundant. Similarly, under such conditions, if the isoprofit line were parallel to the line AE in Fig. 8-2, the same profit would be made for all production arrangements and no best selection (optimum for maximum profit) would exist.

In setting up the constraint equations care must be taken to ensure that the dimensions of the constraint limits and the other terms agree. For example, if the constraint is total hours, the equation terms for item A must be x_A times the hours per item. However, where the production data are given in the usual terms of items of A per hour, the constraint equation coefficients must be in proper units. Thus, the coefficient times x_A in items will make the equation dimensionally correct. Where small numbers (that is, less than one) result for the coefficients, it may be desirable to adjust the equations by using *lots* of each item to get larger numbers for the time per lot. Then, the profit function should be expressed in terms of profit per lot of item A, etc., to keep the computations straight.

Where the criterion equation is set up on the basis of costs rather than profits, the optimum program is selected for minimum costs. For such situations, the isocost lines move outward from the origin, decreasing values for the daily costs being at the greater distances. The analysis follows the same form as given above. Since a minimum cost is a maximum negative cost, the linear program analysis for minimum cost is sometimes treated as a maximization one by changing all signs for costs to negative in the criterion equation. This simplifies the computation procedure in complex situations by permitting the analysis to be made in the same manner for all problems.

Note that the preceding discussion assumes no slack variables, since each additional variable complicates the solution. However, this does not mean that no slack time is possible, for as was shown at point A unused time on machine II would be slack time since none of B was made. As pointed out previously, the slack variables are introduced as a necessity where more formal mathematical analytical procedures are required because of the large number of equations involved.

MULTIPLE PROCESSES

Further constraints may be placed on the operation, such as the addition of a packaging operation (machine III) to the previous example. This merely adds another equation to plot on Fig. 8-1. The plot of this equation may or may not reduce the feasible area of operations, depending upon the location of the new equation. If the profit lines are not affected (same profits per unit as before), their positions and slopes will remain the same. For example, assume that a packing machine runs for 1300 minutes per day, that item A uses 10 minutes per item, and that each item of B absorbs 4.34 minutes of packaging machine time. Thus, another constraint equation is added to Eqs. 8-1 and 8-2, or:

$$\text{MACHINE III:} \qquad 10x_A + 4.34x_B \leq 1300 \qquad (8\text{-}4)$$

$$10x_A + 4.34x_B + x_3 = 1300 \qquad (8\text{-}4a)$$

Neglecting the slack variable yields:

$$x_A = 130 - 0.434x_B \qquad (8\text{-}4b)$$

As in previous discussions, the slack variable is omitted here, since it is required only for more complex presentations. Equation (8-4b) may now be plotted on Fig. 8-1 as the line HI. The immediate effect is to reduce the previous area of feasible solutions in this case, since the new area is $OHJCD$. Two new extreme points appear at H and J, and the one at A is no longer feasible because of packing-capacity limitations. For some situations where the new constraint equations lie outside of the area $OACD$, additional processes will not affect the previous result in any manner. In the present case, however, the introduction of the third restraint results in machine II not being used to its full capacity, if point J represents an optimum schedule. On the other hand, if point C represents an optimum schedule, machine III will not be utilized to its full capacity.

Another result of the packing-process limitation is that the maximum profit cannot now reach \$560 because the extreme point J at $x_A = 118$ and $x_B = 27$ determines the profit maximum; this amount is:

$$Z' = 4(118) + 2(27) = \$526 \qquad (8\text{-}5)$$

By observing that the slope of the profit line is greater than that for line HI, one could reason intuitively that the intersection at point J would be the point of highest profit, where an isoprofit line touched the feasible area. Thus, the intersection at J would indicate the optimum program for maximum profit; that is, when all extreme points and the slopes of the line segments, HJ, JC, and CD, connecting them, are considered, the intersection at point J must represent the isoprofit line of highest feasible value.

Additional processes could be added to the plot in the same manner employed for including the packing machine in the analysis.

There are a large number of isoprofit equations [Eq. (8-3a)] that pass through any extreme point such as C of Fig. 8-2. Each different equation has different coefficients for profit per unit of A and B and gives a different total profit, depending upon how the profit line is rotated. For example, the isoprofit line may be pivoted around point C within the limits of where it superimposes upon line AE and upon line BD. For these arrangements, the unit profits for items A and B are different from the values in the original problem. However, the isprofit line could be pivoted around point D (profit per unit of B would remain constant), until it coincided with line BD. In this case, the unit profit of item A must increase. Similarly, the isoprofit line can pivot around point A with a decrease in unit profit of item B without changing the optimal solution at point C. Thus, a range of values, which have the same optimal program at point C, may apply for unit profits of A and B.

MULTIPLE ACTIVITIES

The addition of more items of production (new activities) to the set of equations previously discussed is quite a different matter from merely adding processes (constraints). For each new variable, or activity, the chart in Fig. 8-1 must be expanded in a new dimension. With a three-activity program, the feasible area of Fig. 8-1 becomes a feasible space in three dimensions and boundary points become boundary lines. This leads to the solution of three simultaneous equations for extreme points in the space, two equations for those boundary lines being on the coordinate base planes when algebraic solutions are employed. The detail involved in keeping the results in order is obvious, and graphical solutions are indicated unless formal iterative procedures are available.

The diagram for more than three different items or activities requires vector analysis. Certain techniques (simplex method, modi method, etc.) for solving these multiple activity problems are available, and the bibliography at the end of this text may be consulted for them. When 20 products and 5 or 10 processes are involved, even these special techniques are tedious; however, the use of digital electronic computers provides solutions at a reasonable cost.

Including slack variables to allow for the inequalities in the constraint equations is the same as adding activities. Therefore, slack variables were omitted in the previous discussion and in Fig. 8-1. However, in iterative methods for solving the problem, slack variables must be considered in order to take up the slack when all available capacity is not employed for some optimum program. These slack variables are required for the mathematical manipulations. This type of solution is demonstrated later in the chapter.

The smaller number, of either activities or processes, is a controlling factor in fixing the number of processes to employ at full capacity or the number of activities to be produced. Thus, for two activities and three processes, only two processes are utilized fully; however, for three activities and two processes, only two activities are manufactured at full capacity. Where the number of activities and processes are the same, then all activities *can* be made and all processes *can* be allowed to operate at capacity.

ALGEBRAIC SOLUTION OF THE TRIPLE-PROCESS EXAMPLE

The graphical solution of the previous problem shown in Fig. 8-2 provides a visual aid; however, graphical solution is not necessary, since the problem can be solved by algebra. The available equations are:

MACHINE I: $\quad x_A = 140 - 0.8x_B$, units per day $\qquad$ (8-1b)

MACHINE II: $\quad x_A = 240 - 3.0x_B$, units per day $\qquad$ (8-2b)

MACHINE III: $\quad x_A = 130 - 0.434x_B$, units per day $\qquad$ (8-4b)

CRITERION: $\quad x_A = 0.25Z' - 0.5x_B$, units per day at a
$$\text{given profit} \qquad \text{(8-3a)}$$

DAILY PROFIT: $\quad Z' = 4x_A + 2x_B$, dollars per day $\qquad$ (8-3)

The procedure is to locate the extreme points, H, J, C, and D by algebraic solution without the graph and then to determine which gives the maximum profit. This may be done as follows: The maximum feasible boundary point for x_A is the lowest value of x_A when x_B is zero, and this maximum point is obtained by inspection from Eq. (8-4b) to be 130 units, which checks point H. The maximum boundary point for x_B is the lowest value of x_B when x_A is zero; x_B is obtained by solving from Eqs. (8-1b, 8-2b and 8-4b) or:

MACHINE I: $\qquad x_B = 140/0.8 \quad = 175$ units

MACHINE II: $\qquad x_B = 240/3.0 \quad = 80$ units

MACHINE III: $\qquad x_B = 130/0.434 = 300$ units

The lowest, or controlling, value is obtained on machine II at 80 units of item B, which checks point D on Fig. 8-1.

The other possible extreme points are obtained by solving simultaneously the process equations in pairs to determine their points of intersection, which must have positive values. Thus:

MACHINES I AND II:

$$140 - 0.8x_B = 240 - 3.0x_B$$
$$2.2x_B = 100$$
$$x_B = 45$$
$$x_A = 104$$

These are the coordinates of point C of Fig. 8-1.

MACHINES I AND III:

$$140 - 0.80x_B = 130 - 0.43x_B$$
$$0.37x_B = 10$$
$$x_B = 27$$
$$x_A = 118$$

These are the coordinates of point J of Fig. 8-1.

MACHINES II AND III:

$$240 - 3.0x_B = 130 - 0.43x_B$$
$$2.57x_B = 110$$
$$x_B = 43$$
$$x_A = 111$$

These are the coordinates of point K of Fig. 8-1.

At this stage in the solution without the graph it is not known which points are feasible. Without a graph such as Fig. 8-1 each of the points of intersection must be checked for feasibility as shown in Table 8-1.

TABLE 8-1. Determination of Feasibility at Intersection Points in Example Problem (see text, page 208)

Point Coordinates		Point	Machine I $10x_A + 8x_B$ Limit: 1400	Machine II $5x_A + 15x_B$ Limit: 1200	Machine III $10x_A + 4.34x_B$ Limit: 1300	
x_A Units	x_B Units	Fig. 8-1	min.	min.	min.	Feasible
118	27	J	1400	995[a]	1300	yes
111	43	K	1454[b]	1200	1300	no
104	45	C	1400	1200	1235[a]	yes

[a] This process not used at capacity. [b] Capacity of this process is exceeded.

Feasibility is checked algebraically by substituting any two sets of values obtained from the solution of two equations into the third equation.

The next step is to determine the optimum solution by finding the maximum profit for all feasible solutions. These are computed from the results shown in Table 8-2.

TABLE 8-2. Profit Calculations for Extreme Points in Example Problem (see text, page 208)

Point, Fig. 8-1	x_A Units	x_B Units	$4x_A$ Dollars	$2x_B$ Dollars	Total Dollars
H	130	0	520	0	520
J	118	27	472	54	526
C	104	45	416	90	516
D	0	80	0	320	320

The optimum program for maximum daily profit is a production of 118 units of A and 27 units of B, which will use up all the availability capacity of machines I and III. However, machine II will be idle part of the time. A practical result of this analysis is to show that the machine II operation is not a bottleneck, and certainly no attempt should be made to increase its capacity.

Of course, the graphical solution is considerably more clear than the algebraic method, since it permits visual observation of the feasible areas and may reduce the amount of necessary calculations.

THE SIMPLEX PROCEDURE

The algebraic procedure has limitations and is complex to use with more than two or three variables. Accordingly, other methods of solution have been developed. However, only the rudiments of the *simplex* method will be presented, although it has many refinements. Reference to Metzgar in the bibliography is valuable for further study. Although the procedure here is quite similar to that author's, others employ numerous variations of the exact form.

In the previous section, the problem was to determine the level of production of items A and B that will maximize the profit according to Eq. (8-3), subject to the constraints of Eqs. (8-1) and (8-2). This can be accomplished by the simplex method of iteration, or step by step solution, for any number of activities and processes.

The simplex procedure is a formal technique based on matrix analysis, which can be followed readily without a background knowledge of the mathematical theory involved. Furthermore, because it is a stepwise, or iterative, procedure, the model can be programmed readily on digital computers.

In general, after the basic equations have been established, the simplex method consists of performing certain operations according to the subsequent procedure. The operations are performed utilizing the coefficients of the variables in each equation. The terms mentioned in the following general problem are defined later when employed in an illustration that clarifies the procedure (page 215). The present problem is given to outline the steps; these steps will take on meaning when they are applied to the specific example.

Problem

What is the procedure for the simplex method of solving linear programming?

Solution

Step a. An *initial matrix*, or table, is set up based on a starting point of no production. This table consists of all of the equations converted into equalities by the use of slack variables, such as given by Eqs. (8-1a), (8-2a), (8-3), and (8-4a). As pointed out previously, the slack variables correspond to idle time on each process. (Some starting point must be selected, and since certain information is known at the point of no production, it is utilized as the first matrix).

Step b. The *index row* is computed as the last row after the initial matrix is set up by inspection. The numbers for the index row are then computed as follows:

$$\text{Index row number} = \sum (\text{column number in row } i \times \text{objective}$$
$$\text{column value in row } i) - \text{column heading value} \qquad (8\text{-}6)$$

These index numbers then determine the direction of the next step. The most negative value for the index row under the body or the identity of the matrix generally indicates the variable to consider for obtaining the fastest (minimum number of iterations) approach to an optimal solution. The column containing this number is designated as the *key column*, and the variable heading of this column is then placed in the stub by removing the variable in the key row.

Step c. The *key row* is then determined in order to obtain the *key number* for the performance of the next iterative steps. The key row is determined by the row with the lowest nonnegative value for all quotients of the number in the constant column divided by the number in the key column in the same row. The key number is the number in the preceding matrix that appears at the intersection of the key column and key row. The variable in the stub for this key row is then replaced by the variable determined in step *b* above.

Step d. A *new matrix* is set up similar in form to the initial one by first determining the *main row* of the next table and then filling in the other entries of the matrix. The main row appears in the same position as the key row of the preceding tabulation, and is obtained merely by dividing the values of the key row in the preceding matrix by the key number.

Step e. The *remaining numbers in the new matrix* are obtained from the following relation for the constant column, the body, the identity, and the index row.

New =

$$\text{old} - \frac{(\text{key row number in same column}) \times (\text{key column number in same row})}{\text{key number}}$$

$$(8\text{-}7)$$

Step f. *An inspection is made* to determine if the solution is complete. When all the numbers in the index row are positive in the columns below the body and identity, an optimal solution has been reached. If a negative value appears, all procedural steps starting with step *b* are repeated until a positive value for all entries in the index row is obtained.

NOTE: This analysis has considered only the most elementary type of simplex analysis in order to introduce the reader to the technique. There are many refinements, approximations, and short-cut procedures that apply in special cases. The reader is referred to Metzgar, previously cited, for an elaboration, and to other references in the bibliography at the back of this text.

APPLYING THE SIMPLEX PROCEDURE

Applying a two-process–linear program example (page 203), using only Machines I and II, and omitting packaging, the problem may be stated as follows. It is desired to maximize Eq. (8-3):

$$Z' = 4x_A + 2s_B, \text{ dollars per day} \qquad (8\text{-}3)$$

subject to constraint with slack variables x_1, and x_2 as given by Eqs. (8-1a) and (8-2a).

$$10x_A + 8x_B + x_1 = 1400 \text{ minutes} \qquad (8\text{-}1a)$$

$$5x_A + 15x_B + x_2 = 1200 \text{ minutes} \qquad (8\text{-}2a)$$

The first row in Table 8-3 is the same as the criterion equation given by Eq. (8-3), and is called the *criterion row,* or the *objective row.* This is the same in all matrices.

The second row shows the variables for which the numbers are the co-efficients, and is called the *variable row* which remains the same for all matrices.

The third and fourth rows are the same as Eqs. (8-1a) and (8-2a). Any other equations containing constraints would be added as specified in the problem. Each equation adds an additional *problem row.*

The last row is called the *index row* and is computed according to step b of the simplex procedure. (For the case where the *objective column* con-

TABLE 8-3. Initial Matrix for Example Problem

	Objective Column	Stub	Constant Column	Body			Identity		Check
OBJECTIVE ROW			0	$(4)^a$	2	0	0		6
VARIABLE ROW				(x_A)	x_B	x_1	x_2		
	0^b	x_1	1400	$(10)^c$	8	1	0		1419
	$\overline{0}$	x_2	$\overline{1200}$	$\overline{(5)}$	$\overline{15}$	$\overline{0}$	$\overline{1}$		$\overline{1221}$
INDEX ROW			0	(-4)	-2	0	0		-6

[a] This is the key column indicated by parentheses. [b] This is the key row as underlined.
[c] This is the key number.

TABLE 8-4. First Iteration for Example Problem (cf Table 8-3)

		0	4	2	0	0	6
		x_A	x_B	x_1	x_2		
4^a	x_A	140	1	0.8	0.1	0	141.9
0	x_2	500	0	11	-0.5	1	511.5
		560	0	1.2	0.4	0	561.6

[a] This is the main row put in the table to replace the key row of Table 8-3. It may be any one of the problem rows and not necessarily the first row in the table.

tains all zeros as in this problem, the index row is merely the negative of the criterion row for the initial matrix.) The index row is the final computation and determines the next steps.

The first vertical column is the *objective column,* which has entries varying with the nature of the problem. The numbers are the coefficients of the variables in the criterion equation that appears in the stub. In this example, the coefficients are zero. This column is important because it is necessary in the determination of the index row.

The second vertical column of Table 8-3 is called the *stub,* and in the initial matrix this column includes only the slack variables. This will change as the solution proceeds. In more sophisticated problems, there may be other considerations not taken up here for this elementary case.

The third column is the *constant column* and lists the constants of the problem equations. The value of zero appears in the objective row because no profit is made in the initial matrix.

The columns containing real variables are called the *body* of the matrix. The *identity* is that portion of a matrix where all numbers on a diagonal line have a value of $+1$ and all other numbers are zero. This occurs in the initial table.

The last column shown in Table 8-3 is the check column and is not part of the matrix. Its purpose is to verify the iteration calculations, as will be demonstrated. It is the algebraic sum of the numbers composing the constant column, the body, and the identity. In any subsequent iteration, the numbers in the column are computed just as any other number in a column. A verification of correct computation is obtained when these calculated values agree with the algebraic sum for the row.

There are several other points to note which aid in the efficient solution of simplex programs, since a number of entries can be made by inspection.

(1) For any variable in the stub, the column with the same variable has a value of $+1$ in the same row as for the variable, and all other entries in that column are zero.

(2) A value of $+1$ occurs at the intersection of the main row of the succeeding table (Table 8-4) and of the key column of the preceding table (Table 8-3).

(3) A zero appears in the index row for any column that shows the variable in the stub.

(4) A zero in the key column at row i of Table 8-3 means that row i is identical to it in Table 8-4.

(5) A zero in the key row of Table 8-3 results in that column being unchanged in Table 8-4.

(6) The numbers in the index row in all matrices may be checked by the simplex procedural step b (page 214). However, this is only a partial check, since a zero in row i of the objective column will mask any errors in other elements of that row.

The index numbers for Table 8-3 are computed from the simplex procedural step b to give, for example, the value of -4 in the fourth column.

$$(10 \times 0) + (5 \times 0) - 4 = -4$$

An inspection of the index row indicates that the key column (by the criterion given in step b) is the one that contains x_4, which now will be placed in the stub, and a new table for the next iteration will be constructed (Table 8-4).

The key row is obtained from the procedural step c, thus:

$$x_1: \quad 1400/10 = \underline{140}$$
$$x_2: \quad 1200/5 \ = 240$$

Thus, the key row is the one containing x_1 because it is the lower non-negative value in the stub, which therefore will be removed and replaced by x_4.

The physical significance of this determination is that one of the constraint equations has a greater influence on the amount of the most profitable product, A, that can be made. However, the first equation is more restrictive on x_A, and therefore the first equation will be the controlling factor, because fewer items of A can be made.

Since the key number is now known as 10, the main row, or iteration, of Table 8-4 may now be determined, as discussed in procedural step d. The results are shown in Table 8-4.

The remaining rows in the table are obtained by following procedural step e. Thus, for the second problem row of the constant column the calculation would be:

$$1200 - \frac{1400 \times 5}{10} = 500$$

For the index row, the calculation would be:

$$0 - \frac{1400 \times (-4)}{10} = 560$$

For the fourth column, one would obtain:

THE FOURTH ROW: $\qquad 5 - \dfrac{10 \times 5}{10} = 0$

THE INDEX ROW: $\qquad -4 - \dfrac{10 \times (-4)}{10} = 0$

For the fifth column, the calculation would be:

THE FOURTH ROW: $\qquad 15 - \dfrac{8 \times 5}{10} = 11$

THE INDEX ROW: $\qquad -2 - \dfrac{8 \times (-4)}{10} = 1.2$

For the sixth column, the following would be obtained:

THE FOURTH ROW: $\qquad \dfrac{1 \times 5}{10} = -0.5$

THE INDEX ROW: $$0 - \frac{1 \times (-4)}{10} = 0.4$$

For the seventh column, the calculation would be:

THE FOURTH ROW: $$1 - \frac{0 \times 5}{10} = 1.0$$

THE INDEX ROW: $$0 - \frac{0 \times (-4)}{10} = 0$$

This completes the table. Since there are no negative numbers in the index row of Table 8-4, this also completes the analysis for best operations. The check column also verifies the calculations.

The variables in the stub indicate what the production should be for maximum profit. In this case, x_A indicates that only product A should be made, and x_2 indicates that any available time for machine II above that needed for product A will be slack time.

The numbers appearing in the third column (the constant column) of Table 8-4 show the magnitude of the corresponding variables in the stub. Thus, 140 items of A will be employed, and the slack variable, x_2 has a value of 500.

The index-row value of the constant column equal to $560 is the maximum profit that can be obtained and is the solution to the criterion equation for the model. This number was obtained in the original solution of the example problem (page 206) with only two processes.

A word of caution should be mentioned in connection with simplex program solutions; one should avoid rounding errors of numbers, since all subsequent calculations depend on the preceding numbers. For this reason, many simplex tables are carried through with the numbers appearing as fractions rather than decimals. Also, using fractions helps in the locating of errors by inspection if the denominators are not least common multiples.

The preceding discussion of the simplex procedure is called the *primal*. An alternative method in solving linear programs is the *dual* of the previous procedure. In theory, this is a mathematical property of the set of equations. The dual procedure assumes that the constraint is the available profit per unit produced. Then, the highest values per unit time on each machine are imputed or assigned as found by the problem solution. These imputed values must be such that individually and collectively they do not produce a smaller profit. If one example on page 208 is reconsidered, one finds that with three processes the constraint equations are:

ACTIVITY A: $10x_I + 5x_{II} + 10x_{III} \geq 4$ (8-8)

ACTIVITY B: $8x_I + 15x_{II} + 4.34x_{III} \geq 2$ (8-9)

where x_I, x_{II} and x_{III} are the imputed values for machine time in dollars per minute for machine I, machine II, and machine III, respectively.

The criterion equation for the dual is the total cost in dollars which is minimized and is equal exactly numerically to the maximum feasible profit found previously. For example, the criterion equation would be:

$$C = 1400x_I + 1200x_{II} + 1300x_{III}, \text{dollars} \qquad (8\text{-}10)$$

Since there are three variables, a three-dimensional feasible space is indicated. Thus, this solution is actually more difficult than the primal, which was worked on a two-dimensional area. The dual can be programmed readily by the simplex method, and often is more simple than the primal; thus, both attacks may be considered in solving a linear program which aims to determine which will give the desired result with the greater expediency.

For certain types of problems, the production, or value, of an activity must not be less than a certain amount; this constitutes a *requirement* as distinguished from a constraint. An example of this would be the case where activity, or item, A must be equal to or *exceed* 500 items. Thus, a slack variable is necessary to make an equality of the problem equation, or:

$$x_A - x_3 = 500 \qquad (8\text{-}11)$$

Since the slack variable, x_3, has a negative coefficient of -1, it cannot appear in the variable column of the identity because of the definition of the matrix identity as noted above; therefore, it appears in the body as a variable. Also, since the constant column cannot have any negative values (that is, all real solutions require the variables to be zero or positive), an *artificial* variable must be introduced to complete the identity.

The addition of an artificial variable to Eq. (8-11) results in its being considered in the objective equation also. However, since Eq. (8-11) is already an equality, any artificial variable complicates the result and must not appear in the optimum solution. To prevent its appearance in a final solution, any artificial variable is assigned arbitrarily a very large negative coefficient, $-M$, to be used in the objective equation. In effect, this $-M$

represents a large negative profit, thereby making its appearance in a solution unrealistic.

Thus, in the simplex procedure, the identity and stub include the artificial variables; their coefficients of $-M$ appear in the objective column. The slack variables with negative coefficients do not appear in the stub, since as noted above the constant column may contain no negative volues.

For problems involving artificial variables, certain preliminary algebraic manipulations often can be used to simplify the matrix; this has been discussed by Metzgar who has been previously cited.

COMMENT ON ITERATIVE PROCEDURES FOR MULTIPLE ACTIVITIES AND PROCESSES

As illustrated in the section on the algebraic solution of a simple example (page 211), the step by step (iterative) procedure is somewhat tedious even for a problem with only two products. Where any great number of products and/or processes are involved, some formal mathematical method is required to effect a solution to the problem. This is accomplished by the use of space vectors to define the feasible space and the equations which are set up in a matrix or table in a certain order. The mathematical analysis consists in rearranging the entries in the table according to a prescribed procedure based on matrix algebra and tensor analysis. Each rearrangement corresponds to the checking out of an extreme point, and eventually a best final arrangement gives the optimum solution. One general method is the *simplex* method. For special cases where the same activity can be exercised by different processes with the total of each activity fixed, then simplified procedures are available. The *transportation* method is an illustration where, for example, different plants at separate locations may supply three products (activities) to different sales points (processes). The total amount of each product is fixed, and the total capacity of each plant for different products is also fixed. These conditions set the "rim" requirements or constraint values and permit certain shortcuts in general procedures. Manufacturing and transportations costs vary for different arrangements, the program for minimum cost being sought. Other variations are available in more advanced texts, covering such problems as scheduling of fixed production for different machines, assignment of operators to machines, etc.

Where more than first order equations are required for the mathematical relations among capacities and costs, higher order programming is indicated. Although some progress has been made in applying such procedures, no further comment on them will be given here. The present chapter has tried to demonstrate only the rudiments of linear programming, thus giving the interested reader some background to proceed to more advanced techniques.

The present model is certainly a powerful tool for confirming or initiating management decisions. However, in many cases, the linear program model is only an approximation to the real world situation: considering the true situation more accurately requires higher than first order equations. Nevertheless, this does not negate the use of the linear model. It is of value as long as it performs a service for management and its limitations are known and recognized.

PROBLEMS

8-1. A furniture manufacturer is making fixtures for bureaus and tables in two styles A and B (activities) which require three operations (processes). Available times and required times for lots of each style are shown in the tabulation together with the profit per lot.

	Product A, hrs/lot	Product B, hrs/lot	Available Time for Process, hrs
PROCESS:			
1	30	17	170
2	20	10	120
3	150	60	270
Profit per lot, $:	30	23	

Determine the optimal program by (a) graphical analysis and (b) algebraic solution. (c) Solve by the simplex procedure also.

8-2. (a) Compute analytically and confirm graphically the optimal solution for maximum profit with a linear program of two activities A and B on machines 1 and 2 with the following data. Compute the profit using constraint equations based on available machine time.

Activity:	A hrs/unit	B hrs/unit	Constraint, hrs
MACHINE:			
1	2	2	150
2	1	4	100
Profit, $/unit:	1.5	3.0	

(b) Repeat the problem using constraint equations based on profit per unit where the maximum contribution to profit per hour for each machine is established. These inputed or assigned profits rates per unit of time represent machine charges that must be made to secure the same total profit in (a). They also represent the loss of contribution to profit when the machine is not used to its limit. Such an analysis is known as the dual of the solution in (a) and gives the same total profit. (c) Repeat (a) using the simplex procedure.

8-3. This problem is a comprehensive discussion dealing with the properties of the graphical analysis of a two-activity–three-process operation, with data as given in the tabulation.

		x_1, hrs/unit	x_2, hrs/unit	Constraints, hrs
PROCESS:				
	a	2	3	120
	b	2	2	100
	c	2	3	200
Profit per unit, $:		12	15	

Additional requirements are that x_1 must be made in amounts of not less than 6 units (this is a vertical line on a feasibility diagram) and x_2 in amounts of not more than 90 units. (This is a horizontal line on the diagram.)

(a) Show which constraints are superfluous or redundant and solve for optimal activities.

(b) Show how much the slope of the profit lines may vary from that given and still not affect the optimal answer for units of x_1 and x_2 (but the total profit will vary.)

(c) Explain how much the range in unit profit for product x_1 may vary without affecting optimal solution. Repeat for product x_2. What are the total profits for all four cases (that is, the limiting unit profit values for x_1, with with that for x_2 as given; and for the limiting unit profit values of x_2, with that for x_1 as given.)

8-4. In a manufacturing plant, product A requires 0.03 hour per lot to form and 0.02 hour per lot to finish. Product B requires 0.02 hour to form and 0.05 hour to finish. The forming and finishing capacities are each 40 hours per week. The profits are $0.03 per lot for A and $0.04 for B. What is the optimum program? Use both algebraic and simplex solutions.

8-5. A wholesaler handles regular orange juice selling for $0.40 per 46 ounce can and frozen juice at $0.25 for a 6 ounce can. The profit on the former is $0.28 per dozen cans and $0.14 for 24 cans of the frozen product. The maximum number of large cans that he can handle is 20,000 and the maximum number of small cans is 30,000. The costs for the large cans are distributed in a proportion of $\frac{5}{8}$ for purchase cost and $\frac{3}{8}$ for handling and storage. The costs for the small cans are proportioned between ¾ purchase cost and ¼ handling and storage cost. The wholesaler has $10,000 available to take care of orange juices, of which $3000 must be available for handling and storage. What is the optimal program?

8-6. Analyze the example problem (page 211, in the text) with three processes (machine I, machine II, and machine III) by the simplex procedure. HINT: In this analysis there will be three problem rows and three slack variables.

8-7. Analyze problem 8-5 by the simplex procedure.

ECONOMIC BALANCE ANALYSIS

As pointed out in Chapter 1, the over-all management economy model, consisting of a series of steps, includes an equation requiring analysis before the making of a decision. Very often the criterion for the equation is some type of maximum monetary return in the form of maximum profit or minimum cost (that is, minimum loss). For such problems, the procedure is straightforward, provided sufficient data are available and reasonable simplifying assumptions can be utilized.

Where management having only a few alternatives must select the most feasible one, then a simple tabulation of the cost data for each alternative often is sufficient. However, where management being able to vary the controllable variable(s) at will has many alternatives, then the rigorous mathematical model is best, and usually more informative. Nevertheless, since the rigorous equation is sometimes difficult to manipulate mathematically, trial and error solutions may be required. For some situations, the utilization of incremental procedures provides a relatively simple solution, although the incremental equation may be more difficult to formulate than the rigorous equation. This chapter proposes to demonstrate the principles pertaining to the over-all model. In later chapters, the application of these principles in decision making will be illustrated by employing models and accounting data presented in earlier chapters. Thus this chapter serves as a transition between the formulation of over-all models and the use of the model as a basis for a decision. In some situations the model can be formulated on a basis of reasonable certainty for the data. In others, however, uncertainty may enter and the probability for certain data must be considered. The elementary probability situation will be considered in a later chapter.

CRITERION CONSIDERATIONS

The criterion that management seeks in helping to make a decision is some number or combination of numbers that can be employed as an

a	= constant	min	= abbreviation for minimum
b	= constant	n	= production rate
c	= constant	opt	= abbreviation for optimum
d	= constant	p	= sales price, dollars per item
C_F	= annual fixed costs, dollars per year	Q_A	= annual production units
		Q_B	= production units per period
C_0	= annual operating costs, dollars per year	S	= sales dollars
		V	= variable costs, dollars per item
C_T	= annual total costs, dollars per year	x	= variable
		y	= variable
C'_T	= unit costs, dollars per item	Z	= profit, dollars
M	= number of mechanical units in warehouse problem		

index. This index serves as a guidepost, or dividing point, for decision making. If the quantitative numbers from the analysis are such as to indicate positive action, management can take that action with more confidence than it can on the basis of "hunches" or unsubstantiated intuition.

In general, the criterion equation states the relationship of a set of conditions (parameters of the equation) to some variable that can be controlled or manipulated by management. This variable might be the number of units to be produced, the number of machines to produce the product, the method to be used, the cost of an item from a distributor, the cost of a raw material, the interest rate for an investment, the reorder-stock level for maintaining an inventory, etc. When all the costs involved can be related to this controllable variable, an economic analysis is readily made. Usually, the objective of the analysis is to maximize the profit or minimize the costs. When all of the costs are related to the controllable variable, an "economic balance" may be found to exist. Under such conditions, an *optimum* value of the controllable variable meets the conditions set for the criterion, such as maximum profits (or minimum cost), as illustrated in Fig. 9-1. This optimum value becomes a decision or policy-making basis for management.

ECONOMIC BALANCE

Economic balance means a balancing of economic factors to give an optimum result for operation. In Chapter 7, the economic analysis was concerned with the balancing of decreasing revenues at higher sales rates against the change in costs at higher production rates. This is a special kind of economic balance where the controllable variable was specifically

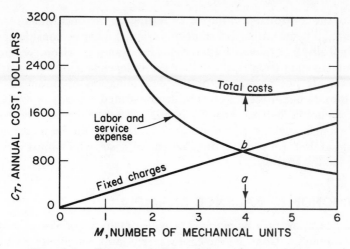

Fig. 9-1. Economic balance for warehouse problem.

limited to production rate. In a more general sense, economic balance may be considered to be a detailed study of any and all revenues, costs, expenses, savings, etc., that pertain to a given operation. Those items that increase are balanced against those that decrease, as illustrated in Fig. 9-1 where the manipulated variable is the number of machines involved. Thus, here the annual costs are evaluated in terms of a method of operation rather than a rate of production. For this special case of a fixed annual production (with a given amount of revenue), the maximum profit is made at minimum annual costs found from the model in Fig. 9-1. The annual costs may serve as the criterion for this situation, which is a common one in the real world where equipment is selected or operated at a fixed output.

To make the analysis, certain operation information ordinarily is required to relate the costs with the variable that can be controlled or manipulated by management. Thus, establishing the proper operational relation and the economic relation that approximates the real world situation are important parts of the analysis. These relations are developed in one of three ways: (a) a rigorous theoretical mathematical relation is known, (b) an empirical relation is available from past performance, or (c) in the absence of real data, if a reasonably accurate estimation of the desired relationship can be established, it may be utilized as a good approximation.

After the operational and economic relations are combined, management may make a step by step analysis in a tabular form and plot a graph, or a rigorous mathematical analysis may be used to solve for the optimum value of the controllable variable. In setting up the operational economic relation, all items (either constant or variable) must be considered to obtain final absolute values (of total costs, for example). However, any costs that are constant (that is, those that are independent of the controllable variable) are not necessary for the determination of optimum values, since they do not affect it, as will be demonstrated.

An economic balance analysis is not limited to the algebraic sum of only two lines such as shown in Fig. 9-1. Instead, the total costs may be the sum of three (or more) individual sets of costs, which must all be considered in the analysis.

AN ECONOMIC BALANCE EXAMPLE

The operation of a warehouse provides many examples of economic balance analysis. Consider the case where a large warehouse with a large labor force has certain operations which can be handled by mechanical units, such as mechanical conveyors and fork-lift trucks. The use of more mechanical units will reduce labor requirements. This reduction in labor will reduce direct labor costs, but the capital required for the mechanical units will be increased, thus also increasing the annual fixed charges. An economic balance analysis will determine the optimum number of mechanical units.

Problem

A warehouse handles a large number of items which have to be moved often for various reasons. The work can be handled by hand labor or by a combination of mechanical aids and hand labor. What is the best combination to use?

The following data are available. The cost of a fork-lift truck is $2400 and is depreciated over 10 years. The annual fixed charge for each mechanical handling unit installed, therefore, is $240. A study of operations in another warehouse at different times has shown that the average operating labor costs, including service costs on the mechanical units, vary as $3840/M$ dollars per year, where M is the number of mechanical units. It will be observed that the operating costs decrease as the number of machines increases. This is readily explained by the fact that less labor is required with more machines. However, service charges for fuel and repairs, as well as other maintenance requirements will increase with a larger number of machines. Nevertheless, the lowered labor costs are a greater influence on the operating costs. In the real world, the person making the economic analysis is required to obtain this operating-cost data by work-study procedures and to fit them as best he can into a mathematical relation. This relation can then be incorporated into the total cost model, as will be explained.

In this problem, there are three points to note: (1) the objective is to determine the minimum cost for the operation—this is the criterion, (2) all costs can be related to a common variable—the number of mechanical units, and (3) the available technical data for a mathematical relation consists of information from another similar operation; this data is assumed to be valid for the new application.

Solution

Since the objective is minimum cost for the operation, the simplest approach is to set up an equation for total annual costs, C_T, or:

$$C_T = C_0 + C_F, \text{dollars per year} \qquad (9\text{-}1)$$

where C_o = operating costs, dollars per year
and C_F = fixed charges, dollars per year
From the data of the problem:

$$C_T = 3840/M + 240M \qquad (9\text{-}2)$$

where M = the number of units installed.

With this equation, values for C_T can be computed for assumed values of M, as is shown in Table 9-1. Here, the results show a minimum total cost at 4 mechanical units, which is confirmed graphically[1] in Fig. 9-1.

The same result can be obtained through calculus by differentiating Eq. (9-2) with respect to M and setting the derivative equal to zero in order to find the minimum point for the equation, or:

$$dC_T/dM = -3840M^{-2} + 240 = 0 \qquad (9\text{-}3)$$

$$M_{\text{opt}} = \sqrt{\frac{3840}{240}} = \sqrt{16} = 4 \text{ units}$$

It will be observed in this analysis that any additional *constant* costs that might be added to Eqs. (9-1) or (9-2) would merely raise the total numbers in Table 9-1 without affecting the optimum M for minimum cost. Similarly, any constant in Eq. (9-2) would disappear in the mathematical differentiation to obtain Eq. (9-3) and thus would not affect the optimum number of mechanical units.[2]

---

[1] In Fig. 9-1 the minimum total cost and cross-over point b occur at the same value of M because of the special form of the cost equations for C_o and C_F in this problem. This special case results when only two equations are involved and one type of cost varies as a linear (straight-line) relation starting at zero and the other varies as a rectangular hyperbola. In other cases, the costs may vary as either positive or negative power functions of the controllable variable, and the optimal solution may occur at other than the cross-over point. In some cases, the economic balance equations may all be set out separately, resulting in more than two equations; thus more than one cross point would appear. This does not affect the analysis as presented.

[2] In this analysis, the example used a continuous relation for the number of machines, which obviously must be in discrete steps of 1, 2, 3, etc., machines. The example was selected to illustrate a real world application. The continuous relation must at best be only an approximation, since any answer must be an integer number of machines. The use of the continuous relation model simplifies the model and the presentation of the concept, and can be considered as a good approximation of the plot that would result for a step-function plot, using discrete numbers of machines.

TABLE 9-1. Calculations for Minimum Cost (Warehouse Problem)

Number of Mechanical Units, M	Annual Fixed Charges in Dollars, $240M$	Annual Operating Charges in Dollars, $3840/M$	Total Charge, Dollars per year
1	240	3840	4080
2	480	1920	2400
3	720	1280	2000
4	960	960	1920
5	1200	768	1968
6	1440	640	2080

THE INCREMENTAL PROCEDURE

Inspection of Eq. (9-3) indicates that it may be written as the derivative of Eq. (9-1), or:

$$dC_T/dM = dC_0/dM + dC_F/dM = 0 \qquad (9\text{-}4)$$

Each differential term of Eq. (9-4) is the slope of an individual curve of Fig. 9-1, and thus represents the incremental cost for an infinitesimal change in the number of units. (As discussed in Chapter 7, this is also the marginal value for the respective costs under study.) Accordingly, if the incremental cost equation [Eq. (9-4)] can be set up, it is unnecessary to set up a total cost equation, such as Eq. (9-2), and differentiate it. By merely taking the algebraic sum of the average incremental costs and equating them to zero, a direct solution is possible for the optimum value.[3]

The basic concept in an incremental analysis is as follows: for an optimum condition to be present, the incremental gains (profits, savings in costs, etc.) are exactly equal to the incremental costs required to obtain the gain. This means that these incremental gains and costs must be known in terms of one additional unit of equipment, input, output, time, manpower, distance, inventory, etc., whichever is the variable controllable by management. Thus, in the preceding warehouse example, the incremental cost for one more mechanical unit was $240 at any level of units involved. However, the incremental operating cost was $-3840M^{-2}$, as

---

[3] Note that the more general form of Eq. (9-4) is:

$$dC_T/dM = \sum_{i=1}^{n} (dC_i/dM) = 0 \qquad (9\text{-}4a)$$

where C_i is any individual cost that varies in some manner with M. Thus, Eq. (9-4a) may have any number of terms and is not necessarily limited to two cost relations.

had been estimated from available data; the cost is found to depend upon the specific value of M at which the calculation is made. Furthermore, the cost is shown to decrease (a negative cost change is a gain) rapidly at first, and then more slowly with the addition of a greater number of mechanical units. Presumably, as more machines are added, the savings in warehouse labor cost begin to be offset somewhat by increased service charges on the mechanical units. However, this is all included in the operating cost relation. This decreasing operating labor cost is assumed as a continuous function *for each mechanical unit added;* that is, the actual available costs (real or estimated) are plotted for each unit added, as shown in Fig. 9-2. This plot is then fitted empirically by a curve to give the cost equation that approximates the data; this is demonstrated in Fig. 9-2. The final cost equation may then be used in the economic analysis.

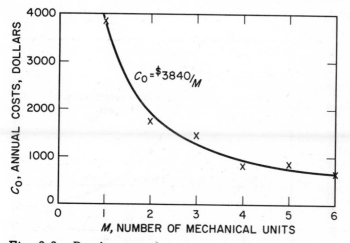

Fig. 9-2. Development of empirical relation for operating costs in warehouse problem.

The reader will recognize here that for each unit increase in the controllable variable some gain should result for the increased cost. When the incremental cost just exceeds the incremental gain, no further units should be added. At the point of equality, it is immaterial whether a unit is actually added or not, since the gain exactly balances the cost; that is, the algebraic sum of all gains and all costs for a unit change in the variable is exactly zero at some optimum value of the controllable variable. This optimum value is what is sought in order to meet the condition set up for some decision criterion (such as minimum cost, maximum profit, etc.)

With the incremental procedure, it is often more difficult to set up the equations equating the incremental expenses and incremental gains; where it can be done readily, extra effort may be saved. As a guide to setting up the analysis the following relation may be useful as a form of Eq. (9-4).

$$\text{Sum of all incremental gains, or savings} = \text{sum of all incremental costs, or expenses} \qquad (9\text{-}5)$$

With this relation, each side of the equation may be a complex equation with different powers of the controllable variable. In such a case, trial and error solutions should be employed by assuming values of the variable and substituting into the equations until both sides give the same numerical result. To expedite the solution, a plot of the incremental equations may be utilized to give a point of intersection, which also would be a solution, as demonstrated in Fig. 9-3, for the simple warehouse problem.

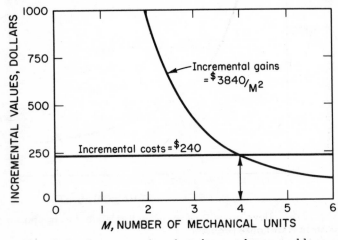

Fig. 9-3. Incremental analysis for warehouse problem.

In the warehouse problem, the incremental gains (rate of change of operating costs) were the cost savings resulting from mechanizations. These incremental gains were estimated at $3840M^{-2}$ dollars, which can be plotted for different values of M, as shown in Fig. 9-3. In addition, the incremental expense for each additional mechanical unit was known to be $240 at any M and, therefore, plots as a horizontal line in Fig. 9-3. The intersection of these plots represents the value of M at which both are equal according to Eq. (9-5); thus an optimal solution occurs at $M = 4$ units, as was found in the total cost analysis.

COMMENT ON THE INCREMENTAL METHOD

Although the incremental method is often quite readily solved after the mathematical relations are set up, certain implications must be clearly understood. For example, in the warehouse problem the question might be raised as to why it is necessary to install any more than one mechanical unit. When M is equal to one, the operating charge is decreasing at a rate of $3840/(1)^2$, or \$3840 per added unit; whereas, at M equal to two, the decrease is only $3840/(2)^2 = \$960$ per unit added. Thus, the answer is that the absolute cost of using only one mechanical unit is higher than that of two mechanical units, despite the difference in the rates of change in the operating cost at the two levels. It might also be noted that the benefits from the second unit (\$960) are still greater than the additional costs (\$240).

In other words, the relationship among the various rates of change (or any other pertinent expression for the incremental gain or cost) with respect to the common variable determines the optimum for the sum (or differences) of their total values, rather than any individual cost equation.

COMMENTS ON SELECTING A CRITERION

The economic balance analysis is a model which may take the form of charts, tables, mathematical equations, or combinations of them. A criterion of minimum annual cost, minimum loss, maximum profit, or some other basis may be selected to determine the optimal value of the controllable variable. From this result, management can establish a policy and specify a decision rule. In the example presented, the criterion was minimum annual cost for the warehouse operation. The decision rule was that for this minimum annual cost, four mechanical units should be installed.

The criterion employed in economic analysis varies with different problems, and in some specific ones can lead to complications. In Chapter 7, it was pointed out that maximum profit is often the proper criterion. However, there are situations where other criteria may be employed as will be discussed. If a general equation for profit, Z, is written as

$$Z = S - (C_0 + C_F), \text{ dollars per year} \qquad (9\text{-}6)$$

where S = annual sales dollars
C_0 = sum of all operating variable expense
C_F = sum of all fixed expense

then a criterion of minimum cost as a decision criterion will apply for all of the following situations under the assumption that all of the production or service generated is sold. For example, where annual sales dollars, S, are constant, obviously the maximum annual profit is obtained when $(C_0 + C_F)$ is a minimum. Similarly, if the unit sales price is constant, the annual sales dollars are constant for a given *constant annual production,* and minimum total costs at that production will also give maximum profit. Where the annual sales, S, and the costs, C_0 and C_F, are all expressible as some function of a controllable variable such as annual production, these relations are substituted in Eq. (9-6); the analysis for maximum profit is then determined by differentiation. In general, the maximum profit dollars are sought; however, where the constant components of Eq. (9-6) can be disregarded, their omission does not affect the optimal solution for maximum profit per unit of time. In certain cases, where continued operation at a loss is desirable, Eq. (9-6) is valid, since a loss can be considered as a negative profit. This loss might occur under conditions where a minimum loss is possible. The minimal loss operation is found by differentiation as before, since the same mathematical procedure is followed for either a maximum or minimum analysis.

If both sides of Eq. (9-6) are divided by the total number of units involved, the unit equation results. This is sometimes used for a criterion equation:

$$Z/Q_A = S/Q_A - (C_0 + C_F)/Q_A, \text{ dollars per unit} \qquad (9\text{-}7)$$

where Q_A = annual units of production.

Inspection of this equation shows that *for a constant value of Q_A* the analysis will give the same results as an analysis using Eq. (9-6), since all terms are divided by a constant, Q_A. Accordingly, the criterion may be based on profit per unit or costs per unit, employing Eq. (9-7), as was discussed in Chapter 7. Where all the costs are expressed in terms of the annual production, Q_A, the latter may be the controllable variable; either Eq. (9-6) for maximum annual profit or Eq. (9-7) for maximum profit per unit may be utilized, depending upon the information sought. The results obtained from using these two equations could be quite different, depending upon how annual sales, Q_A, and the unit sales price, or costs, vary.

Although, in general, for the short run a criterion of maximum annual profit is desirable, for the long run maximum profit per unit may be a better policy. This is particularly true where input resources are limited in availability. (Consider such items as petroleum oil, gold-bearing minerals, etc.) Here, the problem of optimum yield and recovery enters to complicate the analysis.

Economic balance analysis covers a wide field of management economy, with many and varied problems amenable to mathematical treatment. To establish the best criterion and decision policy, each new problem requires a separate analysis according to the principles developed in this chapter. There is no one solution or formula into which all problems can be sorted. Many follow the same pattern, but the analyst ordinarily must set up his own solution based on a knowledge of the basic principles involved.

DIRECT SOLUTIONS BY MATHEMATICAL EQUATIONS

In an analysis for economic balance, if the complete equation with real numbers, such as Eqs. (9-6) and (9-7), is set up, then the relative magnitude and importance of the individual terms can be evaluated. To simplify the procedure, terms having a minor influence may be observed by inspection and dropped from the analysis before differentiation for the optimum evaluation. This is the preferred procedure rather than a blind substitution in an equation. However, it is interesting and valuable to check results by both methods, where a direct-solution equation is available. Special cases where the data may fit directly are now considered.

* * * (*Optional*) * * *

In a general form, the profit per unit time (usually one year) is given for the difference between revenues and expense by Eq. (9-6). A more useful equation, where the terms are in values of units of production (or fractional capacities), is as follows:

$$Z = np - (nV + an^b + C_F), \text{ annual dollars} \qquad (9-8)$$

where n = the numbers of items made and sold
  p = the sales price per item; it also may vary with n
  V = the variable cost per unit, which may be constant or vary with n
  C_F = annual fixed charges, which may vary with n
  a and b = constants and the term an^b represents costs that change (usually increasing) with production particularly at high production rates, where decreased efficiency results from overloads. In some cases, these costs and V may be included together as one term.

Equation (9-8) is a basic equation, which may be manipulated mathematically to give certain generalized expressions useful on occasion for checking other solutions made in a stepwise manner. Several forms will be shown for maximum annual profit, minimum annual costs, and minimum unit costs. Note that for production rate any time basis such as one hour, one day, or one year may be employed in these equations without affecting their form.

Maximum Annual Profit

The critical production rate of Chapter 7 is found by differentiating Eq. (9-8) with respect to n and equating to zero:

$$dZ/dn = d(np)/dn - d(nV)/dn - abn^{(b-1)} - dC_F/dn = 0 \qquad (9\text{-}9)$$

When p, V, and C_F are constant, the following equation results for the optimum n:

$$p - V - abn^{b-1} = 0$$

$$n_{opt} = \left(\frac{p - V}{ab}\right)^{\frac{1}{b-1}}, \text{ items per year} \qquad (9\text{-}10)$$

When p, V, and/or C_F are functions of n, these functions must be incorporated in Eq. (9-8); the differentiation is performed as indicated in Eq. (9-9), from which the optimum annual production rate can be computed. When a fixed annual production rate and a constant sales price apply, the maximum annual profit is found to occur at minimum annual cost.

Minimum Annual Costs

The annual costs from Eq. (9-8) are:

$$C_T = nV + an^b + C_F, \text{ annual dollars}$$

which when differentiated with respect to n and equated to zero gives:

$$dC_T/dn = d(nV)/dn + abn^{b-1} + dC_F/dn = 0 \qquad (9\text{-}11)$$

If V and C_F are independent of production,

$$V + abn^{b-1} = 0$$

$$n_{opt} = \left(-\frac{V}{ab}\right)^{\frac{1}{b-1}} \qquad (9\text{-}12)$$

It is assumed in this analysis that a minimum exists at a positive value of n.

Minimum Unit Costs

The unit costs are obtained by dividing the term in parentheses in Eq. (9-8) by the units of production, n, or:

$$C_T' = V + an^{b-1} + C_F/n, \text{ dollars per unit}$$

Upon differentiation with respect to n, this leads to:

$$dC_T'/dn = dV/dn + (b - 1) an^{b-2} + d(C_F/n)/dn = 0$$

For the case where V and C_F do not vary with n, this yields:

$$(b - 1)\,an^{b-2} - C_F/n^2 = 0$$

which reduces to the following for the optimum n:

$$n^b = \frac{C_F}{a(b - 1)}$$

$$n_{\text{opt}} = \left[\frac{C_F}{a(b - 1)}\right]^{1/b}, \text{ items per year} \qquad (9\text{-}13)$$

It is assumed in this analysis that a minimum exists.

ANALYSIS WITH MORE THAN ONE CONTROLLABLE VARIABLE

In many economic analysis problems, there is often more than one controllable variable to consider. This arises where technical relations because of design, variable-lot size, or other economic considerations produce a basic equation such as Eq. (9-6) with two or more variables. For example, production capacity (and therefore costs) may vary with both the speed of a machine and its size. If both variables are subject to selection and control by management, the best policy could be a combined optimum speed and optimum size. The mathematical relations for two (or more) variables may be of the following type:

$$C_T = ax + \frac{b}{xy} + cy + d, \text{ annual dollars} \qquad (9\text{-}14)$$

where x and y are variables and a, b, c, and d are constants. The annual cost thus will go through a minimum when either x or y is held constant, as indicated by partial differentiation:

$$\partial C_T/\partial x = a - b/yx^2 \qquad (9\text{-}15)$$

and $$\partial C_T/\partial y = -b/xy^2 + c \qquad (9\text{-}16)$$

A plot with x as the abscissa, C_T as the ordinate, and y as a third coordinate will show a curved surface as idealized in Fig. 9-4.

In the economic analysis, the minimum point on this surface is desired for optimum values of x and y. These may be found graphically by plotting Eq. (9-14) for assumed constant values of x, which corresponds to passing vertical planes through Fig. 9-4 in the north and south direction or perpendicular to the C_Tx plane at the assumed values of x. The line made where any one constant x plane intersects the surface will show a minimum cost, C_T, and an optimum, y. This is also obtained analytically by setting Eq. 9-16 equal to zero and solving for the optimum, y, at the assumed value of x. An actual example is illustrated in Fig. 9-5, where the dotted lines are the projections on the C_Ty plane at $x = 0$ of the

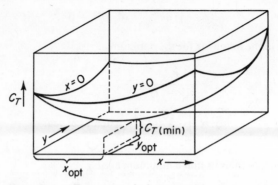

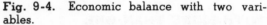

Fig. 9-4. Economic balance with two variables.

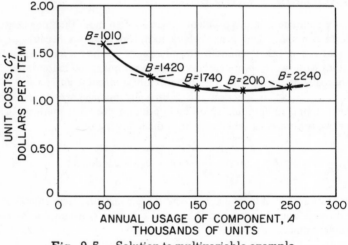

Fig. 9-5. Solution to multivariable example.

intersection of the north-south planes and the surface at different values of x. The continuous line of Fig. 9-5 connects the minimum value of each projection for the unit costs from which the best minimum can be determined. The optimum values of both variables for the minimum annual costs also may be determined analytically by equating both partial derivatives to zero and solving the two resulting equations simultaneously. This simultaneous solution may be carried out analytically, or graphically, by plotting x versus y for each differential equation and determining the values of x and y at the point of intersection of the two equations.

A MULTIVARIABLE EXAMPLE

Problem

A plant is producing a variable number of finished products per year selling for a fixed price. The final product is made from two components A and B, with some of each component being required. However, the costs have been found to vary, depending upon the number (or amounts) of each component required for the total production. Expenses for processes employing component A vary as $2A/10^6$ dollars per unit of product, where A is the annual number used. Costs related to processes employing component B are $36B/A + 720/B$ dollars per unit of product, with B being the number required for the year. Thus, the total costs per unit of product are:

$$C'_T = 36B/A + 720/B + 2A/10^6, \text{ dollars per unit}$$

All other costs are constant regardless of the relative amounts of A and B used.

The problem for this multivariable example may be considered in three parts.

(a) Make an economic balance analysis to determine the optimum number of B units to use when 50,000, 100,000, 150,000, 200,000, and 250,000 units of A are utilized in the product manufacture.
(b) Compute the unit costs for each answer in part a and determine the minimum of the minima to find the optimal quantities for both A and B components.
(c) Confirm the results in part b by an analytical solution of the annual costs, employing partial differentiation.

Solution

Part a. If the number of A components is fixed at some amount, the optimum number of B items can be determined by differentiating the annual cost equation at constant A to give:

$$dC_T/dB = 36/A - 720/B^2 = 0$$

$$B^2 = 720A/36$$

$$B_{opt} = 4.48A^{0.5}$$

Thus, for any annual usage of A the corresponding required units of B can be determined as follows:

A, Annual Units	B_{opt}, Annual Units
50,000	1010
100,000	1420
150,000	1740
200,000	2010
250,000	2240

Part b. Direct substitution of B_{opt} into the unit cost equation will give:

$$C'_T = 36(4.48A^{0.5})/A + 720/(4.48A^{0.5}) + 2A/10^6$$

$$C'_T = 324A^{-0.5} + 2A/10^6$$

Thus, for the rates requested in part a the absolute costs per item can be expressed in terms of A; then A can be computed as in Table 9-2 and plotted as shown in Fig. 9-5.

TABLE 9-2. Computation Table for Minimum Unit Costs In Multivariable Example

A	50,000	100,000	150,000	200,000	250,000
$324/A^{0.5}$	1.45	1.04	0.83	0.72	0.65
$2A \times 10^{-6}$	0.10	0.20	0.30	0.40	0.50
$C_{T_{min}}$	1.55	1.24	1.13	1.12	1.15

From inspection of the graph (Fig. 9-5) it is found that the minimum unit cost occurs at an annual usage of about 192,000 units of A. This result could also be obtained directly by mathematical solution for the optimum A, using the preceding unit cost equation from which the tabulated calculations were made. This production thus will give the maximum profit per finished item since they all sell at a fixed price.

Part c. The same answer as obtained in part b is found by partial differentiation of the general unit cost equation for C'_T or at constant A as before:

$$\partial C'_T/\partial B = 36/A - 720B^{-2} = 0$$

At constant B the result would be:

$$\partial C'_T/\partial A = 36B/A^2 + 2/10^6 = 0$$

The two resulting equations may now be solved simultaneously for A_{opt} and B_{opt}. Starting with the second equation gives:

$$\frac{36B}{A^2} = 2 \times 10^{-6}$$

$$A_{opt} = \sqrt{36B \times 10^6/2} = 4250 \sqrt{B}$$

This value of A may now be substituted in the first equation to give:

$$\frac{36}{4250B^{0.5}} - \frac{720}{B^2} = 0$$

Multiplying through by B^2 yields:

$$0.00846B^{1.5} = 720$$

$$B^{1.5} = 85,100$$

$$B_{opt} = 1940 \text{ units per year}$$

From the equation for A_{opt} above;

$$A_{opt} = 4250 \ \sqrt{1940} = 192,000 \text{ units per year}$$

This confirms the previous graphical solution of part b.

* * * * * * *

EXTENSIONS OF ECONOMIC BALANCE ANALYSIS

Many aspects of economic analysis have been omitted previously in order to avoid confusing main principles by the bringing in of modifications. However, elaborating some of these details is desirable at this point.

First, it should be emphasized that the curves in Fig. 9-1 are by no means general. In other words, in some economic analyses the fixed charges may decrease while other expenses increase, depending upon the variable selected as the abscissa. The actual shape of the curves depends upon the controllable variable and the technical and economic relations involved.

Second, in the search for the decision equation, obtaining a rigorous mathematical expression may be difficult or impossible. For such cases, the individual points for a table or chart must be computed separately in a step by step (iterative) procedure. A decision can then be rendered, based on visual inspection of the iterative calculations rather than by mathematical analysis as was done for Eqs. (9-1) and (9-4). Graphs or charts have some advantage in that they demonstrate exactly how the criterion changes as the controllable variable is altered. Thus, where the total cost goes through a flat minimum, management can see readily that costs will not change much for wide variations from the true optimum. The reverse is true of course for a sharp minimum cost or maximum profit.

Third, another advantage of charts over the analytical procedure is that the latter may sometimes give a fractional answer such as 4.3 units when a real answer must be in an integer unit. A plot of the data would indicate whether it is better to use 4 or 5 units. The use of tabulated procedures have two very important advantages from a communications viewpoint. Stepwise tabulated calculations usually involve the most simple arithmetic, thus making it easy to explain the principles to another person; also, there is less possibility of distorting or hiding the importance of certain expense or profit items. Such a distortion can occur where averaging the data is employed to reduce the work and effort required in making the calculations. Another advantage is that stepwise computations when plotted will sometimes locate an error which might not be so apparent in a single point calculation by differentiation.

Fourth, on the other hand, the advantage of the mathematical analytical procedure consists in the speed with which the result can be obtained; sometimes (but not always) there is also a simplification of the calculating procedure. In this respect, it is often reasonable to make simplifying assumptions for the mathematical relation, thus permitting machine programming.

Fifth, another advantage of the mathematical analytical procedure is that it makes possible the consideration of all alternatives; this is more inclusive than just assuming a few values of the controllable variable and then tabulating the results to find the optimum. Furthermore, for more advanced economic balance problems where multiple equations of the same form as Eq. (9-1) are applicable, the analysis requires a mathematical evaluation to obtain an answer with a reasonable effort of time and manpower.

All models at best are approximations and every possible effort should be made to reduce the mathematical equations to a simple form by the use of assumptions or by the elimination of factors having little or no effect on the final criterion. Quite often after analysis it becomes obvious that certain factors will exert only a minor effect, although this may not have been apparent in setting up the model. This then is another advantage of the analytical method: to ascertain the relative importance of minor factors so they may be removed from further consideration.

A very important point in connection with the mathematical procedure deals with testing to make certain that the "optimal" values of the controlled variables yield a proper criterion value. When a differential equation is set equal to zero, as in Eq. (9-3) or (9-4), the criterion may be either a maximum, a minimum, or a point of inflection for the total-cost curve. A test for a maximum is that the value of the *second* derivative is negative at the maximum, whereas for a minimum the value of the second derivative is positive. If the analysis is made on an incremental analysis basis, then the derivative or the incremental cost equation as shown in Eq. (9-4) will answer the question of a maximum or a minimum, since the incremental cost equation is already the first derivative. For complicated relations that are difficult to differentiate, an alternative test can be applied by direct substitution of values for the controllable variable; these values are substituted above and below the indicated optimum value and then the results are compared.

For nonlinear equations (exponents of greater than one on the variable), there may be several derived answers for the optimum, among which are some absurd results. For example, in the warehouse problem where the value of $M_{opt} = \sqrt{16}$, one solution would be minus 4 units, which of course is unreal. Such possible limitations of the mathematical model must be kept in mind when using these methods.

In many cases management can approximate the incremental values for certain operations where it may be difficult or impossible to obtain the absolute total values. For these circumstances the incremental procedure can be employed to determine the optimal controllable variable. The absolute total values may never be known, but this does not invalidate the use of the optimum value in a decision rule. The important point is for management to recognize that decisions based on such analyses are just as valid as those based on total absolute values.

For certain purposes the incremental method can also assist in determining absolute values from empirical data. For example, if the incremental equation is known, it can be integrated to give the total value equation in terms of integration constants. The latter then can be evaluated empirically, where data are available.

In the mathematical analysis for either total values or the incremental procedures, whether the mathematical expressions are theoretically derived or estimated empirical equations is of no consequence. One assumes that they are valid over the real world range of the controllable variables considered by management. Furthermore, the controllable variable must always appear in the differentiated equation (and in the incremental analysis relation); otherwise, an optimum policy cannot be found.

SUMMARY

Problems in management economy often involve a policy whereby the increasing of one type of cost can produce either a decrease in some other expense or an increase in profits. A criterion model representing a balance of these costs can frequently be set up. From this model a best policy can be established for controlling or manipulating the variables that control the costs.

In economic balance analysis, any costs that are constant or independent of the variables being considered may be neglected, where only optimum values are desired. For example, for a plant already in operation, any truly fixed costs are fixed and constant for a given time period; they only affect the magnitude of the costs and not the optimum operating point. However, the analysis may be concerned with both the design and operation of a proposed venture. In this situation, if the design affects the controllable variable and thereby the annual fixed costs, then the latter are a function of the variable(s) of interest. Since annual fixed costs will influence optimum value, these costs must be included in the analysis.

The procedure for setting up the model follows the principles outlined in Chapter 1 and consists of two major steps. First, the economic principles involved are expressed as an economic model, or criterion equation. Second, the operational or technical terms in the criterion equation must be evaluated in a real world form from known theoretical or empirical relations and data. The resulting relation is then analyzed by appropriate techniques.

The model may be solved by graphical and stepwise solutions, or an analytical procedure may be followed. The latter has many time-saving advantages when approximations to the real world problem can be utilized. Both graphical and purely analytical procedures have their own respective advantages in individual situations.

Economic analyses made on a basis of total costs may be differentiated to obtain optimum values, or the basis for solution may utilize an incremental cost approach. Either method can be utilized with each having its advantages and disadvantages.

A major item to consider in economic analysis is the selection of the proper criterion. A generally useful criterion is minimum annual costs, which will cover a wide variety of situations under a given set of conditions. In special cases, management may be required to set an optimum policy, using two (or more) controlling variables. Where an analytical equation is available, the optimum policy can often be obtained by partial differentiation; this is more expedient than trial and error.

Thus, economic analysis covers a very broad area involving a correlation of technical and economic factors into some type of model that can be analyzed for a best solution. The model often is an approximation to the real world, since minor effects are omitted to simplify the analysis without affecting greatly the over-all end-result. Where approximate mathematical models are utilized, they provide a powerful tool to validate management decisions. In the final analysis, management relies on judgement and experience to set policies; however, if these intuitive methods can be confirmed or supplemented by objective mathematical results, the policies chosen can be put into practice with greater confidence.

PROBLEMS

9-1. Develop a total cost equation and an optimum policy for management in the following case where an operation produces a constant number of units per year. Production machinery has annual fixed charges of $420 per machine but the addition of more machines reduces operating charges according to the following relationship: annual operating costs equal $42,000/M$, dollars, where M is the number of machines.

9-2. A plant manager has a problem of insulating a steam line to prevent heat loss and save fuel costs. His engineers have advised him that each additional inch of thickness of insulation will cost $25 per year. It is also estimated that the incremental savings in fuel will amount to $100/S^2$ dollars per year for each inch of insulation, S. Determine the optimum thickness to use from these incremental costs.

9-3. The manager of a small store is considering remaining open after regular closing hours if it is profitable. Based on certain trial openings, it is established that the average sales dollars return is according to an empirical equation $S = 72 \sqrt{H}$ in dollars, where H is the additional hours open. The manager estimates the average hourly cost of remaining open to be $30. (a) Make a graph of sales, cost, and profit dollars versus additional hours open, and confirm the graphical solution for a "best" policy by an analytical solution for maximum profit. (b) Repeat the analysis using the incremental cost procedure and solve with step by step and graphical analysis; then solve by the analytical procedure.

9-4. A manufacturer of electronic parts produces them in batches, where the "get ready" cost is $300 each time a batch is to be made. This cost has to be balanced against the carrying charges (for interest on capital tied up and storage charges). If the latter amount to $14,700/N$ per year, where N is the number of batches per year, how many batches should be made? (NOTE: this demonstrates the principle of the economic lot size, which is a classical problem in management economy.)

9-5. A furniture manufacturer makes sofas that sell for $138 each. Daily fixed costs are $4800, and variable costs per unit are $36 + 0.12n^{1.3}$ dollars, where n is the number of sofas made per day. What is the optimum management policy and daily profit for (a) maximum daily profit? (b) maximum unit profit?

9-6. An entrepreneur can purchase 100,000 items at a distress sale for $0.75 each, these items may be reprocessed and packaged to have a fixed resale value of $2.00, but at different preparation costs, depending upon the recovery. (a) What is the best policy for maximum profit? (b) Is this the same as for maximum profit per original unit? Per recovered unit? Discuss. Additional data are as follows: to obtain the sales price of $2.00 per finished unit, the reprocessing costs increase in order to get maximum recoveries. The relation between costs and recoveries is $C_p = \$1.10\ R^{2.5}$ dollars per original unit, where R is the fraction of original items finally available for resale. (HINT: draw a diagram for sales dollars, costs, and profits for 100,000 units as a function of recovery.)

9-7. An electronics company produces a radio which can be made using two interchangeable types of tubes X and Y. However, some of each of the types of tubes must be used in each radio. An analysis of costs indicates the following: Processes employed in the use of tube X have costs of $22X^2/Y + 70,000/X$ dollars per unit of product, with X being the annual number used. Costs of processes using tube Y are $1.8Y/10^4$ dollars per unit of production, where Y is the number required for the year. Make an economic balance analysis by both graphical and analytical procedures to utilize the optimum number of X and Y tubes for minimum costs per unit of product.

ECONOMY OF CONVERSION AND YIELD

10

The economics of conversion and yield of a final product from various raw materials and resources is quite complicated unless certain simplifying assumptions are made. These assumptions permit the setting up of simple mathematical models that management can utilize to obtain quantitative answers to questions about the influence of certain variables that are subject to managemental control.

TYPES OF CONVERSION ANALYSIS

Although many economic analyses concern productive operations involving the conversion of a given amount of material to a final product, here the analysis differs in some respects from the regular economic production charts (breakeven charts) discussed in Chapter 7. This difference can be more sharply delineated as follows: Breakeven charts generally pertain to increased production as function of increased input (that is, they are a function of the *over-all magnitude* of operations). This chapter, however, considers those models that make possible the most efficient operation dollar-wise for handling a *fixed quantity* of raw material. There are two main types of problems: (1) those in which varying amounts of a product of uniform quality are made and (2) those in which the manufactured product varies in quality, the amount produced being some function of the conversion of an input to the product. As pointed out in Chapter 6, the conversion cost is the value added by manufacture.

In a diagrammatic form, the operations to be discussed can be represented as follows:

Raw Stock, $Q \rightarrow$ Conversion operations $\nearrow$ Product P

$\searrow$ By-products, waste, unused raw stock, etc.

NOMENCLATURE

b_i = the value of an individual by-product, waste material, or any side product of value, dollars per period

B = composite credit per unit of production for any items having recovery value

F = fixed cost for a given period

G_f = grade, or per cent, of key input component expressed as a fraction

G_p = grade, or per cent, of key component in product expressed as a fraction

m_i = the value for an individual raw material entering the process, dollars per period

M_r = composite cost of all assigned raw materials entering operation per unit of Q

P = product, items per period

Q = quantity of raw material entering the operation

R = conversion defined as the ratio of P/Q, which may be greater than unity, depending upon the units of P and Q

R_r = recovery or ratio of key component in product to key input component

S = unit value of any given product

S_a = average unit sales price

V_c = variable conversion cost per unit of product

V_f = variable cost per unit of input component

Z = profit, dollars per period

Thus, there are many variations of the problem including consideration of the by-products, waste, reusable materials, and unconverted raw stock as they are affected by the quality and amount of the main product. The variations in the analysis also depend upon the criterion selected for the management model and the number of variables to be considered. In general, the criterion is profit dollars, or profit per unit of key starting material. The basic relation is total profits dollars for a specific time period for a given quantity of feed:

$$\text{Profit} = \text{total revenues} - \text{total costs}$$
$$Z = PS_a - (PV_c + F) + \Sigma b_i - \Sigma m_i, \text{dollars per period} \quad (10\text{-}1)$$

where Z = profit dollars per period

P = units of saleable product per period

S_a = average unit sales price, which might vary with both the quality and quantity of product

V_c = variable conversion cost per unit of product, P, and includes any consumable supplies

m_i = the value in dollars of any individual raw material entering the process, which appears as part of the product. (There is some flexibility of interpretation here as to distribution of of materials cost, either here or in V_c.)

b_i = the value in dollars per period of any and all by-products or waste materials which might be used for fuel or raw materials in any operations

and F = the fixed costs in dollars for the period

Because of the summation terms in Eq. (10-1), a rigorous treatment requires a step by step calculation of all costs at each production or conversion level in order to determine the optimum operation for maximum profit. Where it is possible to replace the summation terms by a single composite function of the operation level and where S_a, V_c , and F may also be expressed as continuous functions of the production level, P, then a straightforward mathematical analysis can be made for optimal operation. A proposed model equation might appear as follows:

$$Z = PS_a - (PV_c + F) + PB - QM_r, \text{dollars per period} \quad (10\text{-}2)$$

The terms in parenthesis are often called conversion costs

where M_r = the composite cost of all raw materials entering the operation per unit of Q

Q = the composite amount of raw materials so that the product, QM_r, represents the materials cost at each level of operation

and B = the composite credit per unit of production for any items that have a value for reuse at each level of operation. Thus PB is the total credit allowed for by-products, reworkable material, etc.

In this analysis certain items of supplies may be considered either as a conversion cost in V_c or as a materials cost in m_i. However, it is the intent of this analysis to include in m_i only those basic raw materials constituting the major portion of the product.

UNIFORM PRODUCT QUALITY

The amount of product, P, of a given quality obtained in an operation and used in Eq. (10-2) to compute the profit is a function of the conversion, R, of raw stock, Q, to that product[1]; that is, $P = RQ$. Therefore a

---

[1]As pointed out in Chapter 6, the value added in manufacture results from the conversion of an input, or raw material, into another product which is more useful to the consumer. The degree of this conversion can vary depending upon the type of operation, its efficiency, and other factors. Thus, the conversion, R, is a variable to be included in certain models.

new equation may be written in terms of R and the quantity of raw stock, or from Eq. (10-2):

$$Z = RQS_a - RQV_c - F + RQB - QM_r, \text{ dollars per period} \quad (10\text{-}3)$$

If the analysis is made on a basis of profit per unit of raw stock, Eq. (10-3) becomes:

$$Z/Q = RS_a - RV_c - F/Q + RB - M_r, \text{ dollars per unit of raw stock} \quad (10\text{-}4)$$

Equations (10-3) and (10-4) are rigorous, and are mathematical models generally applicable to conversion and yield problems. If the quantity, Q, of raw stock is fixed, then both equations will optimize at the same value of conversion, R.

The components of Eq. (10-3), when divided by R to give a basis of one unit of product, may be plotted as in Fig. 10-1 where the independent variable (controlled by management) is conversion; revenues and costs relations are somewhat as shown.

The average sales value, S_a, may decrease with increased conversion because more product is made, which may depress the market price. The variable costs per unit of product, V_c, and the fixed costs, F, increase at different conversion levels because more expensive equipment is needed

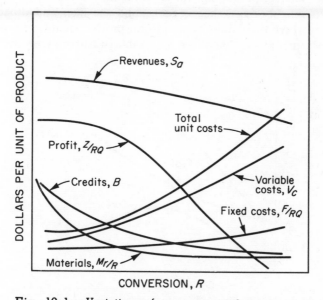

Fig. 10-1. Variation of revenue and cost components with conversion at a constant raw stock.

and higher direct costs are chargeable to the operation. Credits, B, on the operation per unit of product for reworkable material, by-products, waste, etc., may either decrease or increase with changes in the level of operation. In general, the unusable product decreases with conversion. However, if a by-product is made in proportion to the main product, its increasing credit as conversion improves may offset the decreasing credit for less waste.

Where all economic data may be expressed by mathematical relations in terms of conversion, R, these may be inserted into Eq. (10-3), and this equation can be differentiated at constant, Q, for the optimum conversion. Where the conversion, R, may vary with the quality of the input or there are any other expenses that cannot be related to conversion alone, then the mathematical relations will include more than one variable, making necessary a multivariable analysis as shown in Chapter 9 (page 237).

Where the mathematical relationships are not available, a step by step calculation is needed at each conversion, the best one being determined from tabulated values. Curves may be plotted from the calculated data to give the results diagrammed in Fig. 10-2. From such a plot, the optimum policy can be found by inspection.

Examples of this type of analysis are as follows: those for conversion of thread into cloth; cloth into garments; sheet paper stock into bags, containers, etc., for packaging and shipping; the fabrication of steel ingots and billets into forgings, sheets, gears, and the like; and other similar operations where there must be a uniform product quality. In short, the product is either acceptable or not acceptable (waste or by-product).

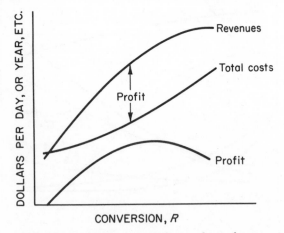

Fig. 10-2. Economic analysis chart for conversion at constant raw stock.

CONVERSION EXAMPLE

Problem

A paper box manufacturer uses raw material equal to 100 tons per day, and can operate at three conversions of 75, 85, and 90 tons of product per day. These levels require daily fixed costs of $2250, $2350 and $3500 at the three levels, respectively. All raw material not appearing as finished product is considered recoverable at $60 per ton. The variable conversion costs in dollars per ton of production are $V_c = 29.8\,R^{0.7}$, where R is the conversion in units of product per unit of raw stock. The raw materials costs are $150 per ton. Because of market conditions at this level of operations, the average sales price decreases with production as follows: $S_a = 260 - 10R$, dollars per ton of product. Using one day as a basis for analysis, what is the optimum or critical production level for maximum profit?

Solution

Since all of the data are not given in terms of conversion, it will be necessary to use a stepwise calculation of all components of either Eq. (10-3) or Eq. (10-4) at each conversion and tabulate the results as shown in Table 10-1.

To solve the problem, one typical calculation will be made to illustrate the procedure, using Eq. (10-3) and a one day basis for operation at 75 tons per day of product. The complete results are given in Table 10-1. For $RQ = 75$ tons:

The revenue dollars are $RQS_a = 260\,RQ - 10R^2Q$ or:

$$RQS_a = 260(75) - 1000\,(0.75)^2 = \$18,938$$

The daily variable conversion costs are $RQV_c = 29.8QR^{1.7}$, or:

$$RQV_c = 2980(0.75)^{1.7} = \$1825$$

TABLE 10-1. **Calculations for Conversion Example at Constant Charge of Raw Materials (Basis: $Q = 100$ tons per day)**

Conversion, $R = P/Q$	0.75	0.85	0.90
Daily production, $P = RQ$, tons	75	85	90
		Dollars	
Daily revenues, RQS_a	18,938	21,378	22,950
Daily costs			
Variable costs, RQV_c	1825	2280	2520
Material costs, QM_r	15,000	15,000	15,000
Fixed costs, F	2250	2350	3500
Total costs	19,075	19,630	21,020
Credit for recoverable material, RQB	1500	900	600
Net total costs	17,575	18,730	20,420
Daily Profit, Z	1363	2648	2530

The fixed costs are stated in the problem for each level of production. The material costs are QM_r, which are constant at all levels of operation and equal to $100(150) = \$15,000$.

Credit for recoverable material is \$60 per ton and is taken for $100 - RQ$ in amount for this problem. For other conditions, the analysis of credit must be considered individually. For this problem at $RQ = 75$ tons per day, there are 25 tons of reusable material worth \$60 per ton; thus, RQB of Eq. (10-3) = $60 \times 25 = \$1500$. The value of B is $1500/75 = \$20$ per ton of product at this conversion, but it varies with conversion. Although this is given as a matter of information, B is not needed to solve this particular problem.

The net total costs are $(RQV_c + QM_r - RQB + F)$. This amounts to \$17,575, which when subtracted from the total sales of \$18,938 gives a profit of \$1363. Profits at the other two levels of conversion are shown in Table 10-1, which demonstrates that the optimum policy will occur at the 85 tons per day level of operation.

The same type of analysis applies to the exploiting of a natural resource such as petroleum, metallic ore deposits, a tree forest, etc., where a fixed quantity, Q, is available for producing a given product at economic advantage. The long-run total profit often might be at a recovery or conversion level different from one that would give a maximu profit for a short period. In the interests of conservation, the operating level may be fixed by governmental agencies to obtain maximum recovery. Equation (10-3) provides a basis for economic analysis, where Q is the total quantity available.

THE ISOPRICE ANALYSIS

One variation of the constant quality conversion analysis is where price is determined; at this sales price, the profits returned at one conversion are equal to the profits returned at a higher alternative conversion. In the isoprice analysis, the sales price is determined in terms of a unit of raw stock (or pure component) as a basis of comparison. At the isoprice point (the value where either alternative returns the same profit) the profit from the low conversion operation equals that for the higher conversion because of the greater expense incurred. By employing the sales price per unit of raw stock as a common basis for comparison of the alternatives, one can find the total revenues readily.

Problem

A clay product costing \$15 per ton in the raw condition is processed with an 85 per cent conversion to finished product at a cost of \$2 per ton of product. Conversion can be increased to 95 per cent at a process cost of \$4 per ton recovered. Determine the sales price at which the increased conversion will be warranted based on: (a) a criterion of total profit for an amount Q of raw stock available; (b) a criterion of profit per unit of raw stock available.

Solution

Part a. The analysis for total profit follows Eq. (10-3) which is simplified because all of the cost items are combined.

At 85 per cent conversion (that is, for each ton of raw clay 0.85 tons of product are obtained) the profit is:

$$Z_1 = 0.85QS_a - 0.85Q(2) - 15Q, \text{dollars}$$

This profit must equal the profit for the same Q tons of raw clay at 95 per cent conversion, or:

$$Z_2 = 0.95QS_a - 0.95Q(4) - 15Q, \text{dollars}$$

Equating Z_1 and Z_2 yields:

$$0.85(S_a - 2) = 0.95(S_a - 4)$$

or: $\qquad S_a = \$21, \text{per ton of product}$

Thus, when a sales value of \$21 per ton of product prevails, the same profit is made by either process. At a lower sales price the low conversion gives more profit; however, if a higher sales value is desired, the high conversion method is more profitable. Note that all cost items that are the same in both converting operations vanish from consideration when Z_1 and Z_2 are equated.

Part b. The unit profit for a given quantity, Q, of raw stock occurs at the same price level as for part a, since both Z_1 and Z_2 are divided by Q, which therefore is factored out in the solution for S_a. This same result is obtained by using Eq. (10-4) as a basis for solution.

VARIABLE PRODUCT QUALITY

The second common management problem in conversion operations is the selection of the best product quality for maximum profit. In these situations, a better quality product (and therefore a better price) can be obtained at additional expense, and an economic balance exists with some optimum operation being possible. For this analysis, the basic Eq. (10-2) applies, and its modification in Eq. (10-3) will be discussed and illustrated.

Where the product quality or grade, G_p, is variable, the revenues and costs will vary with the grade somewhat as shown in Fig. 10-3. Each item of Eq. (10-2) must be considered separately and the credits and materials cost might change differently. The diagram is drawn for a given raw stock

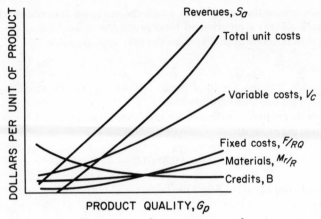

Fig. 10-3. Variation of .revenues and cost components with quality for a given raw stock.

so that it is not a variable in the analysis. Under these conditions, by means of Eq. (10-3) an economic analysis can be made for a constant amount of raw stock, Q, which will give RQ units of product for the different qualities. Where all the revenue and cost items (such as could be determined empirically from Fig. 10-3) are known in terms of G_p, an analytical solution is possible for the optimum grade. Where the economic data are not related mathematically, a graphical plot such as Fig. 10-4 can be made by tabulating the costs and profits. Note that this type of chart can be used for a variable or a constant production level of saleable product, depending upon how R varies for products of different quality. Where R is a variable along with product quality, the solution of the problem requires a multivariable analysis.

Problem

A garment manufacturer can produce an item in three qualities A, B, and C which sells for 15, 20, and 25 dollars, respectively. Daily fixed costs are $600; other costs for the different products are as follows:

Product	A	B	C
Variable Conversion Costs, V_c, dollars per garment	4.50	7.00	9.50
Credits for Cuttings, B, dollars per garment	1.30	1.20	0.50

The conversion is essentially constant at 200 product units per 1000 square yards of raw stock. The raw-stock cloth costs $1.20 per square yard, and this operation is carried out using 2000 square yards per day. Which product should be made?

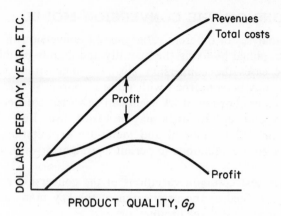

Fig. 10-4. Economic analysis chart for product quality at constant raw stock.

Solution

The solution follows Eq. (10-3) for optimization of quality that should be selected. A basis of a fixed amount of raw stock, Q, is used; since the conversion is essentially constant, this produces a fixed amount of product of each quality. Where the quality of product for this type of problem cannot be expressed as a continuous function, the indicated solution is to tabulate all values for costs and revenues as in Table (10-2).

TABLE 10-2. **Tabulated Data for Cloth Conversion Example** (Basis: Q = 200 square yards of cloth per day)

Product	A	B	C
Conversion, R, garments per 1000 square yards	200	200	200
Daily production, RQ	400	400	400
Daily sales, RQS_a	6000	8000	10,000
Daily costs:			
Variable costs, RQV_c	1800	2800	3800
Material costs, QM_r	2400	2400	2400
Fixed costs, F	600	600	600
Total costs, dollars	4800	5800	6800
Credit for recoverable material, RQB	520	480	200
Net total costs	4280	5320	6600
Daily profit, dollars	1720	2680	3400

For this operation, the best policy is to produce the highest quality product.

EXTENSIONS ON THE CONVERSION MODEL

Previous analyses of different programs on conversion of a raw stock have been simplified by fixing the quantity and quality of the raw stock, and in some cases the conversion also. In some situations, however, management may be concerned with more elaborate programs in which the production level or raw-stock levels can be varied. Moreover, in some cases, the grade of the raw stock may be varied also. For these types of problems, common in mineral and metal-ore exploitation, constraints are necessary on the equations to permit a straightforward mathematical analysis.

One of the most common variations of the conversion concept is the recovery, R_r, defined as the proportion of the key material in the raw stock that is recovered in the product, or:

$$R_r = PG_p/(QG_f) = RG_p/G_f \tag{10-5}$$

where G_p = grade or quality of product P in terms of the key input component

 G_f = grade or quality of raw stock in terms of the key input component

and R = the conversion, P/Q, previously defined

Where a pure input item or pure product is involved, then G_f and G_p may separately or both be equal to unity. Thus in the previous paper bag problem, the total raw stock and total product were involved so that both G_f and G_p were unity and $R_r = R$. However, the relation from Eq. (10-2) can be more generally written for $P = R_r QG_f/G_p$.

Repeating Eq. (10-2) gives:

$$Z = P(S_a - V_c + B) - F - QM_r, \text{ dollars per period} \tag{10-2}$$

For problems of conversion only where product or input quality is not involved:

$$Z = RQ(S_a - V_c + B) - F - QM_r, \text{ dollars per period} \tag{10-3}$$

The general relation, therefore, from Eq. (10-2) is:

$$Z = (R_r QG_f/G_p)(S_a - V_c + B) - F - QM_r, \text{ dollars per period} \tag{10-6}$$

This multivariable equation requires that certain conditions be fixed or constant in order to avoid too complex an economic analysis. For example, by fixing the amount of raw stock and its grade, G_f, the raw material item becomes constant. An analysis then may be made for the optimum recovery or product quality when the costs and revenues are known as a function of R_r and G_p. If either of these is made constant, the analysis becomes quite simple as previously demonstrated. If conversion and product quality are both variables, the cost data must include the influence of these simultaneous variations on solutions by partial differentiation.

In general, Eq. (10-6) is used for the analysis under one of the following arrangements with the basis for calculation indicated.

(1) Where production level only is involved and revenue and cost data are known in terms of production level with $P = Q$, Eq. (10-2) may be employed.

(2) Where conversion is involved with $P = RQ$, where quality of input or product is not a variable, and where all revenue and cost items are known in terms of R, then Eq. (10-3) is used at constant raw stock, Q.

(3) Where variation in product quality is considered with revenues and costs known in terms of product quality, G_p, the analysis is best made with Eq. (10-6) at constant values of QG_f, that is, a fixed quantity of key material in the raw stock is employed as a basis for all comparisons. The quantity of key material in the product is $PG_p = R_r QG_f$ and for pure product, $G_p = 1$. Thus, the cost data are known as functions of R_r with $P = R_r QG_f$. This is similar to Eq. (10-3).

(4) Where variations in raw stock quality, G_f, must be evaluated for selection of the best input stock, then the analysis should be made at a constant amount, P, of a given quality, G_p. The variation of revenues and costs is known in terms of, G_f, and the conversion or recoveries permit use of Eq. (10-6) as the model for the analysis. A variation of this problem, where conversion costs are essentially constant at all input qualities, is to process constant amounts, Q, of each input for a given product and quality with a maximum profit.

In all of the foregoing analyses, the conversion term, R, may be greater than unity, since an analysis made on one key raw stock of quantity, Q, may be combined with other items to produce more units of P than there was stock consumed. Also, the dimensions of P and Q may vary as they did in the garment problem. These complications are automatically allowed for in the basic analysis, but care must be taken that the costs are related in the proper manner to the respective base item and that all composite items are calculated correctly.

In such problems, however, desirable simplification often results if the subordinate-materials cost can be included in the variable-conversion costs, so that only the main raw-materials costs must be considered in M_r.

As with all optimization problems, any cost items that are constant in the analysis can be omitted without affecting the optimal value although all cost data must be considered if real cost or profit dollars are to be found. Also, the incremental procedure for economic balance may be applied here, where incremental revenues are equated to incremental costs at the optimization value for the variable controllable by management.

SELECTION OF THE CRITERION

In conversion and recovery models for economic analysis, management has a variety of criteria it may select as a basis: These are as follows.

(1) Profit per unit time period (per day, per year, etc.)
(2) Total profit available for a total elapsed time which may vary with conversion level
(3) Profit per unit of key material available for input item
(4) Profit per unit of key material in final product

Each of these possible criteria serve specifically under certain conditions. The most generally useful probably are (1) and (3) because the meaning of the results can be visualized quite easily in day to day operations. For example, management usually desires the maximizing of daily or yearly profit. However, since in the long run a smaller profit per unit of time for a longer period of time may be more economical, the time element should be considered, as under criterion (2) above. Where maximum possible economic benefit is desired, conservation considerations indicate the use of criterion (3). Also, conservation of natural resources with legal restrictions may be considered in connection with item (2), since governmental authorities often limit production rates to avoid waste.

Criterion (4), although perhaps the least useful, could be employed in certain management decisions where different products and manufacturing processes are being compared.

All of the criteria above are available from the basic Eq. (10-6). Thus, Eq. (10-6) may be divided by any quantity (product per period, raw stock per period, key material per period, etc.), and the conversion relationships R_r or R may be stated in the appropriate units to provide a mathematical model.

One further comment on criterion selection relates to the use of costs instead of profits. Where minimizing of costs is desired, the same equations used for profits may be employed; the operation is based on the last terms, with omission of the revenue portions of these equations. The comments that are pertinent to any analysis with economic production charts also apply here. In selecting a criterion, management must consider each situation separately; selection depends upon the controllable variables(s) being studied, the manner in which revenues vary, and the objective of the analysis, that is, whether profits or costs are being studied.

CONSIDERATION OF MULTIPLE PRODUCTS

Where a large number of joint products are being made from a given raw stock such as petroleum, management must consider the relative market values and recoveries for each product.

For a given breakup of a barrel of crude oil into gasoline, into fuel oil, into lubricating oil, etc., the over-all cost and over-all profit per barrel will not change, regardless of any values assigned to the individual products for unit costs and profits. Thus, if certain expenses that are part of the gasoline operation are charged to lubricating-oil manufacture, the gasoline operation will show a greater profit than it should and the lube oil operation will have a lower than real profit. However, since the products are all made from the same barrel of oil, the total profit on each barrel is merely the revenues for the products made therefrom minus the costs for processing the barrel of crude petroleum. Thus, total profit is the same, even though absolute values for costs and profits on each component are in error.

However, the necessity for obtaining absolute cost figures is apparent when the economic analysis involves different procedures that give a variety of breakups for a given barrel of oil. In these circumstances, the desired goal is optimum recovery or conversion of the feed to those products giving the maximum over-all profit. The economic analysis must consider variations in volumes with both unit costs and over-all profits. With two products (such as one major product and one by-product), the analysis is relatively simple. When three or more products are produced by alternative procedures, many complex technical process operations may be involved so that evaluating the optimum conversion of feed to each product may result in an extremely complex economic analysis. In such a case, the best course is to set up an individual profit equation for each product in terms of actual costs and sales values and to weigh each equation according to different amounts of products made, the sum of the equations being the over-all profit per barrel of feed. As many as possible of the technical variables and economic variables (such as equipment sizes, annual production, etc.) should be fixed to limit the variables in the over-all equation and permit a solution by simple analytical or graphical procedures. When the relations among yields, costs, and profits are linear, the analysis is amenable to linear programming, using automatic calculators for the solution.

Where all costs are expressed in terms of feed units, a form of Eq. (10-2) that can be applied for multiple products is as follows:

$$Z = Q(S_1 R_1 + S_2 R_2 + \cdots + S_n R_n)$$

$$- (QV_f + QM_r + F), \text{profit per period} \qquad (10\text{-}7)$$

where Q = the annual quantity of feed material
S = the unit value of any individual product
R = the recovery of any product per unit of feed
M_r = the price for raw feed per unit of feed
V_f = the variable cost per unit of feed for all processing
F = the annual fixed cost for operations

When V_f and F vary either with the feed material or with the particular recoveries (R_1, R_2, etc.) of products desired, these variations may be incorporated into the general equation. If technical relations among recoveries can be related mathematically, an analytical economic analysis can be obtained for maximum profit as a function of the recovery of any one product.

MULTIPLE PRODUCTS

Problem

A manufacturer of a certain chemical A, as its main product, also makes products B and C in the process. A table of data available for the process shows the following:

Product	A	B	C
Recovery per unit of feed, tons per ton	0.7	0.2	0.1
Value per ton of product	$16	$8	$5

What is the annual profit per 100,000 tons of feed material if the investment is $200,000 with a 10 year life (or annual fixed charges of $20,000) if the costs for raw feed materials are $8 per ton of feed, and if the variable costs per ton of raw feed are $4.3 per ton? Consider losses as negligible.

Solution

If one has a knowledge of the available information, this problem can be analyzed quite simply, since Eq. (10-7) may be employed directly.

$$\text{Profit,} \ Z = 10^5[(16)(0.7) + (8)(0.2) + (5)(0.1)] - 10^5(8 + 4.3)$$

$$- 20{,}000, \text{dollars per year}$$

$$Z = 10^5(13.3 - 12.3) - 20{,}000 = \$80{,}000 \text{ per year}$$

In some situations with joint products one key product may be considered critical for the operation; where all other items are known, the limiting price of the key product may be determined by means of Eq. (10-7). Where Z is equal to zero, the limiting value of the key product is at a breakeven price. An alternative use of Eq. (10-7) is to establish the price, M_r, to pay for the feed at which some specified profit is to be made. If the profit basis is zero, the raw-material price for breakeven is a minimum. Such analyses can of course be made by dividing through Eq. (10-7) by Q so that the basis is profit per unit of feed rather than total profit. For a given amount of raw stock, the same result is obtained by either criterion.

PROBLEMS

10-1. A manufacturer of automobile parts uses 10 tons of steel per day on a machine that uses three alternative procedures A, B, and C which operate at 88, 92, and 96 per cent efficiency, respectively, in turning out parts; the lost efficiency is in terms of waste which is recoverable at $80 per ton. Raw material costs are $220 per ton. Fixed costs per day are $200, $300, and $500, respectively, variable costs being a function of conversion, R, as follows: $50R^{1.2}$, dollars per ton of production. Sales price per unit is constant at $350 per ton of product. Which is the best procedure to use?

10-2. A firm is considering a prospective investment and operation utilizing a mineral deposit of 100,000 tons which can be purchased for $200,000. Operations can be at a 75 or 90 per cent recovery, with costs of $3 and $7 per ton of product, respectively. What process should be used if the product can be sold for (a) $30 and (b) for $25? Why?

10-3. A manufacturer has a choice of producing three plastic products of increasing quality, which he sells for $24, $39, and $45 per dozen. Conversion costs are $5, $7, and $14 per dozen items. Conversions of raw material are essentially constant at 4 dozen items per 10 pounds, the raw material being worth $4 per pound and all waste being of same value and essentially constant for each product. The operation uses 1000 pounds daily. Which product should be made?

10-4. Repeat problem 10-3 but assume that the conversion is 1.00, 0.95, and 0.88 for the three product qualities, respectively, the recovered unconverted raw material being worth $1.60 per pound.

10-5. In the analysis of a plant operation starting with a given raw stock, the following models were found for the relations of Figure 10-1 as a function of conversion, R, where R is the ratio of units of product per unit of raw material:

Dollars per Unit of Product

Revenues	$S_a = 1000 + 100R$
Variable costs	$V_c = 180R^{-0.5}$

Dollars per Unit of Product (Cont.)

Fixed costs	F/RQ	$= 400 + 320R$
Credits	B	$= 200(1 - R)^2$
Materials	M/R	$= 200$

What is the optimum conversion? Discuss several possible answers.

10-6. A petroleum crude oil yields 28 per cent by volume of gasoline worth 14 cents per gallon, 52 per cent of fuel oils worth 10 cents per gallon, and the remaining yield of asphalt worth 8 cents per gallon. If the variable costs for processing are 2 cents per gallon of feed and the fixed costs are 0.8 cents per gallon of feed, what is the breakeven price that should be paid for the raw crude oil, assuming that one half per cent of the feed is lost in processing. HINT: Make the analysis on a basis of no loss, and divide the price by 0.995 to obtain the real world price.

PROFITABILITY MODELS 11

Evaluation of profitability involves the use of a model for the quantitative determination of the economic feasibility of a venture in which capital is invested. The venture may be of any form, and may include the manufacture of products, the investment of capital, the performance of services, as well as other functions in which capital is employed for profit. The model may be a chart, a table, or a mathematical calculation that assists management in making an objective study in a quantitative manner of the financial condition of a given operation.

Any venture includes revenues and associated expenses for the services performed or the other outputs produced. The excess of revenues over expenses provides an economic return on assets, which may then be evaluated either with or without considering the time value (interest) of the capital employed. Profitability measures this economic feasibility in numerous ways; however, the major procedures are: (a) to evaluate the present worth of the excesses of revenues over expenses at a given interest rate on capital for a fixed number of future periods, (b) to estimate the earning rate, or interest rate, of the net cash in-flows for a fixed number of future periods that gives exactly the present worth of the invested capital at some datum time, or (c) to estimate the number of time periods required for the future net revenues to repay the value of the investment at some datum time.

Procedure (a) is a special case where the result is a function of the specific venture and the magnitude of the excess present worth may depend on the amount of capital invested. In (b) and (c) the time value of money can be considered or neglected.

The analysis for profitability is complicated by the fact that two types of capital—depreciating and nondepreciating—are involved in the real world situation. Accordingly, the basic analysis must allow for both. In addition, the real world problem must allow for profits tax and for possible variations in annual revenues, expenses, and depreciation.

NOMENCLATURE

C = annual cost or expense dollars excluding depreciation in some cases, C_n is in the nth year, C_0 is for initial rate, $C(N)$ is for the terminal year

s = the compound interest factor equal to $(1 + i)^n$ for discrete periods or e^{in} for continuous compounding; values are listed in the Appendix or Fig. 2-1

D = annual depreciation dollars; D_n is in the nth year, $D(N)$ is a function of N

d = the fraction of investment that is charged annually for depreciation

e = the transcendental number, 2.718

I = initial investment; I_x is its equivalent at some datum year, x

i = interest rate expressed as a decimal

i_c = a special interest rate called capitalized earning rate and defined by Eq. (11-7)

i_m = marginal profit rate defined as dY/dI

k = a constant; k_s for annual sales rate, k_c for annual cost rate

L = recoverable value in dollars of any type (working capital, land, salvage value, etc.); $L(N)$ is a function of N

n_c = a special type of age called capitalized payout time, defined by Eq. (11-8)

n = end-of-year age, total life, or total periods

N = total life or total periods

P = present worth; P_x is at some datum year, x; or P is an initial investment

$a_{n/i}$ = a present worth factor equal to $(s - 1)/is$, which may be found in Appendix C-2 or C-4, or Fig. 2-2; it is equal to P/R

R = a uniform end of period payment

S = annual sales dollars; S_n is in the nth year, S_0 is for initial rate, $S(N)$ is for the Nth year

SL = straight-line depreciation

t = profits tax rate expressed as a decimal

x = a datum time or year, or fraction of a year

Y = net profit dollars after taxes

THE BASIC RELATIONS

Management may avail itself of many proposed models for profitability, involving various procedures. At present, it seems that names of individual methods have proliferated to such an extent as to be confusing. However, actually these numerous methods are merely modifications of previous basic procedures. In reality, they all refer to the basic theoretical economy equation for discounting a series of future returns to obtain their present worth. These future payments represent net revenues in the form of profits, depreciation deposits, and salvage items. Furthermore, any or all of these items may vary from year to year. Accordingly, a generalized equation incorporating the time value of money may be written to consider all variations as follows:

| Discounted amount of all future values | = | present worth of future excess of sales over costs excluding depreciation | + | present worth of tax credits for depreciation[1] | + | present worth of salvage value, land and working capital |

$$P_x = (1 - t) \left[\sum_1^N \frac{S_n}{(1 + i)^n} - \sum_1^N \frac{C_n}{(1 + i)^n} \right]$$

$$+ t \sum_1^N \frac{D_n}{(1 + i)^n} + \frac{L}{(1 + i)^N} \qquad (11\text{-}1)$$

where P_x = present worth in dollars of all items on right at some datum year, x; this is the total discounted cash flow.

S_n = sales dollars in nth year in dollars

C_n = all costs (except depreciation) in dollars incurred to obtain a revenue of S_n dollars in the nth year

D_n = cost in dollars for depreciation in the nth year. Note that where the tax rate is zero that part of the return equivalent to depreciation is automatically included in the bracketed items, since there it is not considered as a cost. Note also Eq. (11-1) permits the use of any form of depreciation procedure desired.

n = end of year age for which computation is made

N = life being considered, years

L = the value of all salvageable items (land, working capital, physical salvage) in dollars available at any time

i = interest rate, fractional

t = the tax rate, fractional

From Eq. (11-1) one observes that the cash flows in future periods after taxes pay off the depreciable investment and pay the interest on the total investment; this is true because these future revenues are corrected to a zero-time datum period allowing for interest. The recoverable assets, L,

---

[1] Although it may not be apparent, the annual return after taxes, consisting of net profits plus depreciation, may be computed in either of two ways which are equal:

(a) Annual return = (net revenues − depreciation) $(1 - t)$ + depreciation
(b) Annual return = (net revenues excluding depreciation) $(1 - t)$ + t (depreciation)

The term, t (depreciation), is called a tax credit. Note also that there must be a profit; otherwise, t is zero. However, Eq. (11-1) is valid.

have a terminal time value equal to L, but the zero-time value of L is less than L because of the time value of money. There is no tax effect on salvage value, since the exact amount of returned value is considered to have the same value as originally assumed. If there is a loss of expected salvage value, a tax credit applies for the lost portion; conversely, if there is a gain in salvage value, the gain is treated as additional revenue in the real world. However, neither of these possibilities is considered in the projected cash flow evaluation.

The basic principles upon which Eq. (11-1) rests were developed in Chapter 2. These may be reviewed for a system of periodic payments equivalent to a present sum of money as follows:

Annual periodic amount equivalent to a starting capital	=	annual sinking-fund deposit on depreciable capital	+	interest on the total starting capital

$$R = (P - L)\frac{i}{(1 + i)^{n-1}} + iP \qquad (11-2)$$

where R = the annual net revenue available to pay the capital charges on depreciable assets plus the interest on the total investment

P = a present sum of capital

L = the value of all recoverable assets (land, working capital, physical salvage) in dollars at the nth (terminal) year

i = interest rate, fractional

t = tax rate, fractional

and s = the compound interest factor equal to $(1 + i)^n$ for discrete values of n. At low interest rates, s is approximately equal to e^{in} for continuous compounding, where n is years and e is the base of natural logarithms. The use of continuous compounding is discussed later in the chapter.

Thus, the annual returns from the operation of an economic venture correspond to the amounts on either side of this equality. The sum of these returns over a number of periods, n, may be employed to evaluate the economic feasibility by establishing an interest rate on the initial capital. This may be done by using the present worth of the sum of the periodic payments as developed in Chapter 2 for uniform payments. In such a case, each annual return must be reduced to the zero starting time by dividing through by $(1 + i)^n$, where n is the corresponding year. This was done in Eq. (11-1), but the corresponding form may be developed from Eq. (11-2). Thus, after all expenses have been subtracted from net sales and taxes have been considered, annual return or net cash flow for the year is determined. This annual cash flow corresponds to the R of Eq. (11-2) if it is a uniform amount each year. The present worth of such a series for n years was shown in Chapter 2 to be $Ra_{n/i} = R[(1 + i)^n - 1]/[i(1 + i)^n]$. (NOTE: the factor $(1 + i)^n = s$; in addition, $(s - 1)/is = a_{n/i}$, the present worth factor, which may be computed or estimated from Appendix C-2 or C-4 or Fig. 2-2.)

Performing these operations on Eq. (11-2) and solving for the indicated present worth yields the following, which may also be derived from Eq. (2-8):

$$\frac{R(s - 1)}{is} = (P - L)/s + P(s - 1)/s \qquad (11\text{-}3)$$

or $\qquad Ra_{n/i} = (P - L + Ps - P)/s = P - L/s$

and $\qquad P = Ra_{n/i} + L/s \qquad (11\text{-}4)$

This equation for present worth is the critical one and is the theoretical basis for Eq. (11-1) to which it has an obvious similarity, since R is the real dollars of revenue (sales minus costs plus depreciation tax credits) available each period to recover the depreciable capital and pay interest on the total capital involved. Where the values of $S, C, D,$ and L of Eq. (11-1) are constant, the amount equivalent to R is constant and may be computed. It may be substituted directly in Eq. (11-4) for calculation purposes.

Where the sales dollars, costs, depreciation, etc., are not constant, recourse to Eq. (11-1) must be made, with summations for year to year values. However, here also certain tools are available for the economic analysis. For example, where nonregular depreciation charges occur according to a prescribed scheme (such as the sum-of-digits method), the summation formulas of Table 3-2 may be utilized. Similarly, if either the sales or the costs vary annually in a geometric or arithmetic series, their present worths may be determined by the summation equations given in Chapter 2 (page 31). These expressions are much less laborious than a stepwise procedure for each year by year value.

The basic relation of Eq. (11-4) has four variables $P, R, i,$ and/or n, any of which may be used as a basis for profitability. For example, a company invests $10 million in a venture at a certain date. The present worth at that date of all expected future returns in real dollars for N years can be computed and then compared with the $10 million. The relative values will be one measure of the profitability of the venture. Obviously, the interest rate used, the type of depreciation selected, and the estimated life will affect the calculated present worth of the real dollars returned. Alternatively, a profitability model can be considered which will give exactly the present worth, when it is at zero time and equal to the investment, some specific interest rate being used for an assumed life of N years. Here then the interest rate is a measure of the profitability, and this specific method is called the economic rate of return.

Still another procedure is to fix the interest rate and solve for the unknown time that makes the present worth of the returns just equal to the investment. This time is the economic payout time and is also a relative measure of profitability.

Thus, management employs numerous models to evaluate profitability; the relative merit of any one method is difficult to evaluate objectively because each has certain advantages. As various methods are presented and illustrated, the reader may judge their applicability to any specific problem. In addition, more sophisticated analyses are discussed.

SOURCE OF DATA

In an existing project, the data in a profitability calculation are obtained from the accounting records. Where the operation is a new or a proposed one, the amount of available exact data will vary, since some figures of necessity are estimates. These estimates are subject to a varying amount of uncertainty. However, in some instances, the uncertainty can be evaluated from probability considerations.

Much, if not all, of the variations among profitability models lies in the specific procedures employed in the basic economic equation. For example the treatment of taxes and tax credits for depreciation and interest payments may be considered or omitted. In other procedures, the time value of money may be completely ignored. Conversely, compound interest may be included with discrete compounding (definite fractions of a year) or with continuous compounding. Investment interest charges prior to plant operations may be optional in other methods. Accordingly, any discussion of profitability demands an exact statement as to what and how the data were obtained and used in the model.

Since the profitability model is defined by Eq. (11-1), profitability can be evaluated readily if the exact items to be substituted for investment, depreciation, earnings, and time (or interest rate) are specified. However, the selection of these items causes the great diversity of proposals for defining profitability, which in some cases unduly complicate the equations. The approach of this book is to use the most simple ones if they are in accord with the required accuracy from a theoretical basis.

In any real world solution for present worth, the most controversial item is probably the value of the interest selected. Here a wide range is permissible, varying from current bond rates to the maximum the industry attains. Any selection is based on a company policy that has been established for such evaluations.

THE EXCESS PRESENT WORTH MODEL

This method is also called the "venture worth" procedure or the "incremental present worth" model. In this method, the present worth of all future returned capital is compared with the initial capital investment plus the present worth of any other subsequent investments. The magnitude of this excess is a measure of the profitability. The procedure followed is to evaluate P_x by a suitable equation and compare its magnitude with the total present worth of all investments, I_x, evaluated at the same datum year, x.

The concept of the profitability model may be demonstrated by a diagram such as Fig. 11-1, where all data are expressed as their value at zero time, x, when operations start. Two years prior to year x, at point a, the project commitment of $1 million was made, the cash being immediately

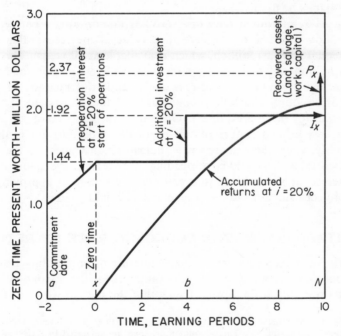

Fig. 11-1. Zero reference time profitability diagram.

available. The investment was for a new merchandizing and warehousing plan expected to provide annual returns of $506,000 for ten years. During the time period, a to x, the warehouses were built and other costs incurred. Thus, the investment, I, had a value then of $I/(1 + i)^n$ at time x, where n is -2 (that is, two years for preoperation interest periods) and i is the interest rate employed by the company. This interest rate may be, for example, at a rate of 20 per cent, whence the value of I_x is $1.44 million. If no further investments are made, the zero-time present worth remains at $1.44 million for the life of the project. However, if an additional non-depreciable investment[2] of $1 million is made at four years, its additional zero-time present worth is $1.0/(1 + i)^4$, which at $i = 0.20$ is about $0.48 million. The project zero-time worth value, I_x, thus becomes $1.44 + 0.48 = $1.92 million for the remainder of its life. This sum is the reference amount to which revenues from the operations should be compared. All data on Fig. 11-1 are values at time zero.

Revenues are adjusted for all deductions of expense, taxes, and interest charges for each year of operation, and the present worth of cash flows is accumulated over the life of the project, which in this case is ten years. The accumulation of new revenues is expressed by Eqs. (11-1) if this accumulation is not uniform. However, if uniform annual cash flows apply, Eq. (11-4) may be used.

In Fig. 11-1, a terminal value of P_x is $2.37 million after a correction of $0.25 million is added for the zero time value of the recoverable assets, $L;$ L is equal to $1.54 million, which is available at 10 years. This value of $2.37 million may then be compared with the $1.92 million for I_x. The excess (or deficiency) of P_x over I_x is one quantitative measure of profitability (or discounted cash flows) compared to the investment at the assumed value for the interest rate, i. Obviously, the selection of the interest rate is a very critical feature of this particular procedure for evaluating profitability, since a low interest rate indicates high values for the excess present worth, which could be misleading. Conversely, the assumption of high interest rates could lead to a conservative policy of approving proposed projects, since the excess present worth would be low.

PROFITABILITY AS THE ECONOMIC RATE OF RETURN

Clearly, the next procedure for the evaluation of profitability from Fig. 11-1 is the determination of the rate of interest for the discounted values of the net revenues from a project such that the present worth of the dis-

---

[2] If this new investment is depreciable, the present worth at plus 4 years for its depreciation must be determined. This amount must then be converted to zero-time present worth in one step by dividing by $(1 + i)^4$, and included in Eq. 11-1.

counted values is exactly equal to the present value of the investment. For the problem shown in Fig. 11-1, this equality occurs after trial and error at about $i = 25$ per cent with a zero-time present worth of $1.98 million. In a simple case involving a single investment point (the present time), constant values of returns each year, no salvage value, and a fixed value of n, the rate i may be computed from Eq. (11-4) or from some equivalent form. This specific interest rate is called the *economic rate of return,* since it is based on the exact economy equation for a constant annual return, R, which is equal to the cash flow.

This method of evaluating profitability is utilized by all methods considering the time value of money. The differences in the methods lie in the values employed for the accounting terms. For example, certain forms of the model may be called the discounted-cash-flow method, the investors method, the profitability index, the interest rate of return, and by other names. The form of the equation may vary from Eq. (11-1) as discussed here—with discounting of all amounts on an annual end of year (discrete) basis—to a form that uses continuous compounding[3] (investors method).

Some models employ straight-line depreciation, while others use the sum-of-digits method or other accelerated procedures. Refinements, such as the sum-of-digits method and other comparable procedures, are unnecessary complications, unless their magnitude has been sufficiently demonstrated to require separate consideration. Usually, this is necessary only if time value of money (interest) is considered. For example, if any return but a constant annual return, R, is postulated, each nonuniform future estimated return must be corrected separately to a present worth value; this holds unless the returns form some type of mathematical series. Furthermore, the correction must be *at an interest rate at which the ultimate for defining profitability is the sought-for-rate.* Obviously, trial and error solutions are laborious unless summation equations are available. However, one or two trials usually are sufficient to cause the accumulated returns curve of Fig. 11-1 to intersect the I_x value at the expected life of n or N years. Computing machines simplify this labor considerably. The question is whether such refinements are warranted. Whenever nonuniform depreciation (and/or earnings) is considered (which is the case with other than the straight-line method), the annual sales or annual expenses must vary to yield constant earnings; otherwise, the earnings will vary for a given annual sales dollars. Thus, in

---

[3] In the case of continuous compounding, the form is the same as for Eq. (11-1), except that $(1 + i)^n$ is replaced by e^{in}, since the limit of $(1 + i/m)^{mn}$ is e^{in} where m approaches infinity and where i is the nominal annual interest rate and m is the number of times compounded per year when n is the number of years or fraction thereof. The discrete and continuous forms can show considerable disagreement at interest rates of 20 per cent for values of n of 10 years.

order to simplify the analysis, it seems reasonable to merely postulate straight-line depreciation and constant earnings for a constant average annual sales (since all projected cash flows in the future are at best an estimate.) The alternative would be to postulate a variation in sales in such a manner that earnings vary in an amount to exactly compensate for nonuniform depreciation charges. This postulation still permits a uniform annual return, R, if such is desired to simplify calculations. Otherwise, each annual nonuniform total of depreciation and earnings must be reduced to present worth separately by trial and error at different interest rates. Errors in estimating annual sales certainly are of the same magnitude as differences in depreciation allowances caused by the use of different methods, and the utilization of an average constant return (depreciation plus earnings) is most expedient. This does not imply that nonuniform values should not be employed. In fact, certain problems at the end of the chapter use nonconstant figures as being more realistic.

ECONOMIC RATE OF RETURN EXAMPLE

Problem

What is the profitability for a firm with an initial total investment of $250,000; a uniform return is made up of depreciation at $25,000 per year, and annual net profits after taxes are $12,000?[4] The firm's operations have an estimated life of ten years with no salvage.

Solution

The solution is readily found by trial and error, using Eq. (11-4) or Fig. 2-2. Thus, from Eq. (11-4) which may be rearranged as follows:

$$a_{n/i} = (P - L/s)/R \tag{11-5}$$

---

[4] Such a net revenue or cash flow, for example, would be generated by sales of $101,670 and costs (excluding straight-line depreciation) of $50,000. Thus, $S_n - C_n = \$51,670$, and at a tax rate of 55 per cent, the net cash flow for any year, n, would be by the usual accounting procedure:

	Dollars
$S_n - C_n$	51,670
Depreciation 250,000/10 =	25,000
Taxable profit	26,670
Profits tax at 55 per cent	14,670
Net profit	12,000

Total annual return = $25,000 + 12,000 = $37,000

The same result is obtained for any year n from Eq. (11-1) as follows:

$$(1 - t)(S_n - C_n) + tD_n = 0.45(51,670) + 0.55(25,000) = \$37,000$$

where P, L, and R are known, where a value of i may be assumed, and where for a life of n years the calculated $a_{10/i}$ of Eq. (11-5) must agree with the tabulated value in Appendix C-2 at the assumed interest rate or with the values in Fig. 2-2. Where there are no recoverable assets ($L = 0$), Eq. (11-5) reduces to $a_{n/i} = P/R$; the values in this equation and the corresponding unknown interest rate may be obtained from the table in Appendix C-2 or from Fig. 2-2.

Thus, $L = 0$, the annual return of real dollars available is $25,000 + 12,000 = \$37,000$. Therefore,

$$a_{10/i} = 250,000/37,000 = 6.76$$

At $n = 10$ years, the tabulated value of 6.76 in Appendix C-2 occurs at $i = 7.84$ per cent by interpolation. Similarly, for the chart in Fig. 2-2, which is entered at 10 years and at $a_{n/i} = 6.76$, the interest rate is estimated from the family of curves.

In this simplified example, the annual returns were considered to be constant. However, where they vary, the summation procedure of Eq. (11-1) must be followed year by year for each term from zero time to the terminal point. Eq. (11-1) is shown as a smooth curve in the idealized Fig. 11-1, but it may take any form (including negative net values in the first years of operation). When the economic rate of return is assumed correctly, the present-worth-level-for-investment line and the accumulated-present-worth-of-returns will intersect exactly at the terminal life at point N on the idealized diagram. At a lower value of money (lower interest rate), the discounted returns have a higher present worth, and vice versa. At any time, n, after the zero point, the vertical distance between the present-worth level and accumulated curve represents the unrecovered value of the capital expenditures. This unrecovered value is *at the zero time reference point,* not at time, n. Its value at time, n, can be obtained; the zero-point value is multiplied by $s = (1 + i)^n$ for the elapsed time, n, that is being considered.

EXAMPLE WITH NONUNIFORM RETURNS

To demonstrate the calculation with nonuniform returns, one may consider a problem where the data are as shown in Table 11-1 for a project having an initial investment of \$40,000, all capital being depreciable by the sum-of-digits method for 10 years. There is no salvage value and a 50 per cent profits tax. The calculations follow the terms in Eq. (11-1) for two assumed economic earning rates of 20 per cent and 25 per cent. The former is too low and the latter too high. However, by interpolation of the two percentages and the calculated present worths compared to \$40,000, the estimated earning rate for the zero time present worth is

TABLE 11-1. Calculation of Profitability for Nonuniform Cash Flows—$40,000 Zero-Time Present Worth. (All Values in Dollars)

Year, n	Cash Flows before Taxes [a]	Net Cash Flows after Taxes [b]	Sum-of-Digits Depreciation	Total Annual Return [c]	Discounted Values = Zero-time Present Worth [d]			
					Trial 1, 20%		Trial 2, 25%	
					$s=(1+i)^n$	Present Worth	$s=(1+i)^n$	Present Worth
1	8760	4730	7270	12,000	1.200	10,000	1.250	9600
2	4540	2450	6550	9000	1.440	6250	1.562	5760
3	7740	4180	5820	10,000	1.728	5790	1.953	5120
4	7240	3910	5090	9000	2.074	4340	2.441	3690
5	6740	3640	4360	8000	2.488	3270	3.052	2610
6	15,480	8360	3640	12,000	2.986	4020	3.815	3150
7	11,280	6090	2910	9000	3.583	2510	4.768	1890
8	16,330	8820	2180	11,000	4.300	2560	5.960	1850
9	15,830	8550	1450	10,000	5.160	1940	7.451	1340
10	17,170	9270	730	10,000	6.192	1610	9.313	1070
Totals	111,110	60,000	40,000	100,000		42,290		36,080

[a] These are gross profits.   [b] These are net profits.

[c] These are annual individual values of $(1 - t)(S_n - C_n) + tD$ of Eq. (11-1) before being converted to present worth. Since $(1 - t)(S_n - C_n) + tD = (1 - t)(S_n - C_n - D) + D$, alternatively, if desired, the zero-time present worth of net profits in the third column may be computed separately, and the values for the depreciation credit may be added by calculating them, using the summation equations of Table 3-2. Thus, the total would be the zero-time present worths of all the net profits and depreciation charges, which are the same as for the annual returns as shown.

[d] Discrete compound-amount factors, s, as read from Appendix C-2.

about 21.9 per cent. From column five, the average annual return is $10,000 for 10 years. If this is considered as a uniform value for R, then from Eq. (11-5) or Table 11-2:

$$a_{10/i} = P/R = 40,000/10,000 = 4.00$$

By interpolation from Appendix C-2 for $a_{10/i} = 4.00$, the value of i is about 23.5 per cent for the computation with average values compared to the more exact 21.4 per cent.

Some authors omit the tax rate. In writing Eq. 11-1 without the tax credit, they subtract the zero-time present worth of the entire investment, I_x. In this way, they obtain the excess of P_x over I_x directly. However, this is unrealistic for the real world because taxes must be paid and also because present worth of the tax credits for depreciation will vary depending upon the method employed. Equation 11-1 allows for these variations.

The model shown in Fig. 11-1 represents a solution where two years before operations started a commitment of capital was made and revenue was received. In the example on page 272 where a $250,000 investment is involved, the commitment point for capital is taken to be identical to the zero point for operations. This, of course, is not true in actual practice, and interest must be charged on capital from the time it is spent. Thus, different interest rates would cause a variation in the present worth level at zero time in the diagram. Note also that the diagram shows the initial commitment of $1 million as occurring all at one time at -2 years. Actually such expenditures would be irregular in timing. However, all such expenditures could be corrected exactly to zero time, and added together to establish the present-worth level by the economic equations. Thus, a capital expenditure before zero time is corrected by dividing it by the compound interest factor, $s = (1 + i)^n$, where n is years before zero time and has a negative sign as shown in Fig. 11-1. Expenditure after zero time such as at b is corrected back by dividing by the s factor, where n is years after zero time. If there is a change in salvage value (or other recoverable assets) because of changes in capital investment, these corrections should be included by making the summations in parts where necessary. For example, if at the sixth year the amount of L is reduced, only a portion of the initial amount is recoverable at the terminal age, the other portion being recovered at $n = 6$ as a cost in that year. This affects only the net revenue line.

As noted previously, either the discrete end of the period factor or the continuous compounding factor, e^{in}, can be employed. The arithmetic averaging of expenditures and revenues, when they are of similar magnitude, does not cause too great an error, and is an obvious simplification over a tedious step-by-step calculation for all irregular items. As illustrated in the previous example, simplification of a lump sum commitment at a fixed initial point permits a quick solution.

There is a terminal point N, the life of the project, at which the lines intersect in Fig. 11-1 at the proper assumed interest rate, which is the economic rate of return. At this point, the operation has no further obligation to pay for the capital costs and associated interest charges. Additional earnings beyond this point indicate further profitable returns (or that a longer life at a higher interest rate could have been predicted). Thus, if too low an interest rate had been postulated, the zero-time present-worth level was reached in less time than the projected life.

The result obtained by the economic-rate-of-return method requires that all returns and investments continue to earn *at the interest rate* obtained by solution of the model. As suggested by other authors, this implication is an important conceptual aspect of this method and its variants.

PROFITABILITY AS THE ECONOMIC PAYOUT TIME

The basic equation Eq. (11-2) contains another variable, n, that can be used as a measure of profitability. If the interest rate is assumed arbitrarily, Eq. (11-6) can be utilized to solve for n in the same manner that Eq. (11-5) was employed for i when n was assumed.

Where $a_{n/i} = (s - 1)/is$, one may write from Eq. (11-4):

$$P = \frac{R(s - 1)}{is} + L/s$$

which gives[5]:

$$n = \frac{\log\left[(R - iL)/(R - iP)\right]}{\log(1 + i)}, \text{ years} \qquad (11\text{-}6)$$

Since this relation is based on the theoretical economy equation, it is called the "economic payout time," and represents the value of n required

---

[5] The derivation of the equation is as follows:

$$iPs = sR - R + iL$$

$$s(R - iP) = R - iL$$

$$(1 + i)n = (R - iL)/(R - iP)$$

$$n = \frac{\log\left[(R - iL)/(R - iP)\right]}{\log(1 + i)} \qquad (11\text{-}6)$$

for the discounted average yearly returns, R, at an assumed interest rate to equal the value of the investment capital, P, referred to zero time. Applicable here is the discussion (page 271) of economic rate of return regarding interest rates, irregular capital expenditures, profits tax, non-uniform annual earnings, depreciation charges, etc. Names for payout time are "cash recovery period" and "payoff period" as well as some others which do and do not include interest.

ECONOMIC PAYOUT TIME

Problem

This problem employs the same information as in the one on economic rate of return (page 272). Thus, if an assumed interest rate of 10 per cent is employed as company policy for profitability evaluations, what is the calculated economic payout time?

Solution

From Eq. (11-6) with L equal to zero:

$$n = \frac{\log [37,000/(37,000 - 0.10 \times 250,000)]}{\log (1 + 0.10)}$$

$$n = \frac{\log (37,000/12,000)}{\log 1.10} = \frac{0.4890}{0.0414} = 11.8 \text{ years}$$

If the previously found interest value of 7.84 per cent for an estimated 10 year life had been used as company policy, the economic payout time would be:

$$n = \frac{\log (37,000/17,400)}{\log 1.0784} = \frac{0.328}{0.0328} = 10.0 \text{ years}$$

The answer agrees with the estimated 10 year life; this is as it should be for the particular interest rate employed.

AN ALTERNATIVE PROFITABILITY DIAGRAM

Although the zero-time present-worth procedure of Fig. 11-1 is preferred by the author, some writers use a modification. This is shown in Fig. 11-2 and might be described as an "unrecovered balance" diagram. It represents the difference between the cash-inflow and cash-outflow balance and relative investment, allowing for interest. It differs from Fig. 11-1 in that it shows the unrecovered balance directly after any

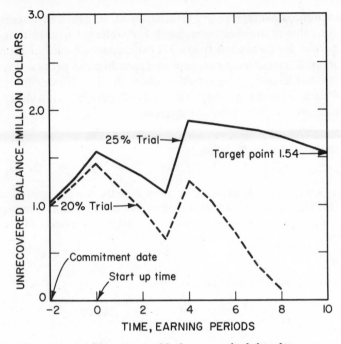

Fig. 11-2. Unrecovered balance profitability diagram.

elapsed time. In Fig. 11-1 this unrecovered balance is shown at a datum of zero time, as represented by the vertical difference between the present-worth-level line and the accumulated curve for revenues. The data in Fig. 11-2 are the same as for Fig. 11-1; these data give a correct economic rate of return when the trial interest rate (of about 25 per cent) results in the unrecovered balance being exactly equal to the target point for the recoverable assets of $1.54 million at the terminal life of 10 years. The calculation data are given in Table 11-2.

Perhaps the zero-time present-worth method is more readily observed as a true economic rate of return than is the unrecovered balance method. This is the case because all revenues and expenditures in the zero-time present-worth method are reduced to a common time (the present or zero age) for comparison. However, the unrecovered balance method provides a more realistic evaluation of the economic state of the venture at any stage of its life. Trial and error computations are required in both methods, each being about equally tedious although both are amenable to machine computing programs. Either uniform, nonuniform, or average annual returns can be considered in both methods.

TABLE 11-2. Unrecovered Balance Method for Profitability[a] (Data same as for Zero-Time Present-Worth Method illustrated in Fig. 11-1)

	Analysis at 20 per cent Interest			Analysis at 25 per cent Interest		
Year	Year End Total	Revenues at End of Year	Unrecovered Balance	Year End Total	Revenues at End of Year	Unrecovered Balance
−2	1.00	—	1.00	1.00	—	1.00
0	1.440[b]	—	1.440	1.560	—	1.560
1	1.728[c]	0.506	1.222	1.950	0.506	1.444
2	1.466	0.506	0.960	1.806	0.506	1.300
3	1.152	0.506	0.646	1.624	0.506	1.118
4[d]	1.775	0.506	1.269	2.399	0.506	1.893
5	1.523	0.506	1.017	2.366	0.506	1.860
6	1.220	0.506	0.714	2.326	0.506	1.820
7	0.857	0.506	0.351	2.275	0.506	1.770
8	0.612	0.506	0.106	2.211	0.506	1.705
9	—	—	—	2.132	0.506	1.626
10	—	—	—	2.033	0.506	1.527

[a] Target value of $1.54 million of recoverable assets at 10th year.
[b] Start up of operations; interest charges for two years added to initial investment of $1.00 million commited two years previously.
[c] Interest for one year added to unrecovered balance for previous year.
[d] Additional investment in this year of $1.00 million.

The theoretical basis for Fig. 11-2 is a rearrangement of Eq. (11-4). Since the latter equation describes the present worth at time zero, the value at n time periods later is s times P, where s is the compound interest factor equal to $(1 + i)^n$ at the proper assumed interest rate:

$$Ps = R(s - 1)/i + L$$

This may be rewritten as:

$$Ps - R(s - 1)/i = L$$

which says that the future value of an initial investment minus the future value of the annual returns must equal the salvage value at the future date. This is a statement of the stepwise analysis shown in Fig. 11-2. For a uniform value of R with P being the equivalent value of investment at zero time, the preceding equation may be solved by trial and error for the value of i that gives the amount L, which is equal to the recoverable assets after n periods. An indicated procedure would be to solve Eq. (11-5) for $a_{n/i}$ and interpolate from tables of $a_{n/i}$ at different trial interest rates to confirm the assumed interest rate.

OTHER PROFITABILITY METHODS

Some writers avoid the trial and error methods of true economic calculations by eliminating the effect of interest. The main argument for eliminating the effect of interest is simplicity; it is also noted that almost any reasonable methods are acceptable for comparative studies provided their limitations are recognized. However, eliminating the effect of interest is controversial. The importance of interest and tax credits indicate that both of these items may be critical for large capital outlays. Under these conditions, it is imperative that they be included in the analysis. Two additional methods (see Table 11-3) for profitability will be discussed, after which an example illustrating all four procedures will be shown.

TABLE 11-3. Comparative Profitability Calculations[a]

1. Economic Rate of Return (Discounted Cash Flow)
Average yearly return using R as sum of depreciation plus annual earnings (profit) after profits tax is computed *as* if none of initial investment capital is borrowed. The $a_{n/i}$ is computed from:

$$a_{n/i} = (P - L/s)/R \tag{11-5}$$

Solve for i.

2. Economic Payout Time
Computed exactly the same as (1) above except that some selected interest rate is assumed and the unknown is n.

$$n = \frac{\log[(R - iL)/(R - iP)]}{\log(1 + i)}, \text{years} \tag{11-6}$$

3. Capitalized Earning Rate
The annual earnings are the net profit after taxes, which includes the tax credit for borrowed funds if any, *and* includes financial costs (interest actually paid out). The total capital is involved and salvage value is neglected unless a different procedure is specifically stated.

$$i_c = \frac{\text{average annual earnings}}{\text{invested capital}} \tag{11-7}$$

Alternatively, the proprietary earning (above annual earnings minus financial costs for interest paid out on borrowed funds) may be used, if desired, in the numerator of Eq. (11-7).

4. Capitalized Payout Time

The investment is divided by the annual return. The annual return is composed of the net profit after taxes, as in (3) above, plus the annual depreciation. Total capital is used and salvage is neglected unless a different procedure is specifically stated.

$$n_c = \frac{\text{invested capital}}{\text{average annual return}} \qquad (11\text{-}8)$$

Note that Eq. (11-8) is *not* the reciprocal of Eq. (11-7), since return (earnings plus depreciation) is not the same as earnings.

Alternatively, the proprietary earning, as in method (3) above, may be employed here instead of net profit after taxes.

[a] In methods (1) and (2), a theoretical concept is emphasized and all calculations are made as if no actual interest is paid out on debts. This is done even if it means a computation of a new income statement by the accountant. In methods (3) and (4), a practical viewpoint is taken and only the real dollars actually recovered are used for the calculation. This will give different results for different methods of financing. Note also that if R is uniform, Eqs. (11-5) and (11-6) may be used directly in methods (1) and (2). If the annual cash flows are not uniform, Eq. (11-1) with a year to year calculation is necessary, trying different rates of interest until the zero-time present worth of the investment is equal to the discounted value of all cash flows.

PROFITABILITY AS THE CAPITALIZED EARNING RATE

In an earlier discussion of Eq. (11-2), it was demonstrated that the earnings part of the annual return corresponds to the iP portion of the theoretical economy relation; that is, annual profits are some fraction of the investment, and this fraction can be defined as an interest rate, or:

$$i_c = \frac{\text{profit}}{\text{investment}} \qquad (11\text{-}7)$$

This equation is analogous to the capitalized cost concept. Sometimes the equation is employed in a different form to estimate the value of a firm from its annual earnings and from an assumed interest rate; such an estimation is a capitalization procedure. For these reasons in this text i_c is called the "capitalized earning rate." The assumptions of straight-line depreciation and constant annual earnings for constant sales dollars apply here also, as noted in Table 11-3.

The investment in the denominator ordinarily refers to the total of invested capital, regardless of source, and includes working capital. As a general rule, any salvage values are neglected for the purpose of simplification. Salvage values also may be neglected for new additions or extensions of a firm's operations, where the calculation is only for that portion of new capital required. Here, in addition, there may be no need to include working capital in the analysis.

The data are taken directly from an accountant's report, and the earnings are the sum of the net profit after taxes including the amounts for interest and other financial costs, since these constitute earnings on the operation. However, the definition of earnings may vary. If desired, the interest charges and other financial costs could be omitted from earnings, the residual earning being called the proprietary earning to be employed in Eq. (11-7). In any case, if interest charges exist, the tax credit is automatically included as part of the earning, assuming that the interest cost is considered in computing the taxes. Clearly, a strict statement of the exact data and accounting methods employed is necessary, where the capitalized earning rate procedure is used to evaluate profitability.

Numerous forms of the capitalized earning rate method have been suggested, and one form is often referred to as the "engineer's method" or "return on original investment." However, as mentioned above, its exact significance depends on the data employed. The engineer's method neglects the time value of money and the influence of this time value on variable earnings, since average earnings are considered constant for the life of the project.

One variation of this method known as the "operator's method" uses the ratio of earnings plus depreciation to the investment. This has little basis in theory and will not be considered further. It is the reciprocal of the capitalized payout time to be discussed later.

Another variation of this method is known as the "accountant's method," where the investment used in the denominator is different from the initial investment. In this variation *the average of the depreciable capital* only plus the working capital (and land) is used as the investment. The basis for this form is that the depreciable capital varies from full value to zero (neglecting salvage value) during the proposed life of the venture. The effect in comparison with the capitalized earning rate will vary with the magnitude of the numbers in the denominator as well as the ratio of depreciable capital to working capital. Where working capital is zero, the accountant's method may be twice (for straight-line depreciation, no salvage value) that given for capitalized earning rate. For this reason, the results may be deceiving to those who do not fully understand the basis for calculation. Furthermore, the results are not comparable from industry to industry where ratios of depreciable capital to working capital are not the same.

An argument that has been advanced against the use of average depreciable capital is that the depreciation reserve is a bookkeeping account anyway; moreover, since the capital it represents is being used in the firm's operations, earnings should be charged against it. In other words, the initial level of investment capital tends to remain the same but is merely present in a different form during the life of the venture.

CAPITALIZED EARNING RATE EXAMPLE

Problem

In the previous example (page 272) an annual profit of $12,000 was obtained on a $250,000 investment. Using the same figures, what is the capitalized earning rate compared to the economic rate of return of 7.84 per cent?

Solution

From Eq. (11-7):

$$i_c = \frac{12,000}{250,000} = 0.048, \text{ or } 4.8 \text{ per cent}$$

The rate here is lower than that which would be obtained using economic rate of return. However, no generalization for differences between the capitalized-earning-rate method and the economic-rate-of-return method can be made, since Eq. (11-2) has a definite theoretical relationship between sinking-fund deposit and annual interest if the annual return, R, is broken down into its components. In accounting practice, it would be fortuitous if depreciation and profit were to fit this breakdown exactly. The net profit, analogous to iP, varies freely and bears no relation to the depreciation, whereas theoretically the annual interest is equal to $(s - 1)$ times (sinking-fund deposit). Thus, in the present case, if the profit had been $20,000 and depreciation $17,000, the economic rate of return and capitalized earning rate would each be about 8 per cent. This is a different problem, however, because the $17,000 for depreciation is not one tenth of the investment required by the problem above.

Note that if the *average investment* had been used (accountant's method) the capitalized earning rate would be 2 × 4.8 = 9.6 per cent, which is in better agreement with theory. This result is fortuitous for the conditions of this problem, however, since calculations with other data can yield results showing less agreement.

PROFITABILITY AS THE CAPITALIZED PAYOUT TIME

As with theoretical economy relations, there is a payout time which complements the capitalized earning rate. This procedure defines profitability in terms of the ratio of investment to annual return. In this text, it will be called capitalized payout time to prevent it from being confused with other methods. It is computed as follows:

$$n_c = \frac{\text{investment}}{\text{annual return}}, \text{ years} \tag{11-8}$$

The data are of the same nature as those used for capitalized earning rate, except here the annual depreciation is added to the profit to obtain the annual return. The discussion of the earnings to be employed for the earning rate applies equally well for payout time, and the use of proprietary earning is optional. When the investment in Eq. (11-8) is the same as the P for economic rate of return, n_c is equal to the ratio of P/R in Eqs. (2-5) and (2-6). This is also the $a_{n/i}$ of Eq. (2-4), provided R is independent of the method of financing.

There is an implied relationship in the capitalized earning rate, which sometimes can be simplified. Thus, when salvage value is not involved, Eq. (11-8) may be rewritten as follows:

$$n_c = \frac{I}{Ii_c + Id} = \frac{1}{i_c + d}, \text{ years} \tag{11-8}$$

where I = the investment

i_c = the profit rate expressed as a fraction of the investment

and d = the *uniform* fraction of the investment that is charged off each year for depreciation.

If salvage value is involved, the analyst must decide whether to include this in the numerator and to allow for it in the depreciation term, Id. The depreciation portion could still be uniform if the ratio of depreciable investment to total investment were constant. However, the d term would not be the same numerically as for straight-line depreciation. On the other hand, the same form of Eq. (11-8) could be used.

CAPITALIZED PAYOUT TIME EXAMPLE

Problem

For the same data in the previous examples what is the capitalized payout time?

Solution

From Eq. (11-8):

$$n_c = \frac{250,000}{25,000 + 12,000} = 6.76 \text{ years}$$

This compared with the theoretical value of 10 years from Eq. (11-6) shows the influence of the time value of money in extending the payout time.

DEPRECIATION TO PROFIT RATIOS

In some financial analyses of operations, the ratio of depreciation to profit is employed as a measure of the status of the firm's earning power. Considering depreciable investment only, a payout time may be written using this ratio.

$$\text{Payout time} = \frac{(\text{original depreciable investment})/\text{profit}}{(1 + \text{depreciation})/\text{profit}} \qquad (11\text{-}10)$$

The numerator of this equation is the reciprocal of the capitalized earning rate based on the depreciable investment. The depreciation to profit ratio as a measure of relative economic feasibility appears to be limited to firms where revenues and operations are expected to give similar values of the ratio. In this respect, the relation may be useful to show how different industries vary in the components of profit and depreciation that make up their annual returns.

CAPITALIZED VALUE AS A MEASURE OF PROFITABILITY

The capitalization of all annual charges of an operation to give a fictitious sum of money, the interest on which will be equivalent to annual operating costs plus annual charges for perpetual renewals of the equipment, is called capitalized cost. This same concept could be applied to a capitalization of the annual return (depreciation plus earning) at the end of each period for a projected venture. Dividing the *annual return* by an arbitrarily selected interest rate yields a fictitious capital sum; this sum is a type of initial investment but can be quite misleading. This is true because such a mathematical operation is not the same as dividing the *earnings* by an assumed interest rate, such as is sometimes done to evaluate the potential value of a going concern. When the *sum* of annual depreciation plus earnings is capitalized at low interest rates, tremendous sums are obtained for capital investment. However, these sums have little or no real significance in relation to the profitability of a complete operation. These large sums result because sufficient interest dollars must be earned to provide for the perpetual life of the operation; on the other hand, this provision is not realistic because of obsolescence.

The concept of capitalized cost for permanent structures has a sphere of usefulness for the comparison of alternatives. Similarly, the concept is practical for comparison of isolated equipment items (such as air conditioning pumps, etc.) whose service is more or less expected to remain continuous, even though the major emphasis of a firm's operations is expected to change in the long run. However, under conditions of change in emphasis, capitalized cost procedures are of limited utility for evaluating profitability. Furthermore, the influence of tax rates and tax credits for capitalized value computations is very important and can lead to entirely erroneous results when this method is employed. In fact, in some cases, a decision without taxes may be reversed when taxes are considered; this holds even for small investments on isolated equipment items.

RELATIONSHIPS IN PROFITABILITY

If there are no complicating factors of debt financing, nonlinear depreciation, salvage value, or profits tax, the various measures of profitability can be compared as shown in Table 11-3.

By employing the nomenclature of this chapter, one can set up certain equations that relate the accounting evaluations and the theoretical ones.

ACCOUNTING: Annual return = profit plus depreciation

$$R = i_c I + dI = I(i_c + d)$$

where I = investment
i_c = some fractional profit rate, actual or required
d = some fractional depreciation rate

THEORETICAL ECONOMY: Annual return = R

$$R = I/a_{n/i} = \frac{i(1 + i)^n I}{(1 + i)^n - 1}$$

where $a_{n/i}$ = the present worth factor

Equating the two preceding relations gives the following form:

$$\frac{i(1 + i)^n}{(1 + i)^n - 1} = i_c + d$$

$$\frac{(1 + i)^n - 1}{i(1 + i)^n} = \frac{1}{i_c + d} = I/R = a_{n/i}$$

Thus, the economic earning rate i that corresponds to $a_{n/i}$ for a given value of n can be computed or estimated from tables when the capitalized earning rate is used for i_c. It is clear that the relationship between the two earning rates depends on the magnitude of the profit rate, the depreciation rate, and the time period used for study.

In a similar manner the capitalized payout time n_c may be computed directly from the values of i_c and d. By definition:

$$n_c = I/R = \frac{I}{I(i_c + d)} = \frac{1}{i_c + d} \tag{11-18}$$

which is the same as $a_{n/i}$ in the previous equation.

This last relation is particularly useful in establishing minimum payout times when d is known and fixed and when i_c is the minimum acceptable earning rate which a company policy dictates. Values for minimum acceptable rate can be selected on a variety of bases by which earnings of comparable ventures are compared. Table 11-4 is typical of the ones employed for approximation purposes.

TABLE 11-4. Earning Rates of Selected Industries

Industry Type	Debt as Per Cent of Total Investment	Return as Per Cent	
		Stockholders Equity[a]	Total Investment
Machinery	10	13	11
Food products	15	9	8
Tobacco companies	30	10	7
Apparel industries	5	5	5
Steel industry	13	12	10
Chemical companies	10	13	11
Paper products	15	15	13
Motor vehicles	4	14	13
Stone, clay, and glass	9	13	11

[a]Current total investment less funded debt, or approximately the same as a net worth basis. Where average debt capital is about x fraction of total investment, the return, as per cent of total investment, is $(1 - x)$ times the rate of return shown for stockholders' equity rate.

EXAMPLE FOR DIFFERENT METHODS

In order to demonstrate actual calculations for the four profitability methods described in Table 11-3, a composite problem will be worked out.

Problem

Using the data in Table 11-5 and Table 11-6, compare the profitability of a proposal for a new venture by the firm when its own capital is to be used (proposal A) and when a portion of the investment is obtained by selling bonds (proposal B).

TABLE 11-5.   Data for Example Problem on Profitability

	Proposal A	Proposal B
Total investment, P	$200,000	$200,000
To be financed by bonds	none	50,000
Annual profit before depreciation	70,000	70,000
Annual financial costs	none	2000
Interim time from commitment to start of operations	2 years	2 years

The depreciable investment is $150,000 for each proposal, and there is an estimated 10 year life. The company uses an assumed interest rate of 12 per cent for payout calculations. The tax rate is 55 per cent for State and Federal taxes, which results in the data shown in Table 11-6.

TABLE 11-6. Calculated Data for Example Problem on Profitability

	Proposal A	Proposal B
Annual profit after depreciation, but before taxes	$55,000	$55,000
Annual financial costs	none	2000
Taxable profit	55,000	53,000
Profits tax at $t = 0.55$	30,250	29,150
Annual profit after taxes, Y	24,750	25,850
Annual proprietary earning	24,750	23,850
Straight-line depreciation (SL)	15,000	15,000
Annual return, $R = [(SL) + Y]$	39,750	40,850
Proprietary return	39,750	38,850

It will be noted from Table 11-6 that the effect of financial costs is to decrease the profits tax and proprietary return, but the annual return for proposal B is larger than for proposal A, since there is a tax credit for the financial costs. This tax credit is tC where C is any cost not included in taxable profit and t is the tax rate. Thus, the tax credit for financial costs is $(0.55)(2000)$, or $1100, which is the difference in profit and annual return for the two proposals.

Some writers determine the annual return directly by using the profit before depreciation and other intangible costs (such as financial costs) as a base. A tax credit for these items is then added to give the same result as illustrated above. Thus, in proposal B the tax base before depreciation and the financial costs is 55,000 plus 15,000, or $70,000, and the calculations are:

		Dollars
Profits before intangible costs and depreciation	=	70,000
Profits tax at 0.55	=	38,500
Corrected profit: $(1 - t)(60,000)$	=	31,500
Tax credits: $0.55(15,000 + 2000)$	=	9,350
Annual return: (depreciation plus earnings)	=	40,850

This final result is the same as in the previous tabulation for the annual return. Although this is a more direct computation, the meaning is not as clear as when the accounting is carried out in steps. The equality was demonstrated algebraically in connection with Eq. (11-1) at the start of this Chapter. The term, (t) (depreciation), is the tax credit for depreciation. Where the tax is computed before subtracting any allowable deductible cost, a correction (for added profit equal to the tax credit for that cost) is made in the same manner.

The computation for profitability by the methods in Table 11-1 are then carried out.

Solution

Economic Rate of Return. By trial and error, Eq. (11-5) is employed with an annual return R of $39,750 and with a P that is the original investment corrected for preoperation interest. For this solution, P_x must equal I_x at the unknown interest rate. The original investment corrected for start up interim time to zero time is $200,000 (1 + i)^2$. By the final trial and error calculation equation, i is solved according to the model of Eq. (11-5):

$$a_{10/i} = \frac{(1 + i)^2 (200,000) - 50,000/(1 + i)^{10}}{39,750}$$

$$a_{10/i} = 5.04 (1 + i)^2 - 1.26/(1 + i)^{10}$$

Various values of i are assumed, and the value of $a_{10/i}$ computed. This is compared with the tabulated values in Appendix C at 10 years until agreement is shown at some assumed i. Various results are in Table 11-7 where the best answer

TABLE 11-7. Trial and Error Solution of Example for Economic Rate of Return

Assumed i	10	12	11.3
Calculated $a_{10/i}$	5.61	5.91	5.82
Tabulated $a_{10/i}$	6.14	5.65	5.82[a]

[a] As read or interpolated from Appendix C at $n = 10$ years.

for i is 11.3 per cent. Note that in this method the assumption is made that no capital is borrowed; therefore, the calculation does not depend on the method of financing and the answers for both proposals are considered to be identical. Note also that if this company believes that its money should earn at least 12 per cent, there is no "venture" profit in this proposal. As demonstrated previously, continuous compounding can be employed here if it is desirable.

Economic Payout Time. In this problem the unknown is n in the basic equation with interest at 12 per cent as given. Employing Eq. (11-6) with the investment corrected to zero operations time, or $I_x = 200,000\,(1.12)^2$, yields:

$$n = \frac{\log\left[(39,750 - 0.12 \times 50,000)/(39,750 - 0.12 \times 200,000 \times 1.12^2)\right]}{\log 1.12}$$

$$n = \frac{\log\left[33,750/9750\right]}{\log 1.12} = \frac{0.528}{0.049} = 10.8 \text{ years}$$

As might be expected, the payout time exceeds the estimated life at 12 per cent interest. There is a theoretical fallacy in this equation which is common to all payout measures of profitability. The fallacy is that the life period is taken as some finite number N in order to obtain a definite value for depreciation in the making of the calculations. This life may bear no relation to the calculated payout time. However, for most cases, this theoretical fallacy may be ignored.

Capitalized Earning Rate. In this computation the actual profits are employed as described in Table 11-3. Thus, there are at least two answers for each proposal that may be of interest. From Eq. (11-7) the results for the annual returns are:

PROPOSAL A: $\qquad i_c = \dfrac{24,750}{200,000} = 0.124$, or 12.4 per cent

PROPOSAL B: $\qquad i_c = \dfrac{25,850}{200,000} = 0.129$, or 12.9 per cent

Thus, proposal B appears to be the better. However, if the proprietary earning is employed, the results are:

PROPOSAL A: $\qquad i_c = \dfrac{24,750}{200,000} = 0.124$, or 12.4 per cent

PROPOSAL B: $\qquad i_c = \dfrac{23,850}{200,000} = 0.119$, or 11.9 per cent

It is also interesting to speculate on the proprietary rate based only on the owner's capital. For proposal A the rate is the same at 12.4 per cent, but for proposal B it becomes $23,850/150,000 = 0.158$, or 15.8 per cent. By this method, the earning rate approaches infinity with increase in the proportion of investment capital financed by bonds. This illustration emphasizes the importance of exact definitions of terms used in profitability calculations.

NOTE: In this method, there is only a slight amount of "venture" profit indicated for either proposal if the company policy takes a 12 per cent earning rate as a criterion.

Capitalized Payout Time. This method employs Eq. (11-8) of Table 11-3 and may yield at least two answers for each proposal depending upon the basis taken for earning.

PROPOSAL *A*:
$$n_c = \frac{200,000}{39,750} = 5.03 \text{ years}$$

PROPOSAL *B*:
$$n_c = \frac{200,000}{40,850} = 4.90 \text{ years}$$

On this basis, proposal B is slightly better. However, if proprietary return is used, the respective payout times are 5.03 years for proposal A, since the calculation is not affected; however, $200,000/38,850 = 5.15$ years for Proposal B. Once again, this demonstrates the necessity for defining exactly the accounting data to be employed.

COMMENT ON THE EXAMPLE FOR DIFFERENT METHODS

In the preceding example, there is an assumption of a constant R and the equivalent of a single total capital amount at zero time. These are logical and reasonable simplifications, even though there might be irregular revenues, costs, and capital investments at times other than in the initial stages. In general, these irregularities may be averaged to give a satisfactory evaluation. If a decision will be affected appreciably by these simplifications, some major influence requiring separate evaluation must be present. At best, future revenues are estimates, and, if desired, their estimated irregularities and timing may be considered by stepwise reduction to present worth. With depreciation the case is similar. Thus, if the sum-of-digits method is employed, the revenues must change, assuming that a constant, R, is desired for calculations. All of these variations affect the annual return after taxes for a given revenue because of the tax credits for deductible costs.

An obvious simplification (where the accounting data are not subject to wide fluctuations) is the assumption that sales will vary in an amount that exactly compensates for variations in costs, profits taxes, and depreciation charges; in this assumption constant average values are regarded as constant for the annual return, R. Where the revenues before depreciation are constant, the method employed for depreciation (which involves a consideration of financial costs) and the profits-tax effects will influence the value of the annual return, R, and accordingly the rate of return.

In connection with depreciation it should be noted that the arithmetic *average* annual amount is the same for all methods. However, the calculations that directly use the time value of money with R in Eq. (11-2) explicitly employ sinking-fund depreciation as part of the annual return, as has been discussed. The remainder of the R is interest on the investment. This breakdown of R into its component parts does not appear separately in the calculations because the accounting and subsequent calculation require only the total amount R.

Whenever profits taxes are considered and an irregular depreciation is computed as a separate component of the annual return, the depreciation method will affect the present worth. For these situations, the accounting may be carried out at zero time by computing the present worths of (a) the net revenues (before deduction of depreciation and financial or intangible costs) corrected for profits tax, and by adding (b) the tax credits for the present worths of any depreciation and intangible costs. This procedure is defined by Eq. (11-1).

INFLUENCE OF DEPRECIATION PROCEDURE

An illustration based on the previous example may be helpful to explain the influence of the depreciation procedure. It was shown on page 289 that when the time value of money is considered, the earning rate for the revenues in proposal A for a zero-time investment of $(200,000)$ $(1.113)^2 = \$247,760$ and for an annual return of $\$39,750$ is equal to about 11.3 per cent. The annual return in proposal A is the sum of the annual profits plus depreciation, or $\$55,000 + \$15,000 = \$70,000$.

Problem

Employing sum-of-digits depreciation, compute the economic earning rate from Eq. (11-1), using a profits tax of 0.55 and the equations for summations given in Tables 3-1 and 3-2.

Solution

$$P_x = (1 - 0.55) \sum_1^{10} \frac{70,000}{(1 + i)^n} + 0.55 \sum_1^{10} \frac{2(10 - n + 1)(150,000)}{10(10 + 1)} + \frac{50,000}{(1 + i)^{10}}$$

This expression can be further simplified, since the summations themselves are evaluated by Eqs. (2-4) and (3-9), respectively.

$$P_x = (0.45)(70,000)a_{n/i} + \frac{(0.55)(2)(150,000)(10 - a_{n/i})}{10i(10 + 1)} + \frac{50,000}{(1 + i)^{10}}$$

$$P_x = 31,500\, a_{n/i} + \frac{165,000(10 - a_{n/i})}{110i} + \frac{50,000}{(1 + i)^{10}}$$

Thus, one must assume values of i and solve for P_x until by trial and error i is exactly equal to I_x for a preoperation time of two years at the unknown interest rate, or $I_x = 200,000(1 + i)^2$. Trying $i = 0.10$, or 10 per cent, results in:

$$P_x = 31,500(6.145) + \frac{165,000(10 - 6.145)}{11} + \frac{50,000}{2.594}$$

$$P_x = 193,567 + 57,824 + 19,275 = \$270,666$$

Comparing this with an $I_x = 200,000(1.10)^2 = \$242,000$, it is apparent that the computed present worth, P_x, is too large and a greater interest rate should be tried. Trying $i = 0.118$ results in \$250,200 for P_x compared to \$250,000 for I_x. Thus, the economic earning rate with sum-of-digits depreciation is about 11.8 per cent as compared to 11.3 per cent for straight-line depreciation. Other methods of taking depreciation, according to management policy for such items, will give additional different results.

As mentioned previously, eliminating the variations in annual returns can be done by assuming that projected net revenues vary exactly in a manner that compensates for any method of depreciation selected. By this assumption of a uniform annual return, the constant value of R may be employed directly for a profitability evaluation. Note that the example above gives 11.8 per cent return, using a complicated sum-of-digits computation, whereas a simplification gave 11.3 per cent. The difference in these terminal results, of course, will be affected by the proportion of the total investment that is depreciable capital. However, in any case, the magnitude of the difference is rather small. The difference is a function of the life period and interest rate.

EXTENSIONS ON PROFITABILITY

An interesting extension of the profitability evaluation involves how the present worth varies with time for varying revenues, S, depreciation, D, salvage value, L, and/or costs, C when continuous functions are considered applicable. For such an analysis, Eq. (11-11) is suggested:

$$P_x = (1 - t)\left[\int_0^N S(n)\,e^{-in}\,dn - \int_0^N C(n)\,e^{-in}\,dn\right]$$
$$+ t\int_0^N D(n)\,e^{-in}\,dn + L(N)\,e^{-iN} \tag{11-11}$$

In this equation, the instantaneous value of sales in terms of dollars per unit of time is a function, $S(n)$, of the year n such as illustrated in Fig. 11-3. The accumulated present worth values, if they are discounted to zero time continuously by the factor e^{-in} (see Chapter 2) and then summed over N total periods, are given by the integral $\int_0^N S(n)\,e^{-in}\,dn$. A similar form results for costs, and the summations may be presented as in Fig. 11-4 for any terminal life, N. The interest rate, i, may be any value that company policy employs for such evaluations. The *net revenues* corrected for taxes thus generate a real dollar present worth; when corrections are added for depreciation and salvage (since $C(n)$ does not include these), a zero-time present worth, P_x, results for a life N.

Depreciation requires some further discussion. When there is no profits tax, the third integral term on the right vanishes because t is zero, and the depreciation credit is in the return $[S(n) - C(n)]$, since no depreciation was deducted in this form. However, for a finite t it is necessary to consider the tax credit for depreciation. The boundary or limiting condition for depreciation is the project life, N, and therefore the instantaneous function $D(n)$ terminates whenever $n = N$. The instantaneous function itself may vary depending upon the depreciation method selected

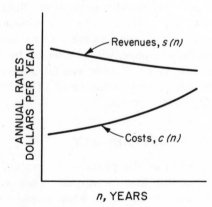

Fig. 11-3. Instantaneous values at n years.

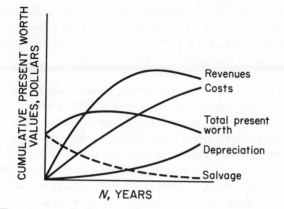

Fig. 11-4. Present worth values after N years at a given tax rate, t.

(see Chapter 3). The salvage value term, L, for nondepreciable assets is not evaluated until the terminal period, N, at the project's completion. Therefore, salvage value also may be shown as a function of N, but it is not an accumulated value as are the other items in Fig. 11-4.

When all of the components of Eq. (11-11) can be expressed in terms of n, a visual concept results as in Fig. 11-4, which shows the summation results after N years. In certain real world cases, the data of Fig. 11-4 may be more readily available than those of Fig. 11-3.

The instantaneous rate of change of present worth with N of any item of Fig. 11-4 is the slope of the respective line or the differential of the terms of Eq. (11-11) with respect to N. For example, for sales it can be shown by advanced calculus that this slope is $S(N)e^{-iN}$. The maximum present worth is obtained mathematically from setting the derivative of Eq. (11-11) with respect to N equal to zero and solving for the optimum, N. This N is then substituted in Eq. (11-11) to determine the numerical value of present worth. This requires advanced calculus because the parameter, N, in the limit of the integral is not the variable of integration. However, a resulting equality[6] after simplification yields:

$$dP_x/dN =$$

$$e^{-iN}\{(1 - t)[S(N) - C(N)] + tD(N) - iL(N) + d[L(N)]/dN\} = 0$$

$$(1 - t)[S(N) - C(N)] + tD(N) = iL(N) - d[L(N)]/dN \quad (11\text{-}12)$$

---

[6] The expressions $S(N)$, $C(N)$, etc., are the same as $S(n)$, $C(n)$, respectively, in Eq. (11-11). They are not integrated functions. N is used to designate a terminal period, whereas n means any period.

Each of the terms in this equation can be evaluated from the corresponding values of Eq. (11-11) or from Fig. 11-4, and the optimum N can be evaluated by trial and error. Note that any function, such as $S(N)$, is equal to the slope of the cumulative line divided by e^{-iN}, since it was noted above that the slope was $S(N) e^{-iN}$.

Equation (11-12) is a form of incremental analysis (see Chapter 9). Thus, the optimum age for the maximum present worth occurs when the additional annual net return on the left side of the equation (incremental gain) at any assumed terminal year, N, just equals the incremental loss for that year. This incremental loss is the right side of Eq. (11-12) and is the interest on the salvage value for the year plus the change (loss) in salvage value for that year. NOTE: Algebraically this change is negative, and the negative sign in the derivative corrects it and makes Eq. (11-12) consistent.

The optimum, N, may be found and the corresponding maximum present worth found for an assumed i by using Eq. (11-11). The excess of $P_{x_{max}}$ over the initial investment of the project can then be compared considering a variety of alternatives if desired. Simple functions for S, C, D and L, although not exact fits for the real world data, often give reasonable and satisfactory solutions for decision policies.

Where the sales and costs are linear with n, then with an initial value of sales rate, S_0, in dollars per year (S_0 is a pseudo value of sales at zero time to permit mathematical manipulation of the equations) the summation for its present worth is $\int_0^N (S_0 + k_s n) e^{-in} dn$ and $S(N) = S_0 + k_s N$ where k_s is the yearly change in sales dollars of income. A similar analysis applies for costs so that a combined form for net revenues excluding depreciation and taxes is:

$$S(N) - C(N) = S_0 + k_s N - (C_0 + k_c N) = S_0 - C_0 + N (k_s - k_c) \quad (11\text{-}13)$$

If sales and costs are constant each year, Eq. (11-13) reduces to:

$$S(N) - C(N) = S_0 - C_0 \quad (11\text{-}14)$$

For straight-line depreciation, the yearly rate, $D(N)$, is a constant equal to $(P - L)/N$. For other depreciation procedures, the $D(N)$ relation is more complex (see Chapter 3).

The analysis for maximum present worth as presented here is the application of a sophisticated mathematical treatment which may not actually be attained in the real world. In other words, a true maximum may not be found in the real case because the costs, revenues, tax credits or salvage value may not vary in a manner to produce a maximum. Furthermore, these variables may not be amenable to simple empirical relations. However, even if such defects are present, the use of approximations and the concept of the mathematical model are powerful tools for management as a guide to formulating policy. Their power lies in a semiquantitative answer that provides guidance in the confirmation of management's decisions. Thus, management can act with more confidence than by the use of intuition alone.

In the evaluation of profitability, some form of a model is a useful approximation for real world data. Accordingly, assumptions are made and accepted as applying to the real world situation. For example, the assumption of straight-line depreciation will produce a yearly rate which must be considered acceptable for tax purposes in order to include the tax credit term in Eqs. (11-1) and Eq. (11-11). Avoiding such complications by omitting the tax consideration from all items in the profitability evaluation merely limits the analysis to the use of artificial dollars that are not really available. Thus, the analysis of profitability may appear excessively complex. However, if a rigorous solution with real cash flow dollars is desired, this complexity if inherent. The model utilized and its complexity depend entirely upon how realistic a result is desired.

At this point, two points should be noted in connection with these types of mathematical models. (1) Empirical equations of orders greater than one will have multiple answers, some of which are absurd and may be discarded. (2) For mathematical relations, the setting of a first derivative equal to zero may indicate either a maximum, a minimum or an inflection point; thus, all solutions should be checked to ensure that the desired result is achieved.

An example will demonstrate the procedures and limitations of the profitability method.

MAXIMUM PRESENT WORTH EXAMPLE

Problem

Determine the maximum present worth for the following problem and compare the optimum, N, by the direct incremental analysis relation. The expected annual sales in any year n of a product is \$4000 with annual costs of $3000 + 200\,n$ dollars. The machine investment is \$1500 with salvage value after N years elapsed time being represented by $1500\,e^{-0.2N}$. Use straight-line depreciation. The profits tax is 55 per cent. Company policy uses 10 per cent interest for economic evaluations.

Solution

The maximum present worth is obtained from a step by step solution using Eq. (11-1); however, the integration of Eq. (11-11) for depreciation credits also could be employed to obtain the result. Accordingly, a table of values is set up for different lives as given in Table 11-8. As shown, the present worth maximizes at about 2 years.

TABLE 11-8. Tabulated Values for Solution to Present Worth of Problem Using Eq. (11-1)

Life, N Years	Net Revenues[a] $0.45 \sum_1^N \dfrac{1000-200n}{1.10^n}$	Tax Credits[b] $\dfrac{(0.55)(1500)(1 - e^{-0.2N})a_{N/i}}{N}$	Salvage Value $\dfrac{1500e^{-0.2N}}{(1 + i)^N}$	Total Present Worth
1	327	136	1118	1581
2	550	236	830	1616
3	685	309	616	1610
4	747	360	459	1566
5	747	396	342	1485
6	696	421	254	1371

[a] The summation for N years is an arithmetic series differing by -200 per year and is given by Eq. (2-12), the first term of the series being 800.

$$0.45 \left[800a_{N/i} - \frac{200(s - 1 - 0.1N)}{0.10^2 s} \right]$$

[b] This is a straight-line depreciation obtained by dividing the accumulated depreciation at any age N and converting these uniform payments to present worth.

This optimum age will now be computed by use of the incremental relation of Eq. (11-12), where $S(N) = \$4000$ and $C(N) = \$3000 + 200N$. Thus, when $n = 0$, the zero-time net revenue is $\$1000$. The value at the end of one year is $\$800$, which is the same value employed for discrete compounding, but a general expression would be:

$$(1 - t)[S(N) - C(N)] + tD(N) = iL(N) - d[L(N)]/dN \qquad (11\text{-}12)$$

Then, solving for Eq. (11-12):

$$0.45\,[4000 - (3000 + 200N)] + 0.55\,\frac{1500 - 1500e^{-0.2N}}{N}$$

$$= (0.10)(1500e^{-0.2N}) - (-300e^{-0.2N})$$

Simplifying this relation gives:

$$450 - 90N + (825/N)(1 - e^{-0.2N}) = 450e^{-0.2N}$$

To solve this expression for N plot the left side of the equation and the right side as separate functions of N, as shown in Fig. 11-5. The point of intersection of

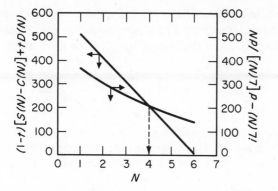

Fig. 11-5. Solution to example problem for optimum N.

these simultaneous equations will give the solution for N. The result of 4 years with a maximum present worth of $2887 may be compared with the answer of 2 years as shown by the method of Table 11-8. It would be fortuitous if both methods agreed because one involves a summation of discrete quantities whereas the other utilizes continuous integrals. In addition, there are four terms in Eq. (11-1) that are affected by the mathematical form of the s factor, which itself differs, depending upon whether discrete or continuous compounding is employed.

INFLUENCE OF INVESTMENT ON PROFITABILITY

The models in the preceding discussion of this chapter involve only one level of investment and expected revenues. However, it is interesting to consider the analysis of profitability of a venture at different investment levels for different plant sizes. This subject, mentioned very briefly in Chapter 7, is now taken up in more detail.

The theoretical models in the preceding part of this chapter are rigorous and can be applied to many real world problems. However, principles must be understood sufficiently so that simplifying approximations can be made with confidence. The remainder of this chapter is concerned with certain simplified real world models that not only are quite useful, but which, in many cases, are the final basis for deciding what level of investment will be committed for a proposed project.

EARNING RATES AS A FUNCTION OF INVESTMENT

The earning rate is defined (item 6 of Table 5-4) as the ratio of profit to total investment, I; the earning rate is given in Chapter 7 as Eq. (7-24) and in Table 11-3 as Eq. (11-7). Accordingly, the earning rate follows the shape of the profit curve in Fig. 11-6 for various plant capacities if a constant investment is involved. The general equation for capitalized earning rate, defined as 100 times the ratio of net profit to investment, is:

$$i_c = \frac{100\,Y}{I} = 100(1 - t)\frac{[nS - (nV + F)]}{I} \tag{7-24}$$

This expression may be plotted on the economic production chart; the maximum for capitalized earning rate is at the same production rate as the rate for maximum profit, since it is simply the profit divided by the total investment.

NOTE: The entire discussion of earning rates and profits in this section applies, whether the calculation is based on profit before taxes, Z (that is, gross profit) or on net profit after taxes, Y. The optimum is unaffected, regardless of which basis is employed. However, the actual numerical values obtained when different bases are used will differ by the ratio of $(1 - t)$, where t is the tax rate expressed as a fraction.

Production at 100 per cent capacities for plants of *different sizes* may not show earning rates and profits optimizing at similar production rates. This is possible because the investment may increase as a fractional power of capacity, as discussed in Chapter 4, but the profits may vary in an

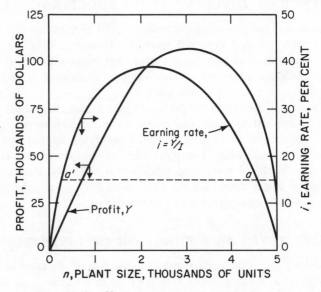

Fig. 11-6. Variation in earnings with plant size.

entirely different manner because of their particular relationship to prices, costs, and capacity.

Consider for example the data given in Table 11-9 for operations of a series of plants of different capacities where the revenues per unit of product decrease as the production becomes larger, resulting in a maximum profit at some critical production. The data in Table 11-9 are plotted in Fig. 11-6; Fig. 11-6 illustrates how, because of varying investment, profit and earning rates for different sized plants may vary separately with *capacity*. The maximum earning rate could be obtained analytically by determining the first derivative of Eq. (11-7) with respect to capacity n and setting it equal to zero. To do this all variables such as, S, V, and I must be known and expressed in terms of plant size n.

TABLE 11-9. Economic and Production Data for Plants of Different Capacities

Plant Capacity, n Units	Profit, Y Thousands	Investment, I Thousands[a]	Capitalized Earning Rate, i_c	True Economic Earning Rate, i
1000	52	160	32.5	41.5
2000	94	242	38.3	48.5
3000	106	310	34.3	43.5
4000	96	368	26.0	34.3
5000	28	420	6.7	10.5

[a] Considering plant investment varies as the six-tenths power of the ratio of size.

Instead of using the capitalized earning rate, i_c, (as given in the fourth column of Table 11-9) one may include the time value of money and thus compute the true economic rate of return. The procedures of choice are those presented in the first part of this chapter. The assumption has been made that the data in Table 11-9 are valid each year for a prospective period of ten years; then from Eq. (11-5) the true economic rates of return have been computed with no salvage value and straight-line depreciation. The results are given in the last column of the table. If nonuniform annual profits or depreciation are involved, Eq. (11-5) can be used with average annual values as a first approximation; otherwise, the more rigorous procedures with yearly values should be employed. These can be programmed on a computer.

If the profit, Y, of Table 11-9 is plotted against the investment, I, as shown in Fig. 11-7, a different type of economic production chart results, which illustrates the maximum that would occur. In addition, this model provides another useful criterion. This criterion is the marginal profit rate, i_m, which is the slope of the profit curve of Fig. 11-7, or:

$$i_m = 100 \, dY/dI \qquad\qquad (7\text{-}25)$$

The significance of Eq. (7-25) is that it describes the additional profit for additional investment. This additional profit is of some concern to management personnel who must make decisions in the allocation of available capital. The marginal rate is the first derivative of the relation between profit and investment which is given by the slope of the line drawn tangent to the curve at any point. As illustrated in Fig. 11-7 at an investment of \$200,000 the tangent has a slope of $\Delta Y/\Delta I = 54{,}000/120{,}000 = 0.45$, or a marginal profit rate of 45 per cent. However, as the investment increases the slope of the profit line decreases, indicating a smaller and smaller marginal profit rate until it reaches zero at the point of maximum profit. Beyond this point a profit is made. However, the costs of operation (or the nearness to market saturation with the reduction in selling price) has brought about diminishing returns.

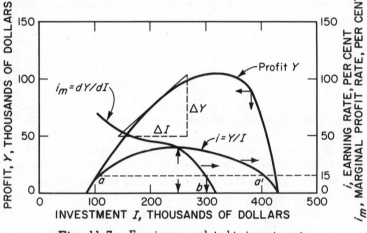

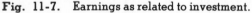

Fig. 11-7. Earnings as related to investment.

CRITERIA FOR OPERATIONS

Clearly, before management selects a scale of operations, a policy must be established. Usually, the controlling factor is investment or earning rate; therefore the curves in Fig. 11-7 are very important. A decision policy may be based on one of several criteria as follows:

(1) *Maximum Profit.* As shown in Fig. 11-7 the maximum profit occurs when approximately 3000 units per year correspond to an investment of about

$310,000. However, this does not produce the maximum earning rate which, as illustrated, occurs at about $245,000. Therefore, if maximum dollar profit is the criterion in this case, some earning rate less than the maximum must be accepted and the necessary capital must be available.

(2) *Maximum Earning Rate.* If a criterion of maximum earning rate is the company policy then about 2100 annual units of production will be made with an investment of approximately $245,000. In this situation, some additional profit is sacrificed to obtain the highest estimated earning rate. This additional profit may be made at some value greater than a minimum acceptable rate.

(3) *Minimum Acceptable Earning Rate.* If a company sets the scale of operations on the basis of a minimum acceptable earning rate, two production levels are possible. For example, if a limit of at least 15 per cent for earning rate is set, the production level could be anywhere between about 250 and 4600 annual units, investments being between about $110,000 and $410,000 as shown by *aa'* on Figs. 11-6 and 11-7. If the capital is available, the production level can be determined from other criteria, such as in (1) and (2) above, that fall in the range described by the minimum acceptable earning rate. If the maximum investment (point *a'*) is selected for the minimum acceptable rate, then those portions of the investment that are above $310,000 are in actuality producing at a loss; that is, no incremental profit is realized from the additional investment.

(4) *Marginal Profit Rate.* The criterion of marginal profit rate may be used in management decisions to limit the amount of investment. When the incremental profit ΔY decreases to some minimum acceptable rate (for example, 15 per cent) as a fraction of the additional investment ΔI, management policy may limit the investment even though more profit dollars could be earned. This limit is reached when management considers it better to invest any additional available capital in ventures that would earn more than the 15 per cent. The figure of 15 per cent as a cutoff point in Fig. 11-7 varies from industry to industry and even among companies in the same industry depending upon company policy.

The capitalized earning rate $i_c = 100\ Y/I$ maximizes when its derivative is zero, or:

$$di_c/dI = 100[IdY/dI - Y]/I^2 = 0 \qquad (11\text{-}15)$$

Thus, $\qquad\qquad 100dY/dI = 100\ Y/I \qquad\qquad (11\text{-}16)$

The left side of this equation is the marginal profit rate, i_m (see Table 7-1). This marginal profit rate is shown to be equal to the capitalized earning rate at the latter's maximum. This point and the zero value of the marginal profit rate (at maximum profit) give two points on Fig. 11-7, which orient the marginal profit curve for quick inspection.

SUMMARY

This presentation has demonstrated the four basic profitability methods, as listed in Table 11-1, for a given investment program. In principle, there are no "newer" methods. In fact, generally all other methods can be shown to be special cases of one of these four, the variations being attributable to the specific manner in which certain accounting data are employed. This statement applies to the Machinery and Allied Products Institute (MAPI) methods as well as all other valid methods.

Where the analysis considers the effect of a change in the investment on the earning rate, either the true economic rate of return or the capitalized earning rate may be used with success, since the results are readily understood. Another criterion, the marginal profit rate, may be employed; however, this case requires further interpretation of how additional profits are a return for additional investment.

Where the time value of money is considered in a correct form, the simplest method is preferred. Complicated refinements may be slightly more accurate. However, if management allows these refinements to be major decision-making factors, its decisions are not on a firm basis and subject to considerable question. This is particularly true for proposed projects where there is uncertainty in the future estimates of the data. In other words, the venture should show a *sufficiently definite* positive (or negative) conclusion irrespective of the particular method employed for evaluation, if the time value of money is included. This may not be true if an analysis is made by two methods, one of which does not consider interest on the investment. Even here, however, if the time value of money indicates a reversal of a previous decision, management should recognize that the interest effect has caused the reversal and should make a careful examination of recommendations. This important point emphasizes the need for a profitability model. The equations, analyses, and interpretations must be clearly understood by management if it is to render a decision in the best interests of the firm.

At best the model and economic analysis of profitability is a complex problem. It involves summations of certain mathematical relations and the consideration of numerous factors, many of which cannot be reduced to absolute values. Accordingly, the purpose here is merely to present certain basic relations and illustrate their use rather than bias the reader toward any one method. In general, some type of earning rate method is the soundest procedure. For the most part, the time value of money is being incorporated in the procedure employed. This is because of its economic and financial advantages. Although its inclusion does complicate the analysis, its importance may outweigh the complication. In addition, profits taxes must be included because they are real dollars not available to the proprietors of a firm. The omission of taxes can contradict a decision made when they are included.

The earning rate methods are criteria of profitability; however, they do not indicate in the form of an easy-to-handle yardstick the order of magnitude of the capital involved or the cash position after a number of years. This may be a disadvantage where a quick appraisal of certain finance considerations is desired. However, all such data may be computed if needed.

The payout time methods, either with or without the time value of money, are the least complicated to compute because of the simplifying assumptions made. However, their significance is somewhat more hidden. Although payout time has the advantage of being related to the number of years required for the average return to equal the investment, the influence of irregular annual returns, the excess returns after the payout period, etc., are difficult to evaluate in making a decision. Moreover, the magnitudes of the capital involved and the cash flows are not apparent when the payout time alone is used as a yardstick. It is true, of course, that such information is available to support the computations.

In the final evaluation of the profitability of an economic venture, numerous criteria are applied by management. Thus, the maximum profit and the maximum earning rates are considered in terms of the investment to be made and the capital available. Some minimum acceptable earning rate usually is employed as a starting point. Furthermore, the marginal profit rate is frequently utilized to ascertain that cutoff point at which additional investment is not productive; in other words, at what point will the capital yield a greater return in an alternative investment?

A word of warning should be issued for analyses made with average annual cash flows. Where different lives and/or nonuniform annual returns are involved, it is possible to find examples where payout times are highest when earning rates are also highest. These contradictions will vary with the life of the project and the time periods in which the annual returns are distributed. However, rigorous economic procedures obviate any discrepancies when the time value of money is included.

Clearly, no one single method or criterion of profitability analysis is preferred. Comparisons and evaluations are best made with several criteria. In any given situation the importance of a critical factor in the mathematical relation may be such as to point out that one particular method is best for that specific case. In another situation, however, some other method may be preferable. The selection of a method will depend on the judgment of management, based on their experience in recognizing critical factors. No one formula, chart, or blind procedure can replace judgment and decision making that is based on an understanding of the principles and limitations of the profitability models.

PROBLEMS

11-1. Management must decide whether it should invest $840,000 in a venture to manufacture automobile parts. It will proceed if the economic earning rates exceed 20 per cent and the capitalized payout time is less than 3 years. Assume straight-line depreciation, a 6 year life, a profits tax of 55 per cent with a uniform annual profit before depreciation of $340,000; neglect salvage value and preoperating time charges. (a) Calculate the earning rate. Should the firm invest in this venture? (b) What would be the comparable numerical results if there were no profits taxes? (c) Would this change the decision?

11-2. A company is considering an investment of $3 million in a venture having an estimated life of 12 years; assume sum-of-digits depreciation and no salvage value. If there is a profits tax of 55 per cent and an annual gross profits before depreciation of $2,000,000, what is the economic rate of return? Use the summation equations of Chapters 2 and 3 where applicable.

11-3. For the data in problem 11-2, assume that the gross profits *before depreciation* starts at $1.5 million at the end of the first year and that there is an increase of $50,000 each year. Using straight-line depreciation, compute the economic rate of return. Repeat the calculation employing a constant annual return equal to the average value. Discuss the results. (HINT: Use summation equations.)

11-4. A large scale operation, with a $42 million investment of which $36.5 million is depreciable, has an annual income of $24 million and annual costs of $7.66 million, excluding depreciation. Life is 10 years and profits tax is 55 per cent. (a) For straight-line depreciation, what is the economic earning rate? (b) If money is worth 6 per cent, what is the economic payout time? (c) What are the capitalized earning rate and capitalized payout time? (d) Is this a profitable venture? Discuss briefly.

11-5. Assume the same data in Problem 11-4, except that the income the first year is $7 million and increases by three million dollars for each succeeding year. Use a diagram to demonstrate the initial negative values for profit. Calculate the economic earning rate only.

11-6. Assume the same data in Problem 11-4, except that the annual costs of operation excluding depreciation increase 3 per cent of the costs for the preceding year, the first year costs being R' and equal to $6 million. (See Chapter 2 for summation equation). Calculate (a) the economic earning rate and (b) the capitalized earning rate, using averaged values for the costs.

11-7. Capital can be borrowed for immediate purchase of cigarette machinery costing $1,400,000. The machinery will start producing immediately after installation. Salvage value is estimated at $140,000 for a 10 year life. Straight-line depreciation is used. Profits tax is 55 per cent; the annual revenue, after all costs and depreciation but before financial costs, is one million dollars. What are the answers for the four basic profitability models when the capital is borrowed at 4.3 per cent interest by a bond issue? Company policy is to use 15 per cent for evaluation.

11-8. A new processing plant for food products is proposed to management. Two years are required to put this plant into operation on an initial capital commitment of $2.4 million. Calculations are made on the basis of an 8 year life, and the depreciable capital is $1.8 million. The estimated future revenue, after all costs but before depreciation, is $400,000 the first year and is estimated to increase $100,000 each year for the remaining period of 7 years. Using straight-line depreciation, compare the economic rate of return and the capitalized earning rate, making corrections for the year preoperation period. Profits tax is 55 per cent. Use average revenues for the capitalized earning rate method.

11-9. For the data in problem 11-4, compute the earning rate, on the basis of a proposed policy of declining-balance depreciation. Assume that at the end of the 4th year of operations an additional $3,000,000 in equipment with no salvage value was installed. (Calculate the residual value at the 10th year for the 4th year investment; then compute its zero-time present worth as an additional cost. Use the same factor as for the major investment and allow for tax credit on the annual depreciation charges from the fourth year.)

11-10. The annual sales rate, S, for a product is expected to vary as $2500 - 50n$ dollars per year, where n is the time from start up. Costs excluding depreciation are expected to vary as $1000 + 150n$ dollars per year. An original depreciable investment of $1500 is involved, but the salvage value varies as $L = 1500\,e^{-0.25N}$. A profits tax of 55 per cent applies, the company policy being to use 12 per cent for such evaluations. (a) After what total years of operation will a maximum present worth occur, employing straight-line depreciation? (b) What is the maximum present worth? It is assumed that the government will allow the annual depreciation based on the optimum life.

ALTERNATIVE AND REPLACEMENT MODELS 12

The profitability principles discussed in the previous chapter can be applied in making a decision among alternatives providing identical services or outputs. There are two major types of alternative decisions. In one type, a choice is available among two or more different policies to be followed for a projected time period of selected duration. In the second type, a projected program is expected to extend into the future long enough to require replacement of the initial installation (defender). The nature of the replacement may be similar to that of the original equipment. In such a case, costs and revenues will be expected to repeat past history or vary in some known manner. On the other hand, the replacement may be an entirely new type of installation.

BASIC MODEL

The basic model rests on theoretical concepts of present worth, which for convenience may be considered either in discrete steps or as a continuous function. The model states that the present worth of the costs for a service shall consist of:

(1) The initial installation cost, I
(2) The present worth of all annual costs, (excluding depreciation) incurred for the service

$$\sum_{n=1}^{n=N} C(n)(1 + i)^{-n} \cong \int_0^N C(n)e^{-in}dn$$

(3) The present worth of any savings or credit attributable to the service. (This is a negative cost.

$$\sum_{n=1}^{n=N} S(n)(1 + i)^{-n} \cong \int_0^N S(n)e^{-in}dn$$

308

NOMENCLATURE

$a_{N/i}$ = the present worth factor equal to $(s - 1)/is$

b = a constant

C = a cost or expense (excluding depreciation) for a period, either uniform or nonuniform; $C(n)$ is for the nth year; C_0 is an initial rate; C_a is an average value

D = the annual depreciation; $D(n)$, is for the nth year, dollars per year

e = the transcendental number 2.718

g = an annual gradient applying to initial cost; may be considered as an obsolescence for a defender

i = interest rate expressed as a decimal

I = initial investment, dollars

I_R = any investment capital added at any time, either in the form of a renewal or major capital rehabilitation

k = a constant in a linear equation representing an annual change or gradient

L = a recoverable value or salvage; $L(N)$ is a function of N

n = end-of-year age, or total life or total periods

N = total life or total periods

P_{cap} = the present worth of all net costs for a perpetual series of renewals of an installation, or capitalized cost

P_{cost} = the present worth of all net costs for one installation having a life of N periods

R = any uniform amount at the end of a period

R_{cost} = a uniform annual cost equivalent to a present worth, P_{cost}

S = an end of period savings or credit value, either uniform or nonuniform, $S(n)$ is for the nth year; S_0 is an initial value

s = compound interest factor $(1 + i)^N$ for discrete compounding or e^{iN} for continuous compounding. Values are given in Appendix C.

(4) The present worth of the salvage value after the terminal year N is reached.

$$L(N)(1 + i)^{-N} \cong L(N)e^{-iN}$$

In these expressions $C(n)$ represents the annual rate for costs in dollars per year; $S(n)$ is the annual rate for any savings or credits in dollars per year; e is the transcendental number 2.718; i is the interest rate as a fraction; n is any year; N is the terminal year; and $L(N)$ is salvage value in the terminal year. The two alternative forms given for the components represent the relations for discrete calculations (that is, calculations performed in integer steps representing time intervals from the end of one period to end of the next period). Discrete calculations can be employed instead of continuous compounding. However, the analysis with continuous functions provides a more useful model where it is necessary to perform certain mathematical operations which are more difficult to handle with the discrete form. The selection of discrete or continuous compounding is often a matter of the personal choice of the one making the analysis.

Utilizing the same principles as in the previous chapter, one finds that the present worth of the net costs, P_{cost} (neglecting taxes), is for the discrete form:

$$P_{cost} = I + \sum_{n=1}^{n=N} [C(n) - S(n)](1 + i)^{-n} - L(N)(1 + i)^{-N} \quad (12\text{-}1a)$$

For the continuous form, the present worth of the net costs is:

$$P_{cost} = I + \int_0^N [C(n) - S(n)]e^{-in} \, dn - L(N)e^{-iN} \quad (12\text{-}1b)$$

Thus, two (or more) alternatives having differing investments, I, with varying costs, credits, and salvage values may be compared by determining their respective present worths according to either of the above equations. The alternative with the lowest P_{cost} is the most feasible.

EXAMPLE FOR COMPARING ALTERNATIVES

Problem

Two manufacturers propose a machine installation to perform a service with the following arrangements. Supplier A has a machine that costs $10,000 and uses $18,600 per year for labor. Other direct annual costs are 8 per cent of the investment, with salvage value of $600. Supplier B suggests a machine costing $30,000, with $11,000 annual labor charge and $3000 other annual charges; the salvage value is $1000. The estimated project life is 10 years. Which is the more feasible installation if money is worth 10 per cent?

Solution

A comparison of present worth for the two installations is quite simple, since the annual costs are constant at C_0 and the $a_{N/i}$ factor of Chapter 2 may be used to convert the summation to present worth[1] by employing Eq. (12-1a) with $S(n)$ equal to zero.

FOR SUPPLIER A:

$$P_{cost} = 10,000 + \sum_{n=1}^{n=10} [18,600 + 0.08\,(10,000)]/1.10^n - 600/1.10^{10}$$

$$P_{cost} = 10,000 + 19,400\, a_{N/i} - 600/2.594 = 10,000 + 19,400(6.145)$$

$$- 231 = \$128,982$$

FOR SUPPLIER B:

$$P_{cost} = 30,000 + 14,000(6.145) - 1000/2.594 = \$115,645$$

Thus, supplier B provides the service at a present cost of $13,337 less than supplier A.

Where a continuous form is used then some initial value C_0 must be employed, and the indicated integration of Eq. (12-1b) is performed for $C(n)$ equal to the annual costs and $S(n)$ equal to the annual credits.

With $S(n)$ neglected and constant annual costs such that $C(n) = C_0$:

$$P_{cost} = I + \int_0^N C_0 e^{-in}\, dn - L(N)e^{-iN}$$

$$P_{cost} = I + [C_0 e^{-in}/(-i)]_0^N - L(N)e^{-iN}$$

$$P_{cost} = I + (C_0/i)(1 - e^{-iN}) - L(N)e^{-iN} \qquad (12\text{-}2)$$

FOR SUPPLIER A:

$$P_{cost} = 10,000 + \frac{19,400}{0.10}(1 - 1/e^{1.0}) - 600/e^{1.0}$$

$$P_{cost} = 10,000 + 194,000\,(0.632) - 600/2.72 = \$132,387$$

FOR SUPPLIER B:

$$P_{cost} = 30,000 + 140,000\,(0.632) - 1000/2.72 = \$118,112$$

Thus, with continuous compounding the present value of the cost of service is higher than for discrete when the nominal annual interest is used in both cases. This difference must be recognized as primarily a comparison of two results obtained by two different methods of computation; this may be of no practical significance.

---

[1] The summation, where R is any uniform amount, is:

$$\sum_{n=1}^{n=N} R(1 + i)^{-n} = R(1 + i)^{-1} + R(1 + i)^{-2} + \ldots + R(1 + i)^{-N}$$

which can be put in the form:

$$\sum_{n=1}^{n=N} R(1 + n)^{-n} = \frac{R}{(1 + i)}\left[1 + \frac{1}{1 + i} + \left(\frac{1}{1 + i}\right)^2 + \ldots + \left(\frac{1}{1 + i}\right)^{N-1}\right]$$

where the sum in the brackets is $\dfrac{r^N - 1}{r - 1}$ and $r = \left(\dfrac{1}{1 + i}\right)$.

the summation becomes:

$$\sum_{n=1}^{n=N} R(1 + n)^{-n} = \frac{R[(1 + i)^N - 1]}{i(1 + i)^N} = \frac{R(s - 1)}{is} = Ra_{N/i}$$

COMMENT ON BASIC MODEL

The basic model, as proposed is a general one that allows for uniform or nonuniform annual expense; the summation may be made in as rigorous a manner as desired.

Where the costs or credits vary from year to year but in some uniform manner, Eq. (12-1a) may be summed algebraically or Eq. (12-1b) may be integrated. Thus, for example if the annual cost rate $C(n)$ varies as a linear function such as $C_0 + kn$, and $S(n)$ varies as $S_0 + mn$, then the summation value for the present worth of all annual net costs up to the year N would be (see summation equations of Chapter 2) for discrete steps (that is $n = 1, 2, 3, \ldots,$) where C_0 and S_0 must be the values at year zero.

$$\sum_{n=1}^{n=N} [C(n) - S(n)](1 + i)^{-n} = (C_0 + k - S_0 - m)a_{N/i}$$

$$+ \frac{(k - m)}{i^2 s} (s - 1 - iN) \tag{12-3a}$$

For continuous-function analysis where n is continuous and C_0 and S_0 are some initial values at zero time:

$$\int_0^N [C(n) - S(n)]e^{-in} \, dn = \frac{(C_0 - S_0)(e^{iN} - 1)}{ie^{iN}}$$

$$+ \frac{k - m}{i^2 e^{iN}} (e^{iN} - 1 - iN) \tag{12-3b}$$

Equations 12-3a and 12-3b will not give the same present values of future amounts at a given value of i, and the present worths so obtained must be considered as separate evaluations according to the respective equation employed (discrete or continuous).

In Eqs. 12-3a and 12-3b, C_0 represents a beginning cost and k an annual change in costs, or an annual gradient; similarly, this holds for S_0 and the gradient m for the $S(n)$ function.

The basic-model concept of comparing present worths of alternatives is a special case of profitability evaluation. Accordingly, if one wishes to take into account taxes in comparing alternatives, complex treatments allowing for tax credits for depreciation (see subsequent discussion, page 329) must be employed. The previous example for the comparison of alternatives in the provision of a given service assumes that the revenues are essentially the same for any alternative. If this is not the case, the difference must be considered as an equivalent savings term). Thus, the revenues would not affect the *difference in present worth* for the alternatives. This would be true for any cost, revenue, or credit that was the same in both alternatives. The absolute values of the total present worths would be different by an added constant amount for each alternative. In all economic analyses for selection of alternative procedures, any items common to all procedures may be omitted without affecting the end result, provided differences only are desired. However, where absolute values are required all items must be included.

As demonstrated in Chapter 11, an optimum time period for future returns from a project can be determined by a criterion of maximum present worth. The analysis on page 297 consisted of a trial and error solution for the maximum present worth or, alternatively, an incremental analysis as described by Eq. (11-12). The same procedures can be employed for determining the optimum life of a project at which the present worth of the annual costs is at a minimum. The uniform annual costs, then, for that minimum present worth is also a minimum. If revenues are constant, they can be omitted from the analysis without affecting the optimum, and the maximum instantaneous annual profit is determined for the year of minimum cost. Such information may be of interest and value. However, since the long run amounts for cumulative profits are more important than the maximum instantaneous annual profit, the present worth of all future values is what is desired where nonuniform annual cash flows are involved.

The subject of minimum annual costs will be reconsidered in a later section.

The interest rate generally used in replacement calculations is a preselected value determined by management. It may vary from a 4 per cent bond rate to a 30 per cent (or higher) rate, which has been selected as company policy for making economic analyses.

COMPARISONS BY ANNUAL COSTS

Management often desires a comparison of the annual costs instead of the present worth of the costs[2]. For this purpose, all costs of a uniform or irregular type can be converted to the present worth. The equivalent uniform annual costs, R_{cost}, can be computed readily (see Chapter 2):

$$R_{cost} = P_{cost}\, is/(s - 1)^N = P_{cost}/a_{N/i}, \text{dollars per year} \qquad (12\text{-}4)$$

where s is the compound interest factor $(1 + i)^N$ and $a_{N/i} = (s - 1)/is$ for discrete compounding; s is e^{iN} for a continuous analysis.

---

[2] The term costs as used in this discussion may mean actual costs only, or it may refer to net costs (that is, the net difference between any costs and any credits for the situation under study.)

This comparison of annual costs is for a limited time period of N years. It is also based on the equivalence doctrine, regardless of the uniformity of the annual quantities, since they are all referred to a single time point as present worth.

Problem

Applying Eq. (12-4) to the previous problem (page 310), what are the annual costs for the two alternatives? At $i = 10$ per cent, and $N = 10$ years, the value of $a_{N/i}$ is 6.145.

Solution

SUPPLIER A: Annual cost $= 128,982/6.145 = \$20,990$ per year

SUPPLIER B: Annual cost $= 115,645/6.145 = \$18,819$ per year

Where the present worth is given by Eq. (12-1a), it can be shown that the annual equivalent uniform cost in dollars per year is in the general form:

$$R_{\text{cost}} = \left\{ I + \sum_{n=1}^{n=N} [C(n) - S(n)] (1 + i)^{-n} - L(N)(1 + i)^{-N} \right\} / a_{N/i} \qquad (12\text{-}5a)$$

The continuous form is:

$$R_{\text{cost}} = \left[I + \int_0^N [C(n) - S(n)] e^{-in} dn - L(N) e^{-iN} \right] \left[\frac{1}{1 - e^{-iN}} \right] \qquad (12\text{-}5b)$$

As Eq. (12-1b) is employed in this problem the bracketed function $[C(n) - S(n)]$ equals a constant value, C_0, and $L(N)$ is a constant. Thus, a special form of Eq. (12-5a) for this case is:

$$R_{\text{cost}} = I/a_{N/i} - L/(sa_{N/i}) + C_0, \text{ annual dollars} \qquad (12\text{-}6a)$$

which also can be written as it is usually shown:[3]

$$R_{\text{cost}} = (I - L)/a_{N/i} + iL + C_0, \text{ annual dollars} \qquad (12\text{-}6b)$$

This equation states that the annual costs result as follows: (1) from converting

---

[3] The general form of Eq. (12-5a) can be written:

$$R_{\text{cost}} = (I - L) \frac{is}{s - 1} + iL + C_0 = I/a_{N/i} - \frac{iLs + iLs - iL}{s - 1} + C_0$$

Therefore,

$$R_{\text{cost}} = I/a_{N/i} - \frac{iL}{s - 1} + C_0$$

Since $i/(s - 1) = 1/(sa_{N/i})$, Eqs. (12-5a) and (12-6a) are identical.

the initial cost less salvage value of the installation to an equivalent annual basis (capital recovery), (2) from adding the interest on the salvage value, and (3) from adding the other annual costs.

Substituting numbers in Eq. (12-6b) gives:

SUPPLIER A: $\quad R_{\text{cost}} = 9400/6.145 + 0.10\,(600) + 19,400 = \$20,990$

SUPPLIER B: $\quad R_{\text{cost}} = 29,000/6.145 + 0.10\,(1000) + 14,000 = \$18,819$

These results are in agreement with the previous ones for discrete compounding. However, note that the annual costs equivalent to the present worths for continuous compounding with a different value of $a_{N/i} = \dfrac{e^{iN} - 1}{ie^{iN}}$ as based on Eq. (12-4), become:

SUPPLIER A: $\qquad\qquad \$132,387/6.321 = \$20,945$ per year

SUPPLIER B: $\qquad\qquad \$118,112/6.321 = \$18,686$ per year

These annual costs are in substantial agreement with those for discrete compounding.

APPROXIMATE ANNUAL COSTS

An approximation to the more rigorous solution of Eq. (12-6a) can be used where nonuniform annual costs apply and/or the lives are different. This procedure combines the approximate capital relation of straight-line depreciation and average interest plus the arithmetic average of all other annual expenses, C_a; thus:

$$R_{\text{cost}} = \frac{I - L}{N} + (I - L)(i/2)\left(\frac{N + 1}{N}\right) + iL + C_a \qquad (12\text{-}7)$$

where $C_a = (1/N) \displaystyle\sum_{n=1}^{n=N} C_n$ with C_n being the expense in any one discrete time period. In using this equation for comparing alternatives, one should employ the respective lives N for each alternative. Equation (12-7) could be applied to the previous problem (page 310) to yield:

SUPPLIER A:

$$R_{\text{cost}} = \frac{9400}{10} + 9400\,(0.10/2)\,(11/10) + 0.10\,(600) + 19,400$$

$$= \$20,917 \text{ per year}$$

SUPPLIER B:

$$R_{\text{cost}} = \frac{29,000}{10} + 29,000\,(0.10/2)(11/10) + 0.10\,(1000) + 14,000$$

$$= \$18,595 \text{ per year}$$

The extent to which this result agrees with the more rigorous Eq. (12-6a) depends upon the interest rate, salvage value, and time involved. However, Eq. (12-7) has many areas of usefulness for quick economic analysis.

As previously mentioned, where the annual operating expenses are uniform, it is not necessary to compute an average value. However, where the expenses vary from year to year, the average expense, C_a, is used.

ALTERNATIVES WITH DIFFERENT LIVES

Where the life of one alternative is different from that for another, it is obviously incorrect to compare the present worth involved. The longer-lived alternative undoubtedly will cost more because it has features built into it to give a prolonged life. These additional features usually increase the initial investment, and comparisons of present worth show such installations at a disadvantage.

Where a present worth comparison is desired, one may consider each installation to be renewed at the end of its life until an elapsed time is the same for both series of renewals. This total life is the least common denominator for the lives of all alternatives involved. Thus, if three alternatives are being considered with 3, 4, and 6 year lives, then a 12 year period for comparison is necessary, with installations of 4, 3, and 2 times, respectively, for the alternatives. The amount of each and all renewals discounted to zero time would be added to Eq. (12-1a) by the addition of a term $I_R/(1 + i)^n$, where n is the specific age at which each new capital in amount I_R, allowing for salvage value, is expended. This procedure would also apply to a continuous-function analysis by adding $I_R e^{-in}$ in Eq. (12-1b).

The problem of different life is circumvented by simply converting the present worth of each alternative to annual costs for each, as was discussed in the previous two sections. The comparison of alternatives is thereby made by observing the respective annual costs of the alternatives. In this procedure, it must be recognized that the alternative with the longer life has an inherent advantage of greater potential longevity not shown by the analysis but which should be considered in a final decision.

THE CONTINUOUS SERVICE MODEL

In certain types of economic analysis, the continuous service model is useful. In this model, the service facility is renewed periodically either at its terminal life or on the basis of some other management policy. In conceptual form, this is a perpetual renewal of the model for a specified life after every N years, as defined by Eq. (12-1a). In effect, this is the capitalized cost for the service (see Chapter 2). This model, thus, gives the present worth of a perpetual series of models, each of uniform life, N; the present worth of each renewal is P_{cost} given by Eq. (12-1a) or Eq. (12-1b).

$$P_{cap} = P_{cost}[(1 + i)^N + (1 + i)^{2N} + \ldots]$$

$$P_{cap} = P_{cost}\frac{(1 + i)^N}{(1 + i)^N - 1} = P_{cost}s/(s - 1) = P_{cost}/(ia_{N/i}) \tag{12-8a}$$

This is the same as the capitalized cost of the annual cost, R_{cost}, as given by Eq. (12-5a).

The continuous function corresponding to Eq. (12-8a) is:[4]

$$P_{cap} = P_{cost}\frac{e^{iN}}{e^{iN} - 1} = P_{cost}/(1 - e^{-iN}) \tag{12-8b}$$

From the concept of capitalized cost as the capital equivalent to the annual costs at some interest rate, i, it then follows that the annual costs for a series of renewals is merely the interest times the capitalized cost. Therefore, the annual costs for comparing such alternatives is:

$$\text{Annual Costs, } R_{cost} = iP_{cap} = P_{cost}/a_{N/i}, \text{ dollars per year} \tag{12-9}$$

EXAMPLE FOR PERPETUAL SERVICE

Problem

Assume the same data used in the problem on page 310. (a) What are the capitalized costs for the two alternatives? (b) What are the annual costs?

Solution

Part a. The capitalized costs are determined from the present worth of one cycle of events corrected by Eq. (12-8) for perpetual renewals. Thus for $i = 10$ per cent and N equal to 10 years with $a_{N/i} = 6.145$ the results are:

---

[4] Note that the present-worth factor for continuous compounding is $(e^{iN} - 1)/ie^{iN}$, which transposes Eq. (12-8a) to (12-8b).

SUPPLIER A: $P_{cap} = \$128,982/(0.10 \times 6.145) = \$209,897$

SUPPLIER B: $P_{cap} = \$115,645/(0.10 \times 6.145) = \$188,194$

Part b.   The annual costs would be merely 10 per cent of the answers for part a, or $20,990 and $18,819, respectively, as was determined in the solution (page 315). The student should recognize the equivalency of these analyses.

If the continuous functions are used in the analysis then Eqs. (12-1b) and (12-8b) would apply.

REPLACEMENT OF THE SAME KIND

One useful application of the continuous function method is in replacement analysis, where an item of equipment or some service function has an expected history which is likely to be repeated continuously.   In this situation, there is often some optimum life for which the annual costs go through a minimum.   A basic economic balance concept applies in replacement policies for such continuous service.   In general, manufacturing equipment requires more and more maintenance with age, which results in greater annual costs each year.   At the same time the annual capital recovery charges decrease with age because the amortization is spread over a greater length of time.   Thus, an economic balance is present to give an optimum life.   When salvage must be considered, it affects the result because it may vary with the age selected for turning in an old machine on a new replacement.

By means of a model similar to Eq. (12-1b) and/or an incremental analysis, the minimum present worth may be determined if the costs can be approximated by continuous functions.   This analysis is similar in form to that presented in Chapter 11 for maximum present worth of net revenues, except that the analysis is made for minimum costs.   The relation for costs with continuous renewal, from Eq. (12-8), is:

$$P_{cap} = P_{cost}/(ia_{N/i}) = \frac{I + \displaystyle\int_0^N C(n)\bar{e}^{in}\,dn - L(N)\bar{e}^{iN}}{ia_{N/i}} \qquad (12\text{-}10)$$

The derivative of this equation when equated to zero will yield the following equation (compare the analysis of Eq. (11-11), page 294, where taxes were included) which gives the minimum annual equivalent cost when taxes are omitted:[5]

$$iP_{cap} = C(N) + iL(N) - d[L(N)]/dN \qquad (12\text{-}11)$$

---

[5]The same comment applies here as for Eq. (11-12). That is, the form $C(N)$, etc., is used where N is a terminal year; it is not the integrated value of $C(N)$. Note that annual costs are equal to the interest on the capitalized cost.

This relation states that the optimum age, N, for any one installation for minimum costs results when the annual costs for a perpetual *series of installations* is equal to the annual costs for the Nth year plus the interest on the salvage value for the year minus the loss in salvage value for the year. This change in salvage value correction is therefore positive, since the derivative involving salvage value is itself negative. The analysis will be recognized as another example of an incremental type; in this case, at the optimum year the annual charges equivalent to the capitalized value are just equal to the incremental costs at the optimum year. The solution of Eq. (12-11) requires trial and error solution for the optimum replacement age, N, which is demonstrated in the following problem.

Problem

A machine installation has an initial investment of $15,000; the salvage value for any terminal life varies as $15,000\,e^{-0.18N}$. The annual expenses pertinent to the operation increase each year and are approximated by the expression $C(n) = 6000 + 700n$ dollars per year. The relation $S(n)$ is zero here. The company uses 15 per cent for economic evaluation. Determine the age at which the installation should be replaced by a new one of the same kind, and estimate the average annual costs.

Solution

The solution is straightforward either by substitution in Eq. (12-10) stepwise year by year or by a more formal solution of Eq. (12-11). Employing Eq. (12-11), one must evaluate the capitalized cost, P_{cap}, as follows:

$$P_{cap} = \frac{15,000 + \displaystyle\int_0^N (6000 + 700n)e^{in}\,dn - 15,000e^{-0.18N}e^{in}}{ia_{N/i}}$$

$$P_{cap} = \frac{15,000}{1 - e^{-0.15N}} + \frac{6000}{0.15} + \frac{700}{(0.15)^2} - \frac{700N}{0.15(e^{0.15N} - 1)} - \frac{15,000e^{-0.18N}}{e^{0.15N} - 1}$$

[see Eq. (12-3b) for the integration, noting that $ia_{N/i} = 1 - e^{-iN}$.]

The annual interest of the capitalized cost is iP_{cap}, which is plotted in Fig. 12-1 as the left side of Eq. (12-11). The right side is also plotted to determine a point of intersection. The left side from above is:

$$iP_{cap} = \frac{2250}{1 - e^{-0.15}} + 10,670 - \frac{700N}{e^{0.15N} - 1} - \frac{2250}{e^{0.33N} - e^{0.18N}}$$

The right side of Eq. (12-11) for this problem from differentiation is:

$$6000 + 700N + 0.15(15,000)e^{-0.18N} - (-2700e^{-0.18N})$$

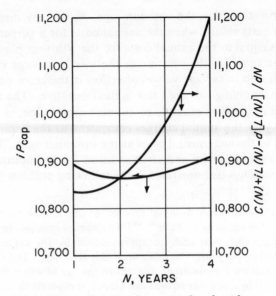

Fig. 12-1. Solution for example of replacement of the same kind.

(Note that the last term is $d(15,000e^{-0.18N})/dN$ without the $e^{-0.15N}$ from Eq. (12-11) or:

$$6000 + 700N + 4950/e^{0.18N}$$

The point of intersection at $N = 2$ years is the optimum life for replacement in kind where the cost history is expected to be repeated. For the optimum, N, at $i = 0.15$, the capitalized present worth of the costs is computed as follows:

$$P_{cap} = \frac{15,000}{1 - e^{-0.15N}} + 71,113 - \frac{700N}{0.15(e^{0.15N} - 1)} - \frac{15,000e^{-0.18N}}{e^{0.15N} - 1}$$

$$P_{cap} = \frac{15,000}{1 - e^{-0.30}} + 71,113 - \frac{1400}{0.15(e^{0.30} - 1)} - \frac{15,000e^{-0.36}}{e^{0.30} - 1}$$

$$P_{cap} = \$72,400$$

The annual equivalent cost is iP_{cap}, or $(0.15)(72,400) = \$10,860$.

Where the cost prognosis (expected future behavior) for a replacement is expected to be different from that for the original machine, then the problem is a replacement of a new kind, which will be discussed in the next section. In this situation, the replacement will have some demonstratable superiority. However, in the present problem, the costs vary as $6000 + 700n$ dollars per year, and the \$700 represents an increasing amount, or annual gradient per year. This concept is utilized in some replacement evaluations.

REPLACEMENT OF A NEW KIND

Replacement of an installation with a new and different one is essentially the selection of an alternative procedure. The model then follows those expressions given in the early part of this chapter, the basic model being Eq. (12-1a) or (12-1b). The comparison of the alternatives may be either on the basis of present worth or annual costs. Other methods of analysis may evaluate interest rate or payout time. For these situations, the profitability models of the previous chapter may be employed.

Replacement studies are somewhat more complex than straight alternative comparisons and require some judicious use of the available information plus an experienced interpretation of the results.

Reduced to essentials this replacement study is a comparison of net costs for continuing operations on a current installation A (often called the defender) with those resulting from replacement of A with a new installation B (often called the challenger). Thus, one may make the analysis by comparing the equivalent annual costs of the respective installations, allowing for the proper values of the variables in the mathematical model employed. Perhaps, the two most controversial variables are the time period to use and the investment value, since these are likely to be quite different for the defender and the challenger.

Where the lives of the alternatives are different, the expenses, credits, and all pertinent items are discounted to their present worth and the equivalent annual cost comparison made by Eq. (12-5a), (12-5b), or directly by Eq. (12-7). In many instances a common life is assumed to simplify the analysis.

In the case of the challenger, the investment value to be used in Eq. (12-1a) or (12-1b) presents no problem, since the challenger's installed cost is known or can be estimated readily. However, for the defender, the investment value, I, is that equivalent capital in real dollars, as represented by the defender in good working condition, which will provide the future service intended in competing with the challenger. This investment value for the defender may have no relation to the book value, since the correct comparison is between a certain number of I dollars at the present time and its associated net costs for some future time period, the result being compared to similar items for the challenger. Thus, the current investment capital for the defender is taken as the realizable value if it were disposed of plus any capital expenditure that would be required to put it in working order to perform the service. For example, a machine may be sold for $2500 with a cost of $500 to dismantle it and prepare it for shipment. Furthermore, if the machine is to be continued in service, it may require $700 worth of major repairs. Thus, if the machine is to remain a defender, the proper current real capital invested in the machine is (2500 − 500) + 700, or $2700. Value of lost production during a changeover would be chargeable to the investment of challenger.

When replacement of a new kind is contemplated, the premiums involved in making an improved product must be given some consideration. Where there is a difference in revenues, opportunity cost effects are important. An opportunity cost is a revenue that is lost when an alternative decision is made. Often some profit for a current operation is sacrificed in a new activity where the total profit in the end is greater than for the alternative already in operation. For example, a defender makes a medium grade product from a base stock at a nominal profit. By changing to a challenger, management may be able to produce a high quality product at a higher profit from the same base. The real gain for the challenger is not the total gain from base product to high quality because the previous profit on the medium quality product has been lost. Thus, these lost revenues, or opportunity costs,[6] must be subtracted from the over-all gains of the challenger if the true benefit is to be evaluated. In other words, it is the net gain of the proposed alternative that must be considered. Furthermore, any savings attributable to the challenger must be realistic. A conservative attitude must be taken by management in estimating the value of unproven claims of the challenger for decreased costs or other savings.

SIMPLE REPLACEMENT EXAMPLE

Problem

Certain equipment in a printing company requires repairs of $500 and major-repair costs of $2000 in 2 years. No future salvage value is expected; however, a turn-in credit today of $2500 can be realized. A dismantling cost of $300 would be incurred. On the other hand, a new machine can be purchased for $10,000 for a 10 year life; there is a salvage value at 10 years, with estimated annual repairs of $100. The book value on the old machine is $5000, and if a change is made, the equivalent of lost production is estimated at $1000. Money is worth 14 per cent for these evaluations. Is the replacement economical based on (a) rigorous analysis and (b) approximate analysis?

---

[6] The above concept of an opportunity cost is not a universal term. For example, an opportunity cost is sometimes defined as the lost revenues resulting from the operation not being carried out at its best alternative procedure. These potential losses are thus considered as a cost, although they really have not been incurred.

Discussion

The solution is a simple replacement procedure for comparison of alternatives based either on present worth or annual costs. First, it should be noted that the book value of the defender is an expenditure of the past. Very often under the conditions of an alternative procedure none of the book value will be recovered; it is a *sunk cost*. The term sunk cost refers to that portion of an investment which cannot be recovered. In most cases, provision is made to amortize an investment. If subsequent events result in the amortization procedure not being carried out, then any unrecovered portion of the initial capital outlay is a sunk cost. (It is true that a portion of this value *might* be recoverable through tax credits, but such allowances complicate the analysis unduly and will be omitted to demonstrate the principles.) Thus, to evaluate the investment capital for the defender, one must consider the specific situation. If the old equipment is turned in, it will net 2500 − 300 or $2200, which is its equivalent value as it is. However, an additional $500 is required to put the old equipment in condition for future service. Hence the total capital involved for the defender is 2200 + 500, or $2700; this sum is compared to the $10,000 installation charge for the challenger plus the $1000 changeover cost in lost production.

The second consideration is the time involved. Although some expected book life remains for the current machine, this data is not considered. For comparative purposes, management assumes that the defender will continue to give good service for a reasonable time in the future, say five years, which is then the basis for calculations on the defender. The basis for calculations on the challenger is its expected life of ten years, and it is recognized that the challenger has an inherent advantage in its longer life. However, this factor would not be considered in the numbers utilized for comparison. Thus, the annual costs are compared on the basis of five year life for the defender and 10 years for the challenger.

Solution

Part a. Essentially, the problem requires evaluation of the present worth by use of Eq. (12-1a), where the individual items are summed up as follows:

	Defender, Dollars		Challenger, Dollars
Investment		2700	11,000
Other costs	$2000/(1.14)^2 =$	1535	$100(5.216) =$ 522
Salvage value (a credit)		0	$500/(1.14)^{10} =$ (−135)
Total present worth		4235	11,387

Clearly, the high initial cost of the challenger will bias the answer if a present worth basis is employed. However, if for the defender a future period of 5 years is considered and this is set in comparison to the annual costs of the challenger for its life of 10 years, the following results using the respective $a_{N/i}$ values:

	Defender	Challenger
Annual costs:	$4235/3.432 = \$1234$	$11,387/5.216 = \$2181$

The answer indicates that the replacement would not be economical.

Part b. As an approximate method of solution, the method of average annual costs as given by Eq. (12-7) is recommended, the future \$2000 overhaul being averaged over 5 years.

FOR THE DEFENDER:

$$2700/5 + 2700(0.07)(6.5) + 2000/5 = \$1166$$

FOR THE CHALLENGER:

$$(11,000 - 500)/10 + 10,500(0.07)(11/10) + 0.14(500) + 100 = \$2130$$

The approximate answers for annual costs are of the same magnitude as for the rigorous solution.

EXTENDED REPLACEMENT EXAMPLE

Numerous articles and several volumes have been written on replacement models based on the change in annual costs or revenues as an operation ages. In essence, the author believes that these procedures simply confuse the reader when comparing the average annual costs of a challenger with those of a defender. It the replacement is of the same kind (with the expressions for costs or revenues being repeated in the future) the average annual costs will be the same for the challenger as for the defender and the optimum age, by analysis of Eq. (12-10), will apply. If a challenger (because of new design, better quality product, or other factors effectively reducing annual costs) becomes available, its average annual costs may be less than those of the defender and its age cycle may also be different. These results can be evaluated numerically for both alternatives by Eq. (12-10). A comparison can then be prepared and a decision policy established based on minimum equivalent present worth or annual costs for each according to their respective optimum lives. Dollar values can be used to evaluate economic superiority.

If management expects the annual costs for the life of the replacement to be more economical than the average costs for the defender, the replacement can be made. Thus, if special circumstances indicate unanticipated excessive costs in a given year for the defender, management may decide to replace it with the challenger. On the other hand, without this unanticipated expense, the replacement may be unwarranted in the long run.

There is a special type of replacement analysis that utilizes the model of Eq. (12-10). This case occurs where improvements in the challenger are in a form that can be expressed mathematically as part of the net cost function. For example, the improvement in design from year to year may result in lower operating costs in some known or estimated manner. Where this improvement is a relation that is linear with each year the question arises: In what year should a defender be replaced *if the same pattern of improvement* will hold for all future challengers? In other words, the annual costs of new challengers are continuously being decreased, and a criterion (of minimum annual costs) can be employed to determine the optimum age for the periodic replacement of obsolete defenders by continually improved challengers.

The mathematical model for this analysis is merely a special case of Eqs. (12-10) and (12-11), wherein the annual costs function, $C(n)$, is decreased by a savings factor, $S(n)$. In connection with the analysis by Eq. (12-10), it should be noted that the expression $C(n)$ should represent net costs. Thus, any corrections for savings, better design, etc., should be included. This corrected value may be written as $[C(n) - S(n)]$, as in Eq. (12-1b), or it may be considered as a single relation designated as $C(n)$.

Problem

A machine installation cost is $15,000 with annual operating costs of $1000 + 300n$ dollars. Salvage value varies as $15,000N^{-1}$, where N is the age at replacement. It is expected that these machines will be improved such that the present annual operating costs will be $150 less for each year that the replacement is deferred. What is the optimum replacement cycle for this type of machine if money is worth 15 per cent? What are the annual costs?

Discussion

This problem involves a special type of replacement, where each new machine effectively produces a different annual cost function. However, this annual cost function is of the same form as that covered by Eq. (12-10). Hence, the solution is to establish the equation for present worth, P_{cap}, for continuous service and to solve for the optimum life so as to give a minimum present worth. Then, minimum present worth is converted to annual cost.

Solution

Where $S(n)$ is the savings, the function for net costs $[C(n) - S(n)]$ must be evaluated as a function of the year in which the periodic replacement is made. Thus:

$$C(n) - S(n) = 1000 + 300n - 150n = 1000 + 150n$$

The working equation then may be written from Eq. (12-10) and Eq. (12-3b):

$$P_{cap} = \frac{15,000}{1 - e^{-0.15N}} + \frac{1000}{0.15} + \frac{150}{0.15^2} - \frac{150N}{0.15(e^{0.15N} - 1)} - \frac{15,000}{Ne^{0.15N}}$$

$$iP_{cap} = \frac{2250}{1 - e^{-0.15N}} + 2000 - \frac{150N}{e^{0.15N} - 1} - \frac{2250}{N(e^{0.15N} - 1)}$$

The incremental analysis of Eq. (12-11) is then applied. The value of P_{cap} is multiplied by 0.15 and plotted in Fig. 12-2. The appropriate right side of Eq. (12-11) is also plotted in the same figure to obtain a solution for the optimum, N, as follows:

$$1000 + 150N + 0.15(15,000/N) + 15,000/N^2$$

These two plots intersect at $N = 1.9$ years, and the present worth is $44,300. From this result the annual average cost, iP_{cap} is $6650.

This analysis indicates that where a pattern of improvement for reducing costs will be repeated with the particular dollar values as stated, a

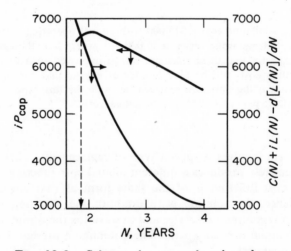

Fig. 12-2. Solution for example of replacement of a new kind.

new challenger should be installed every 1.9 years for minimum perpetual annual costs. The result is the same as for a situation where the net costs for a given machine change as $1000 + 150n$; here the replacement would always be of the same kind.

VARIATIONS IN REPLACEMENT ANALYSIS

As pointed out previously, a replacement analysis is essentially an economic balance between the decreasing annual capital charges for amortization and the increasing costs and reduced salvage values. General cases for annual costs being constant or variable were discussed in previous sections for conditions of continuous compounding. This section discusses certain variations of these procedures.

In Eq. (12-6a) the model was given for annual costs; *constant* annual operating costs, including interest and salvage, were considered. If the annual operating costs vary linearly in some manner with a gradient, k, such that $C(n) = C_0 + kn$, the equivalent total annual costs can be shown to be given by the following for *discrete* values of N [refer to Eqs. (12-3a), (12-3b), and (12-4) for development]:

$$R_{cost} = P_{cost}/a_{N/i} = (I - L)/a_{N/i} + iL + C_0 + k/i - kN/(s - 1) \tag{12-12}$$

This expression may be used with trial and error to find an optimum age, N, at which the annual costs are a minimum[7]. However, the relationship for L as a function of N must be known.

Where a *continuous* function is considered, the analysis corresponding to Eq. (12-12) is given by Eqs. (12-10) and (12-11), as was demonstrated by examples on pages 319, 325.

Another variation of the replacement problem is the determination of the optimum age at which *salvage and interest both are neglected.* If continuous functions are assumed, the equations may be differentiated to determine the optimum age directly without recourse to trial and error. The minimum cost occurs when the change in annual depreciation for capital recovery just equals the change in annual expense, as demonstrated by Fig. 12-3, which is a typical economic balance diagram.

With any type of depreciation, if no salvage value and zero interest are assumed, the *average* annual depreciation is always I/N for different lives, N. Consider I/N to be a continuous function and that the net costs

---

[7]Such a minimum is the *adverse minimum* of the MAPI procedure (Machinery and Allied Products Institute), which makes certain simplifications for a rapid solution.

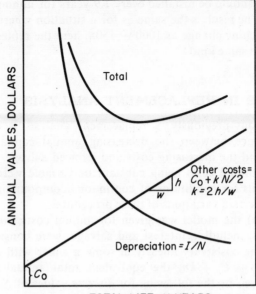

Fig. 12-3. Economic balance for optimum replacement age. (No salvage and zero interest)

for any one year n vary as $C(n) = C_0 + kn$, which may be integrated[8] to determine the average annual costs; then the average annual costs are:

$$C = I/N + C_0 + kN/2, \text{ dollars per year} \qquad (12\text{-}13)$$

Thus:

$$dC/dN = -I/N^2 + k/2 = 0$$
$$N_{opt} = \sqrt{2I/k}, \text{ years} \qquad (12\text{-}14)$$

If this equation is applied to the previous replacement problem of the same kind (page 319) with annual costs showing a gradient k of 700 dollars and an investment of $15,000, the optimum replacement age N is:

---

[8] The integration is as follows:

$$\int_0^N (C_0 + kn)\, dn = NC_0 + kN^2/2$$

Thus, the *average* value for N years would be $C_0 + kN/2$ after dividing by N.

$$N = \sqrt{\frac{2 \times 15,000}{700}} = 6.55 \text{ years}$$

This replacement age should be compared to the 2.0 years found when interest and salvage value were considered. The results perhaps are for an extreme case, but do demonstrate the relative effects of neglecting interest and salvage value.

Note that a direct substitution of N_{opt} into Eq. (12-13) yields the minimum annual cost directly, or:

$$C_{min} = C_0 + (2kI)^{0.5}, \text{ dollars per year} \qquad (12\text{-}15)$$

COMMENT ON TAXES

As demonstrated in Chapter 11, complications are introduced by including profit taxes and tax credits in the replacement analysis. Therefore, if taxes are included, it is suggested that the replacement analysis be made as for profitability (Chapter 11, page 297) but that only one installation be considered. This approach becomes, then, a comparison of alternatives for some given life period. Where taxes are allowed for in a continuous service analysis made for minimum cost, management must refer to Eq. (12-1) as the starting point. Since there is a tax credit available for depreciation but a tax charge on any savings or profits, the basic equation for the *present worth of the net cost* is written as follows:

$$P_{cost} = I + (1 - t) \sum_{n=1}^{n=N} \frac{[C(n) - S(n)]}{(1 + i)^n}$$

$$- t \sum_{n=1}^{n=N} \frac{D(n)}{(1 + i)^n} - L(N)/(1 + i)^N \qquad (12\text{-}16a)$$

For this equation to apply, the net cost of the service must be negative or zero, since only under these conditions does a real revenue necessitate consideration of profits tax. Thus, the terms $[C(n) - S(n)]/(1 + i)^n$ must be greater negatively than the absolute value of $D(n)/(1 + i)^n$ in any one year for the tax to apply.

The continuous function is:

$$P_{cost} = I + (1 - t) \int_0^N [C(n) - S(n)] e^{-in} dn$$

$$- t \int_0^N D(n) e^{-in} dn - L(N) e^{-iN} \qquad (12\text{-}16b)$$

In a real world problem, these present worths are negative, the most negative of several alternatives being the best because the savings or other credits all are negative in sign. This analysis differs from the presentation of Eq. (11-1) and (11-11) in the previous chapter. In Chapter 11 the equations were set up to determine the equivalent interest rate when P_{cost} is made equal to zero; in Eq. (12-16a) and (12-16b) the P_{cost} is calculated for a known interest rate.

For the special case of straight-line depreciation, with constant L at any age, and with zero interest for the value of money, Eq. (12-16) reduces to:

$$P_{cost} = (1 - t)\left\{(I - L) + \sum_{n=1}^{n=N} [C(n) - S(n)]\right\} \qquad (12\text{-}16c)$$

This simplified equation is satisfactory for the comparison of alternatives where the investments are of the same order of magnitude. The one with the most negative value of P_{cost} would be the best.

In analyses using Eq. (12-16a) and its variations, no tax allowance is made for the possibility of the salvage value being different from that originally assumed. As was discussed in connection with Eq. (11-1), a tax credit or additional revenue could occur in a real world situation.

THE RENT OR BUY EXAMPLE

A common economic problem is the one where a service or facility can be purchased outright or rented on an annual or monthly basis. Where a life can be estimated and the time value of money can be established, the solution is straightforward because it is a simple case of comparing alternatives. The comparison can be best made by considering taxes and by using the basic Eq. (12-16a) or Eq. (12-16b).

A separate calculation is made for each alternative. One involves a relatively small investment for preparation or set-up expense plus rather large periodic costs for rent; the other, a large purchase cost plus relatively small periodic costs for services and maintenance. Any basis desired—earning rate or payout time—may be selected.

In this problem where the facility is replaced periodically with a similar machine, the optimum age for operation can be computed and employed in the comparison if desired.

Problem

Consider the situation where a company may (A) pay rent for warehouse facilities at an annual charge of $14,000 plus an initial renovation cost of $6,000 to operate properly or (B) build its own facility at a cost of $150,000 with annual charges for taxes and maintenance of $2000 and straight line depreciation of $5000 per year for a 30 year life with no salvage value.

Solution

If this problem alone is considered there appear to be no revenues or savings to include in the model for taxation, and the present worth of the costs including depreciation may be compared by use of Eq. (12-1a) at some interest rate for the value of money, say 5 per cent. This leads to the following results:

FOR ALTERNATIVE A, where a facility is to be rented:

$$P_{cost} = 6000 + 14,000a_{30/0.05} = 6000 + 14,000(15.37) = \$221,180$$

FOR ALTERNATIVE B, which involves owning facility:

$$P_{cost} = 150,000 + 11,000a_{30/0.05} = 150,000 + 7000(15.37) = \$257,990$$

Thus, the equivalent present-worth cost for renting has a definite economic advantage as would the equivalent annual cost, where time value of money is taken into account.

A different approach in the analysis of this problem neglects the time value of money but utilizes the capitalized earning rate concept and the influence of the tax rate. For alternative B an additional initial $144,000 of investment capital is required, which would result in a savings or increased profit of $7000 of annual costs. At a 52 per cent profits tax, this is equivalent to a real saving of $(1 - 0.52) \times (7000) = \$2660$. Based on the additional investment, this is an equivalent capitalized earning rate of 1.85 per cent, which is an obviously low rate. Clearly, the investment funds will yield a greater economic return for the company if used in some other part of its operation (or outside the company) where opportunities of yields in excess 1.85 per cent are available.

IRREDUCIBLE FACTORS IN ECONOMIC ANALYSES

The preceding discussion covering the mathematical relations involved in economic analyses provides the tools needed to determine the actual advantage gained dollarwise. However, in many instances factors must be considered that cannot be reduced to dollars and cents but which can be of sufficient influence to warrant a greater expenditure. Some of these irreducible factors have been mentioned in connection with the illustrative problems. Among the major factors of this type are the availability of capital for replacements, allowance for growth, anticipation of decrease in demand for products, "going" value of an operation where production will be lost while another process is being installed, the good-will value of nuisance abatement by installation of liquid-waste disposal systems and abatement of air pollution, the effect on personnel of improved working conditions resulting from the provision of continuous jobs and the reduction of potential hazards, and the insurance value of expenditures to allow for the potential risk of future hazards. Some of the above factors can be evaluated partially in terms of dollars. However, the amount of unknown value often is difficult or impossible to evaluate.

The use of annual gradients for changes in revenues, costs, or any other items in the equations represents an attempt to predict future values in a quantitative manner. Obviously, such predictions are at best an experienced guess, but they are certainly better than no guess at all. By making such estimates after considering all potential factors that might influence future values, management can conduct economic analyses that otherwise could not be made.

REAL WORLD PRACTICE

In a survey by MAPI[9] it was reported that the real world methods employed as a basis for replacement policies emphasized that the empirical methods were generally used. The distribution of methods as reported was:

	Per Cent
Payout time	42
MAPI procedure	19
Minimum-average cost	8
Discounted cash flow method	1
Total	70

The report is not clear as to the exact procedures employed in each of the above categories, and evidently the above total does not include a simple rate of return method. Simple rate of return, however, can be estimated at 15 to 20 per cent from the reported data. Apparently, the MAPI procedures and the discounted cash flow methods are the only ones in which the time value of money is considered. However, the importance of interest has received more attention recently, and certainly more firms than reported in the 1956 survey are giving attention to it.

Certain other data in the report were employed to draw Figs. 12-4 and 12-5. Fig. 12-4 indicates that 50 per cent of the firms using an earning rate criterion for replacement require an earning rate (some form of profitability return) of about 18 per cent or less on the initial investment. For those employing a payout time, 50 per cent require a payout of 4 years or

---

[9] "Equipment Replacement and Depreciation—a Survey," Machinery and Allied Products Institute, Washington 6, D. C., 1956.

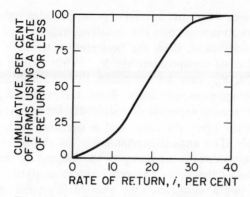

Fig. 12-4. MAPI survey (1956) of distribution of rates of return used for replacement policy.

less (Fig. 12-5). It is not clear from the report, but it is presumed that the methods employed are analogous to the capitalized earning rates and capitalized payout times of Chapter 11.

The fact that this survey did not raise the question of the time value of money once again points up the necessity of a careful explanation of the methods employed if the results are to have meaning for the reader.

In replacement studies, the question of replacement often arises where charges for the next future period have been predicted and a decision as to best policy is required. The theoretical treatment and the incremental analysis indicate that the optimum age for a replacement occurs when the equivalent annual capital recovery charges just equal the net annual

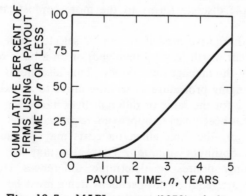

Fig. 12-5. MAPI survey (1956) of distribution of payout times used for replacement policy.

expense; this holds where net annual expense includes repairs and maintenance, interest on salvage, and loss in salvage value.

From this background, then, the preferred criterion is based on the average annual costs for optimum life N_{opt}. Thus, where N_{opt} is known, the annual costs, R_{cost} may be computed; the result obtained represents long run or average expected values. Then, the predicted annual capital costs plus other annual expenses of a defender for any year, N_x, can be compared with the optimum value and a decision made based on the relative amounts. The annual capital costs for the defender constitute the loss of salvage value for the year plus the interest on the salvage value at the beginning of the year. The replacement analysis, therefore, is a comparison of two alternatives. One alternative represents the defender already in operation with certain estimated costs; the second alternative involves a replacement of the same or different kind, having certain expected annual costs for its optimum life. If the predicted next year costs for the defender exceed those expected annually for the challenger, a replacement should be made. For those cases where the predicted costs for the defender are less than the challenger, no replacement is indicated, even though the defender may have exceeded its expected optimum life. Where the salvage value is zero at any age, only out-of-pocket disbursements for the defender need be compared with the optimum annual costs of the challenger.

SUMMARY

Basic profitability functions have been applied to real world situations in which decisions must be made to select proper alternative procedures. The models employed are based on theoretical economy relations, and vary from simple discrete forms to the more sophisticated continuous analysis.

The proper alternative procedure can be selected either on a basis of equivalent present worth (capital involved) or on a basis of annual costs, depending upon the management policy. This also holds for the special case of a challenging procedure as an alternative to an existing defender. Models allowing for the same or different lives have been presented, and consideration has been given to approximate procedures.

All items having the same effect on costs may be omitted from the analysis in an optimum situation. They also may be omitted where management only desires to determine differences. However, where absolute values are required, all pertinent items must be included in the analysis. In a comparison of alternative policies, all costs incurred prior to the time of the decision may be neglected for the analysis. This is specifically true where taxes are not considered. However, when profits taxes are included in the analysis, tax credits and interest on them may influence the cash flows in terms of capital commitments prior to time of analysis. Under these circumstances, the best analyses involve the making of computations of the individual cash flows as proposed in Chapter 11 and the considering of year to year comparisons.

PROBLEMS

12-1. Compare the present worths of the costs and the annual costs by both the discrete and continuous values of $a_{n/i}$ for the following situation. Two bottling machines are offered by different suppliers M and N. Machine M costs $7500 with annual charges of $1000 and a salvage value of $800 with a life of 5 years. Machine N with the same life costs $10,000, with annual expenses of $800 and a salvage value of $600 for the same life. Money is considered worth 5 per cent for this analysis. NOTE: This is not a continuous service model.

12-2. Use the same data as in Problem 12-1, but assume that Machine N will last 10 years. Compare the annual costs for the alternatives by the rigorous and approximate methods.

12-3. Two installations A and B for stamping parts have been investigated and their economic data established for a 10 year life with money worth 10 per cent.

	Machine A, Dollars	Machine B, Dollars
Original installation	12,000	8000
Annual costs, $C_0 + kn$	$1000 + 300n$	$1500 + 500n$
Salvage value at 10 years	2000	1000

Which is the preferred installation and why?

12-4. A manufacturer of electronic parts has a choice between two possible procedures A and B. Costs vary as follows, with money worth 12 per cent:

	Procedure A, Dollars	Procedure B, Dollars
First cost installed	28,000	42,000
Annual costs, $C(n)$	$3000 + 1200n$	$1000n^{1.1}$
Annual quality credit, $S(n)$	0	$600 - 60n$
Salvage value at 8 years	2000	2400

Compare the present worth of the total costs and the equivalent annual costs.

12-5. A textile manufacturer has two possible selections for operations. One, Machine X, has a life of two years with salvage value of $200, whereas

Machine Y has a life of 6 years with $800 recovered. Money is worth 10 per cent, and other costs are irregular as shown by the following schedule:

Year	Machine X	Machine Y
0	initial cost of $12,000	initial cost of $28,000
1	———	———
2	maintenance of $400	———
3	renewal of $12,000; $200 at start of year	maintenance of $100
4	maintenance of $400	———
5	renewal of $12,000; $200 at start of year	maintenance of $300
6	maintenance of $400 and recovery of $200 salvage value	recovery of $800 salvage value

What are the annual costs by a rigorous and an approximate procedure?

12-6. What are the respective capitalized costs and annual costs for the two machines in Problem 12-3?

12-7. (a) Compute the optimum age for replacement of the same kind with money worth 12 per cent for an operation that has an investment of $5000 with costs varying as $1000 + 150n$ dollars per year and salvage value expected as $5000e^{-0.25N}$ where n is any elapsed time period and N is a terminal age. (b) What is the capitalized cost for the series of renewals?

12-8. A challenger having an investment cost of $8000 with expenses of $800 + 50n$ dollars per year and salvage varying as $8000/N$ is available as a replacement for the defender in problem 12-7. Should the change be made and why?

12-9. A report has estimated the costs for the production of synthetic fuels through hydrogenation of coal upon which the following problem is based. An assumed investment of $80 million (of which 10 per cent may be taken as nondepreciable capital, with money to be raised by selling bonds at 2.5 per cent with a life of 10 years) will produce products selling for about $60,000 per day with variable and fixed costs, other than financial and depreciation, of $30,000 per day. Profits tax is 52 per cent. An alternative of financing the investment by selling stock instead of bonds is proposed. Prepare the analysis for selection of the best alternative. How does the procedure for solution differ from that for working problem 11-4 in the previous chapter?

12-10. Using the data of Problem 12-1, compare the P_{cost} of the two machines, they are at zero interest with straight line depreciation and there is a profits tax of 55 per cent when annual revenues for each are $3000. Discuss.

CYCLIC MODELS FOR PRODUCTION AND INVENTORY

13

In the sphere of business, one deals with numerous *cyclic* operating and nonoperating procedures. Cyclic business activities may be defined very simply as activities that repeat themselves over and over again. There are many variations of the economic analysis of these cyclic activities. Among these variations are the classical economic lot size problem, the inventory problem, the periodic renewal of a catalyst in chemical operations, and a large number of others. Although this chapter covers only the elementary problems in this area, it does indicate how elementary models can be applied to more complicated problems.

Since many terms must be considered in discussing cyclic models for economic analysis, an abbreviated glossary is given in Table 13-1. This tabulation will assist the reader in following the discussion.

A continuous process operates 24 hours per day, 365 days per year, and is only out of use periodically for service and repairs. However, although a cyclic process may be at a continuous rate of production during operation, the operation itself may run for short periods. In such a case, the operation is shut down for an interim time, and is later restarted to repeat the cycle. Such an operation is sometimes called a semicontinuous process.

The lead time (or make time), t_m, is the time between the start of the procurement and the receipt of a certain quantity of purchased material, or the time between the start of production and its completion.[1] The

---

[1]Note that there are four production rates where the rate is defined as the production divided by the time in which it is produced. There is (a) an *instantaneous production rate* which may or may not be constant; if constant, it is the same as (b) an *average design rate* based on the time when actually producing, (c) an *over-all production rate* based on the total time required including down time, and (d) an *average demand rate* based on requirements. The average demand rate is the minimum the operation must produce under any conditions and still meet the demand for the problem. When there is no interim time, the rates in (c) and (d) are equal, and when there is no down time the rates in (b) and (c) are equal. The average demand production rate for the year is Q_A, the annual production.

NOMENCLATURE

a = arrival rate, items per unit of time

C = cost of maintaining a stock, dollars per unit

C_c = set-up, get-ready or any costs that are constant per cycle, dollars per cycle

C_F = annual fixed costs that may vary with batch size, dollars per cycle

C_h = annual holding or carrying charges, dollars per year

C_o = other or operating costs specific for one procedure, dollars per year

C_T = total costs in dollars per year

$E(D)$ = expected value of demands in one cycle

f = a deterioration factor

f_h = the fractional value of an item that can be charged as *annual* carrying or holding charge

F = annual production factor, see text

H = annual time available for production

I' = work-in-process inventory, various units

I'' = raw-material inventory, various units

I_a = the average inventory, various units

I_p = protective stock in stores inventory, various units

K = a constant

k = a constant

M = number of machines

m = instantaneous production rate, items per unit of time. This is an instantaneous rate that can never be less than Q_A, which is the required annual rate

N = number of cycles per year

N_{opt} = optimum N for minimum cost

p = product value for inventory purposes

p_p = in-process value of a product, dollars per unit

p_r = raw-material value, dollars per unit

$P(w)$ = cumulative probability of w demands or less

$P'(w)$ = cumulative probability of w demands or more

Q = quantity, various units

Q_A = total production, in time H, various units

Q_B = size of a batch, various units

$Q_{B_{opt}}$ = optimum Q_B for minimum cost

Q_I = inventory, various units

Q_p = a quantity of material held at one station with continuous flow into and out of the station, various units

R = the ratio of units of product per units of raw material, or over-all production rate

S = cost of a stock out in stores inventory, dollars per unit

S_t = a quantity of items

t_d = down time for necessary work, fraction of a year

t_i = interim, or dead, time, fraction of a year

t_m = make time, or lead time, fraction of a year

t_t = total time for one complete cycle, fraction of a year, or other units; it is elapsed time from one point in a cycle to the same point in the next cycle for repetitive operation

V_i = a direct expense applying to a specific procedure, dollars per unit

w = average demand rate or withdrawal rate, items per unit time; if the time unit is one year; w equals Q_A

TABLE 13-1. **Terminology for Cyclic Models**

Term	Definition and Significance
Semicontinuous process	An operation where the production is manufactured at a continuous rate but the operation is stopped or changed at periodic intervals. An example is a production line for making a particular type of chair.
Batch process	An operation where all items of product are finished simultaneously or purchased in a single lot. Examples are purchases of raw stock and production of materials that attain the desired quality or properties after a given production time.
Continual repetitive process	An operation in which a new cycle is started immediately upon completion of the previous cycle —the latter may have down time for cleanup, etc., as part of its cycle.
Lead time or make time	The time required from the initiation of operations to make or procure product until it is available.
Down time	The time period necessary for set up of an operation or the turn around time required to start a new operation. During this period no production is made.
Interim time or dead time	The time period during which there is no operation being performed necessary in any way for production to proceed.
Total cycle time	The time between identical points of the cycle for two consecutive cycles; that is, the time required for the entire operation to repeat itself one time.
Instantaneous production rate	This is the rate of production at any time in items being made per unit of time. It is the slope of a plot of total quantity made versus time.
Average design rate	This is the total quantity made in a given time divided by the time of production excluding the down time.
Over-all production rate	This is the total quantity made in a given time divided by the total production time including the down time. Same as over-all production rate.
Average demand rate	This is the average quantity required per unit of time to meet demands for the product. It is the same as the annual production, if the withdrawals over the year occur at a uniform rate.

interim, t_i, is the time during which there is no necessary operation; it differs from down time, t_d. The down time is necessary time for charging or discharging, aging or curing, etc.; therefore, it constitutes an actual part of the production time.

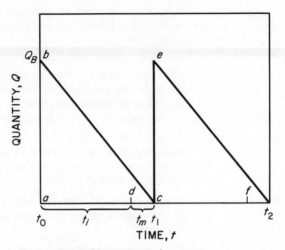

Fig. 13-1. Noncontinual repetitive process.

REPETITIVE OPERATIONS

For the ordinary lot, or batch, problem the quantity involved is obtained all in one batch after a given lead time, t_m. This may be visualized as in Fig. 13-1. Starting at point a one finds that the batch, Q_B, is used up at a constant rate along the line bc until none is left at point c. At point c another batch is available which was started or ordered at point, d, and the distance, dc, is the lead time, t_m. Starting at point c, the cycle repeats and the operation is therefore a repetitive one. There is interim time, t_i, from a to d during which no operations are performed by the producing facility. However, if there were no interim time, the diagram would be as in Fig. 13-2 where dc is the make time; that is, as soon as one lot was completed or received, a new lot would be ordered or started. Where a machine is put in production as soon as available after completing a make cycle, the process is a continual repetitive cyclic operation even though there may be down time. Many economic analyses assume this condition.

Figure 13-3 represents a special semicontinuous procedure where the stock is *produced at a constant rate, m,* and then withdrawn in one lot; thus, in the real world, a batch of stock produced semicontinuously may be sold in one batch. However, Fig. 13-2 differs from Fig. 13-3. Figure

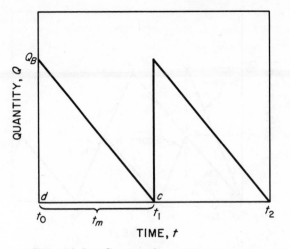

Fig. 13-2. Continual repetitive process.

13-2 represents the practice of purchasing lots which are to be sold at some rate per unit of time (for example, 10 items per day, etc.). In short, the entire lot is available at one time, but is *withdrawn at a constant rate.* Although, there are other variations, the basic economic principles, as presented later, can be applied readily to all cases.

Figure 13-4 may be considered the semicontinuous model where items are received (or produced) at a constant rate, m, during the make period to produce a total batch equal to mt_m. However, product is also with-

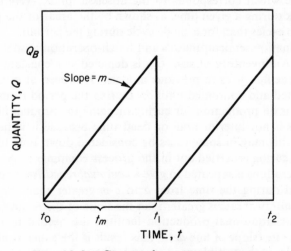

Fig. 13-3. Continual repetitive process.

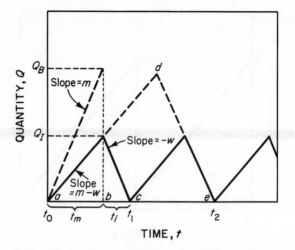

Fig. 13-4. Semicontinuous repetitive process.

drawn at a constant rate w; thus, the net build up of inventory to a quantity, Q_I, is at a rate equal to the difference of these two rates during the make or lead time period. At t_1 the cycle is repeated as shown by the solid lines. The period from a to b is make time or lead time, and since production is stopped at point b, the period from b to c is interim time. The operation, therefore, is repetitive, but it is not continual repetitive. The dashed lines in Fig. 13-4 represent the same type of operation with only one cycle between t_0 and t_2 where the inventory builds up to d with a lead time which corresponds to the abscissa for d. (Here the quantity in stock during a given time, as shown by the areas of the plots, totals less for two cycles than for a single cycle during the period).

A combination semicontinuous and batch-operation model is shown in Fig. 13-5. An inventory of size, Q_B, is depleted at a constant withdrawal rate along the line bc as in previous models. However, at a, new production is started and continued until d. During the period from d to c the equipment is off production for curing or aging the product in the equipment. This is not interim time or dead time because it is necessary for operation; this may in some cases be considered down time for the operation if the curing is carried out in the process equipment. At point c the cycle of operations is repeated to give a *continual repetitive process*. Where the demand during the time from a to c is greater than the batch size, Q_B, the withdrawal rate is greater than the slope of the bc line. To satisfy this demand, additional production facilities are needed to increase the production rate (slope of line ae). This result is the same as having overlapping cycles with multiple production facilities.

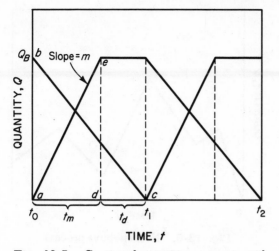

Fig. 13-5. Continual repetitive process with down time.

A nonrepetitive process is demonstrated in Fig. 13-6 where the production rates, lead times, and withdrawal rates are shown to vary from cycle to cycle.[2]

It should be emphasized that the diagrams as shown are simplified analogies of the real world situations. The simplification consists mainly of using straight lines for the relations. Under these conditions the mathematical relations are all linear. Although higher order equations for curved lines can be employed in the analysis, they are more difficult to use for analytical models.

BASIC ECONOMIC RELATIONS

The economic analysis of cyclic processes follows the basic principles presented in Chapter 9; certain costs decrease and others increase when related to some decision variable which management selects to use as a criterion. In general, the model uses the quantity of a lot (batch) or the number of cycles per year as the criterion. Thus, with set-up costs constant per cycle, the annual set-up costs increase as a greater number of cycles are employed per year. However, the utilization of a greater number of cycles means that the average batch size is smaller, the inventory is

---

[2]A variation of Fig. 13-1 consists of a diagram where the inventory is exhausted at periodic intervals because of rapid withdrawal rates. Such an operation would be repetitive if the stockout periods followed a regular pattern.

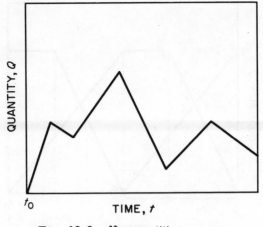

Fig. 13-6.   Nonrepetitive process.

smaller, and the annual interest and storage (holding) charges decrease. As shown in Fig. 13-7, the economic balance involves an optimum number of cycles to give a minimum annual cost.   Instead of cycles per year, the independent variable selected could be quantity per cycle, Q_B.   In this case, however, the holding charge curve increases as Q_B gets larger, and the set-up-costs curve decreases with increase in Q_B, since fewer cycles are used.   The over-all result is the same regardless of which basis for analysis is used.

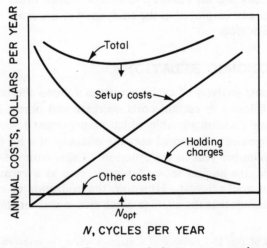

Fig. 13-7.   Economic   balance   in   cyclic process.

The general plan of analysis is to set up the total annual cost equation as follows:

Annual costs = cycle costs + holding costs + other costs

The cycle costs, C_c, are all costs which are repeated each cycle for N cycles. The holding costs, C_h are the costs related to total inventory of product and work in process plus storage charges. This includes space for storage, interest, insurance, and other charges. The operating, or other, costs, C_o, are any other expenses that apply for the particular program under study. As shown in Fig. 13-7 these other costs do not affect position of the optimum value for the point of minimum cost if they are constant. These constant costs are necessary to establish the absolute values for the annual costs where such figures are desired, but they often may be omitted in the analysis of cyclic problem. The algebraic equation is:

$$C_T = NC_c + C_h + C_o, \text{dollars per year} \qquad (13\text{-}1)$$

By expressing all the individual costs in terms of the costs per year in Eq. (13-1), one can differentiate the total cost mathematically and solve for the minimum cost.

The first step in the analysis is to express the number of cycles, N, in terms of the technical information available. Thus, if an annual amount, Q_A, is required for a situation, then $N = Q_A/Q_B$, where Q_B is the quantity for any batch or lot. Similarly, if H hours per year are available for operation, then $N = H/t_t$, where t_t is total cycle time in hours. Note here that all of the quantities Q_A, Q_B, H, and t_t, are variables with known relations or variables which can be estimated. Thus, the optimum operation may be determined in terms of any variables sought.

Second, it is necessary to establish the average inventory quantity size in order to evaluate the carrying costs part of Eq. (13-1). This will be done in the next section.

The analysis as presented by Eq. (13-1) may take the form of an incremental cost analysis, when the analysis does not include all of the costs of the operation. An incremental cost analysis is concerned only with those costs that change with the batch size or number of batches. Any costs that are unaffected do not influence the optimum batch size, since they disappear when the equation is differentiated. All costs other than those shown in Eq. (13-1) that may cause some effect, such as changes in design, special discounts, method of procedure, in-process inventory, etc., must be considered and included in the analysis if the effect is appreciable. Otherwise, they may be omitted for practical purposes even though they are necessary in theory; some statement concerning them, however, should be made in the economic analysis.

The value of holding charges for storage, insurance, interest on capital held up in inventory, etc., can ordinarily be expressed as a fraction, f_h, of the product value, p. Thus, the annual charge for hold up (that is, for carrying an average quantity, I_a, in stock) is $I_a p f_h$. The numerical value of I_a is specific with each real problem, as will be demonstrated. Note that the factor f_h can be made to include spoilage allowances, style changes, obsolescence, etc., merely by increasing its magnitude.

AVERAGE QUANTITY CALCULATIONS

To compute holding costs, one must employ an average value for the quantities held. These quantities may be raw materials, materials in process, or finished products. Referring to Figs. 13-1 to 13-6, one can see that the *average quantity* for any time period represents the area under the respective plots divided by time along the abscissa. Where the total *elapsed* time is one year, the average quantity is for one year of elapsed time, and any annual hold up cost is merely this average quantity times the holding charge per unit. Each average quantity calculation must be considered separately although many types such as the economic size of an order, the economic lot size, etc., fit into a general model. By definition the average inventory is some equivalent average quantity that is held for a period of a year (or some other time basis). Thus, the total inventory is the sum of all quantities in inventory multiplied by the time each item is held therein. The average value is the total inventory divided by the elapsed time involved to obtain a mean equivalent for the year.

Referring to Fig. 13-1, one sees that the average inventory quantity, I_a, for an elapsed time from t_0 to t_2 involves the area of the two triangles (2 cycles); in a general form for all cases, the expression is the sum of the quantities times their time held in inventory divided by the total time period. This is given by integral calculus from the following expression[3]:

$$I_a = \frac{N \int_{t_0}^{t_t} Q \, dt}{N t_t}, \text{(items-year)/year} \qquad (13\text{-}2)$$

where N is the number of cycles in time, t, and t_t is the time for one *complete* cycle in years; t_t is the time along the abscissa of the figures between

---

[3]It is not necessary for the reader to perform the integration at this point since the final result will be given and its use illustrated, but the mathematics are included here for reference.

a point for one cycle and the corresponding point on the next successive cycle in a repetitive process.

For Fig. 13-1 the inventory during time ac is linear starting at Q_B and decreases at a rate w with time, or the value of Q at any t within limits is:

$$Q = Q_B - wt \qquad (13\text{-}3)$$

Therefore, from Eq. (13-2) and Fig. 13-1 for limits of t:

$$I_a = \frac{N \displaystyle\int_{t_0}^{t_1} (Q_B - wt)\, dt}{N(t_1 - t_0)} = \frac{Q_B t - wt^2/2}{t_1 - t_0} \bigg|_{t_0}^{t_1}$$

$$= \frac{t(Q_B - wt/2)}{t_1 - t_0} \bigg|_{t_0}^{t_1} = \frac{(t_1 - t_0)[Q_B - w(t_1 - t_0)/2]}{t_1 - t_0}$$

but since $w(t_1 - t_0)/2 = Q_B/2$, the average inventory is:

$$I_a = \frac{t_1 - t_0}{t_1 - t_0} (Q_B/2) \text{ items} \qquad (13\text{-}4)$$

Therefore[4]:

$$I_a = Q_B/2 \qquad (13\text{-}5)$$

This same relation applies to Fig. 13-3.

Where Fig. 13-4 applies to a semicontinuous operation, a different treatment for the average inventory is necessary because the relation between Q and t is different. If m is a constant production rate and w is a constant usage or withdrawl rate, the net inventory build up during the lead time period is according to the following equation:

$$Q_I = mt_m - wt_m = (m - w)t_m \qquad (13\text{-}6)$$

but it decreases in the interim time according to Eq. (13-3) where $Q = Q_I - wt$ with Q_I being the maximum inventory attained at point b. (Note that Q_I is not the amount produced in a batch, since $Q_B = \displaystyle\int_{t_0}^{t_m} m\, dt$.) The

---

[4]A rigorous proof of this relation can be shown, since the make rate is proportional to the Dirac delta function. The integral of this function is the unit step function.

average inventory then is[5]:

$$I_a = Q_I/2 = (m - w)t_m/2 \tag{13-8}$$

For the general cases represented by Eqs. (13-5) and (13-8), the average inventory is one half of the maximum value attained, since the areas are triangular (that is, m and w are constant with time). However, where the operation has nontriangular areas, rigorous integrated equations must be employed; these are covered in the next chapter. These rigorous equations allow for the time when there is no inventory (noncontinual or stock outs) or where relations other than Eqs. (13-3) and (13-6) apply to the manner in which inventory varies with time. For those patterns of areas bounded by straight lines such as have been illustrated, the areas can be obtained by visual inspection to yield results proven in Eqs. (13-5) and (13-8). For other areas such as in Fig. 13-6 or those bounded by curved lines, it is necessary to use graphical or other procedures to estimate the areas when equations relating Q and t are not available. The values for m and w may be for any time unit (per hour, per day, etc.), but if they are on a yearly basis, then $w = Q_A$ units per year.

THE ECONOMIC ORDER SIZE PROBLEM

A classical application of the principles in the preceding sections is the selection of the economic size of an order to be placed. It costs C_c dollars to make out an order and procure each lot of goods and thus the more orders per year the higher is the ordering cost. However, for a given total quantity ordered each year, more orders mean smaller quantities per order and the interest and other carrying charges C_h will be decreased as illustrated in Fig. 13-7. These carrying charges or holding costs include many hidden items such as tax liability on inventories, insurance, spoilage, pilferage, and other associated costs.

---

[5]If these are substituted in Eq. (13-2), there results during the two times a to b and b to c for the area divided by the elapsed cycle time:

$$I_a = \frac{N \displaystyle\int_{t_0}^{t_m} (m - w)t \, dt + N \displaystyle\int_{t_m}^{t_1} (Q_I - wt)dt}{N(t_1 - t_0)} \tag{13-7}$$

$$I_a = \frac{(t_m^2 - t_0^2)(m - w)/2 + (t_1 - t_m)[Q_1 - w(t_1 - t_m)/2]}{t_1 - t_0}$$

Noting that $w(t_1 - t_m) = Q_I$ and Q_I also equals $(-w)(t_m - t_o)$, the following can be obtained by algebraic manipulation, since $t_0 = 0$:

$$I_a = Q_I/2 = (m - w)t_m \tag{13-8}$$

Problem

What economic order size should a distributor use to purchase automobile tires when it costs him $14 to make up an order and the tires are worth $7? Annual over-all carrying charges are estimated at 12 per cent of product value for insurance, storage, interest on capital, etc., and all other costs are constant. Annual sales are 120,000 tires.

Solution

Since this problem follows the pattern of Fig. 13-2, Eqs. (13-1) and (13-5) are utilized. The lead time is not stated in the problem and may be greater or less than the time from t_0 to t_1, but this variation will be omitted in the analysis of this problem. The total annual costs pertinent to the problem are:

$$C_T = NC_c + I_a pf_h + V_i Q_A \qquad (13\text{-}9)$$

where N = the number of cycles (orders) per year
$\quad C_c$ = the total of all costs that are constant per order for each order
$\quad I_a$ = average inventory of item
$\quad p$ = value of the product
$\quad f_h$ = the fractional value of the product that is chargeable for annual holding or carrying charges, (i.e., storage and interest)
$\quad V_i$ = the cost per unit handled for materials, labor, etc.
and $\quad Q_A$ = the annual quantity

Where the process is repetitive as in this case it was shown that $I_a = Q_B/2$ which yields:

$$C_T = NC_c + (Q_B/2)pf_h + V_i Q_A, \text{dollars per year}$$

This relation has two variables N and Q_B, but for a constant annual quantity, Q_A, the number of cycles (or orders) N is:

$$N = Q_A/Q_B = H/t_t \qquad (13\text{-}10)$$

where H = the number of total hours per year for the operation
and $\quad t_t$ = the total time for one cycle

Equation (13-9) may now be written:

$$C_T = \frac{Q_A C_c}{Q_B} + \frac{Q_B pf_h}{2} + V_i Q_A, \text{dollars per year} \qquad (13\text{-}11)$$

This is a general equation that has many applications, and it may be differentiated readily. With the derivative set equal to zero, the optimum batch size for minimum cost can be obtained; this procedure is found in many textbooks. Note that where the $V_i Q_A$ term is a function of the batch size it must be considered. However, where the analysis is not concerned with changes in variable costs and where the

$V_i Q_A$ term is constant, the term may be omitted, since in such a case it drops out in the differentiation. Thus:

$$\frac{dC_T}{dQ_B} = -\frac{Q_A C_c}{Q_B^2} + \frac{pf_h}{2} = 0$$

$$Q_{B_{opt}} = \sqrt{\frac{2Q_A C_c}{pf_h}} \qquad (13\text{-}12)$$

For the order problem with the numbers substituted in Eq. (13-12), there results:

$$Q_{B_{opt}} = \sqrt{\frac{2(1.2 \times 10^5)(14)}{7 \times 0.12}} = 2000 \text{ tires per order}$$

Thus, the optimum number of orders is:

$$N = \frac{120,000}{2000} = 60 \text{ orders per year}[6].$$

If one wished to calculate the absolute value of the minimum cost, it would be necessary to substitute the lot size of 2000 back into Eq. (13-11) and to know the value of V_i.

It is possible therefore to obtain Q_B or N for optimum operation by direct substitution into equations such as Eqs. (13-12) and (13-12a). However, this procedure has possibilities of error for any real case, since these equations were developed for a certain specified set of conditions.

The best procedure is to establish the appropriate cost equation, analogous to Eq. (13-11), as a model and then perform the proper mathematical manipulations to determine the correct optimum value. This approach ensures that any cost equation is a good representation of the actual problem and that the use of numerical values will produce reasonably correct real world answers that can be employed with confidence.

Thus, the real world cost equation for the preceding problem would be as follows, the $V_i Q_A$ term of Eq. (13-11) being omitted because data of this nature were either not available or not pertinent.

---

[6] Note that the optimum N could be obtained from Eq. (13-10) by expressing Eq. (13-11) in terms of N rather than Q_B, or it may be evaluated directly from Eq. (13-12):

$$Q_{B_{opt}} = \frac{Q_A}{N_{opt}} = \sqrt{\frac{2Q_A C_c}{pf_h}}$$

$$N_{opt} = \sqrt{\frac{pf_h Q_A}{2C_c}} \qquad (13\text{-}12a)$$

$$= 60 \text{ orders per year}$$

$$C_T = \frac{120,000(14)}{Q_B} + \frac{Q_B(7)(0.12)}{2}$$

$$C_T = 1,680,000 \ Q_B^{-1} + 0.42Q_B$$

$$\frac{dC_T}{dQ_B} = -1,680,000 \ Q_B^{-2} + 0.42 = 0$$

$$0.42Q_{B_{opt}}^2 = 1,680,000$$

$$Q_{B_{opt}} = \sqrt{4,000,000} = 2000 \text{ tires}$$

This procedure of establishing a specific model for each problem has merit. In this way, general principles are logically adapted for each specific problem. Moreover, such a model permits an analysis for the relative importance of the individual costs that make up the total. From such knowledge it is often possible to omit certain cost items to simplify the mathematics without affecting the final result appreciably. However, where the same computations are performed at periodic intervals, it is often desirable to prepare charts which provide the mathematical solution. An example is illustrated later in this chapter in Fig. 13-9. Such charts have an advantage in that nontechnical people can be readily taught to use them correctly. The importance of understanding the basis for these charts cannot be overemphasized and that is the purpose of presenting the material under discussion. When the principles are known, then short-cut charts can be established for many purposes in connection with inventory. In fact, the charts are really a technique to permit a rapid and correct solution of a mathematical model.

The economic order size problem is perhaps the simplest cyclic one. The typical analysis applies to many warehousing, distributing, inventory, purchasing, and production operations. However, it becomes more complicated when the purchase price varies with the quantity ordered (quantity discounts). In such cases, the value of p used in the equations will vary. Although the analysis is not difficult, it is best to employ graphical methods and trial and error procedures for the different ranges of quantity unless the cost, p, can be related algebraically to the quantity, $\dot{Q}_B$. More advanced texts cover this type of problem.

When the optimum Q_B is known, the optimum cost can be determined by substituting in Eq. (13-11). However, for the strictly batch process, it is more common to use:

$$C_{T_{min}} = \sqrt{2pf_h Q_A C_c} + V_i Q_A \qquad (13\text{-}13a)$$

Since $Q_A = NQ_B$ and $Nt_t = 1$ year, then:

$$N = 1/t_t \text{ and } Q_A = Q_B/t_t$$

where t_t is a fractional part of a year. Thus, Eq. (13-13a) can be rearranged in terms of Q_A:

$$C_{T_{min}} = Q_A \left[\sqrt{2pf_hC_ct_t/Q_B} + V_i \right] \qquad (13\text{-}13b)$$

Note that the total cost minimum and the minimum unit cost $= C_T/Q_A$; both occur at the same optimum value of Q_B for a fixed cycle time and a constant annual quantity Q_A.

THE ECONOMIC PRODUCTION LOT PROBLEM

A classical cyclic problem is the economic lot size to be produced when "set up" or "make ready" costs are balanced against the inventory charges of finished product. If the production is a strictly batch operation where all of the batch is finished at the same time, then the analysis is identical with the economic order size and Figs. 13-1 or 13-2 illustrate the problem.

However, if the product is made at some constant production rate m for a make time t_m with a uniform withdrawal or usage of finished product, then Fig. 13-4 describes the mathematical relations and Eq. (13-9) will apply with the average inventory I_a being given by Eq. (13-8). (Note that the total size of the batch Q_B differs from the maximum inventory Q_I by an amount equal to the quantity withdrawn during the make period). Where the annual production is Q_A, the number of batches or cycles N is known, since Nmt_m must equal Q_A and $t_m = Q_A/(mN)$. Thus, Eq. (13-9) becomes:

$$C_T = NC_c + \frac{(m-w)Q_A}{2mN} pf_h + V_iQ_A, \text{ dollars per year} \qquad (13\text{-}14)$$

Differentiating with respect to N and equating to zero (note that the V_i term disappears if it does not vary with N) yields:

$$\frac{dC_T}{dN} = C_c - (1 - w/m)(Q_A/2)pf_h/N^2 = 0$$

or
$$N_{opt} = \sqrt{\frac{(1 - w/m)Q_Apf_h}{2C_c}}, \text{ optimum lots} \qquad (13\text{-}15a)$$

An alternate form with the answer in the size of a batch $Q_B = Q_A/N = mt_m$ can be obtained by substituting for $N = Q_A/Q_B$:

$$Q_{B_{opt}} = \sqrt{\frac{2Q_AC_c}{(1 - w/m)pf_h}} \text{ units} \qquad (13\text{-}15b)$$

which is analogous to Eq. (13-12).

The minimum total cost by substitution of the value of N_{opt} into Eq. (13-14) would be:

$$C_{T_{min}} = \sqrt{2(1 - w/m)Q_A pf_h C_c} + V_i Q_A, \text{dollars per year} \quad (13\text{-}16a)$$

However, since $Q_A = wt_t N$ this relation can be rearranged in terms of Q_A as follows with Nt_t equal to unity when $Q_A = w$ in annual units:

$$(C_T/Q_A)_{min} = \sqrt{2\left(\frac{1}{w} - \frac{1}{m}\right)pf_h C_c/(t_t N)} + V_i, \text{dollars per unit} \quad (13\text{-}16b)$$

This form is found in the literature and is discussed in the next section.

AN ECONOMIC PRODUCTION LOT EXAMPLE

An example will be given to illustrate the application of the equations to a real problem. As noted before in the economic order problem, the preferred procedure is to set up the basic equation rather than substitute blindly in the final equations.

Problem

A paint factory manufactures paints continuously but in lots of different kinds. The plant operates 320 days per year and has a withdrawal rate for sales of 1000 gallons per day based on 320 days. The value of the product may be taken as $2 per gallon with an inventory carrying charge factor of 10 per cent. The production rate is 5000 gallons per day and "make ready" charges are $100 per lot. (a) What is the optimum production lot size based on 320 days a year for all calculations? (b) What is the lead time for a batch?

Solution

Part a. Equation (13-14) applies for this case; V_i is omitted, since it is not a factor. The pertinent total costs are for an annual requirement of $320 \times 1,000 = 320,000$ gallons per year.

$$C_T = 100N + \frac{(5000 - 1000)3.2 \times 10^5}{(2)(5000)N} (2)(0.10)$$

$$C_T = 100N + (0.256/N) \times 10^5$$

$$\frac{dC_T}{dN} = 100 - 0.256 \times 10^5/N^2$$

$$N_{opt} = 16 \text{ lots}$$

Thus, $320,000/16 = 20,000$ gallons per lot.

Part b. The lead time, t_m, is related to the batch size through the daily rate, m, or:

$$5000t_m = 20,000$$
$$t_m = 4 \text{ days per lot}$$

The available time per lot is 320/16 = 20 days so there are 16 days between lots. A portion of this time is down time to clean up or get ready for a new batch; the remaining time is dead time or interim time.

The simple economic production lot problem applies to a specific total demand, Q_A, with a fixed production rate, m, and demand rate, w. However, it is not limited to these specific conditions. From Eq. (13-16a) the minimum total cost is given as a function of Q_A, V_i, w, and m, and thus economic analysis can be made for alternative programs from which management may select a best policy. First, however, it must be clearly understood that Eq. (13-15a) was developed on the basis of one procedure in which any variable cost, V_i, was constant per unit of production. However, for different programs, or for scheduling on different types of machines, etc., V_i is a variable to consider, and Eq. (13-14) can be analyzed graphically as illustrated in Fig. 13-8 where Eq. (13-16b) is plotted for different demand rates for three different programs, each of which has a different production rate, m_i, with V_i constant. The points of intersection represent *iso-minimum cost* points for the respective programs. Thus, at point *a* the first and second programs have the same minimum cost. However, below a demand rate of 130, the first program is more economical at a given withdrawal rate; or at a given minimum unit cost a higher withdrawal rate below (130) can be attained with the second program. The situation is reversed above demands of 130 units per unit time. Corresponding conclusions are possible regarding points *b* and *c*

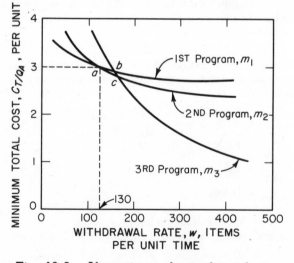

Fig. 13-8. Alternative production lot analysis.

and the other program. With V_i constant, the analysis may be made without V_i being known, since the curves would all be raised or lowered the same amount.

There are many other variations to consider in the analysis of the economic production lot size, which are discussed in the next section. Finally, the general inventory problem will be presented from a store-keeping viewpoint.

* * * (Optional) * * *

CHARTS FOR SOLUTION OF THE INVENTORY PROBLEM

Where routine estimations of batch size or economic production lot quantity are to be made, it may be worthwhile to utilize charts for direct reading of the values. Such charts are particularly valuable where the estimations are to be made by nontechnical staff such as warehousemen, storekeepers, or other clerical labor.

A generalized chart for this purpose is illustrated in Fig. 13-9. This chart is a mathematical solution of Eq. (13-12) or (13-15b) for either the economic batch or order size or the economic production lot quantity. The basis for the chart is that both Eqs. (13-12) and (13-15b) may be written as follows:

$$Q_{B_{\text{opt}}} = \sqrt{\frac{2C_c}{pf_h}} \cdot \sqrt{F}$$

where F is a production factor which varies with the type of problem. Thus, from Eq. (13-12) for the economic order size or batch quantity the production factor, F, is:

$$F = Q_A, \text{items per year}$$

and from Eq. (13-15b) for the economic production lot quantity:

$$F = \frac{m}{(m/Q_A) - 1}, \text{items per year}$$

where m is the instantaneous production rate. This is expressed in items per year to be consistent with w or Q_A when the withdrawal rate is considered on a yearly basis.

Taking the logarithm of both sides of the basic equation gives:

$$\log Q_B = 0.5 \log F + \log[2C_c/(pf_h)]^{0.5}$$

As shown in Fig. 13-9, this equation plots as a straight line with a slope of 0.5 and a value of Q_B equal to $[2C_c/(pf_h)]^{0.5}$ when $F = 1$ (or $\log F = 0$). Thus, a

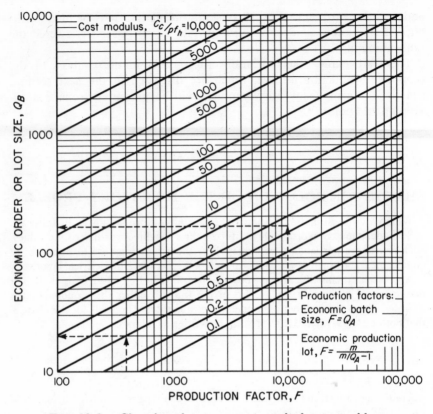

Fig. 13-9. Chart for solving economic cyclic lot size problems.

family of lines all having the same slope may be drawn for different values of the second right hand term, each of which corresponds to a certain cost modulus, $C_c/(pf_h)$, which is defined arbitrarily as this group of terms.

The use of the chart is quite simple. For example, in the previous economic order size problem for automobile tires (page 349), the cost modulus may be computed for units of tires in dozens (or 12 per unit). Thus:

$$C_c/(pf_h) = 14/(7 \times 12 \times 0.12) = 1.39$$

The value for F for the batch problem was 120,000 tires, or 10,000 dozen tires. Thus, the chart may be entered at the dashed line for $F = 10,000$, and Q_B is read at the ordinate, where the dashed line intercepts the cost modulus line at 1.39. The answer is 167 (in dozens of tires) as illustrated. This is $167 \times 12 = 2004$ tires, which agrees with the answer obtained in the previous numerical solution.

For the economic production lot problem previously demonstrated for paint quantity (page 353), the cost modulus for units of 1000 gallons each was:

$$C_c/(pf_h) = 100/(2 \times 1000 \times 0.10) = 0.5$$

The value of the production factor, F, is for $Q_A = 320$ thousand gallons per year with m equal to $(5000 \times 320)/1000 = 1600$ thousand gallons per year:

$$F = \frac{1600}{1600/320 - 1} = 400$$

Entering Fig. 13-9 at $F = 400$ at a cost modulus of 0.5 gives an optimum lot quantity of 20 thousand gallons, which confirms the earlier solution.

It will be noted in these examples that the units of production (dozens of tires and thousand gallons of paint) were adjusted to fit the range of the chart given so the chart is universally useful. Care must be taken that the proper value of p, the cost per unit, is utilized in calculating the cost modulus to be employed on the chart.

* * * * * * * * * * * * * *

ANALYSIS FOR SEASONAL DEMANDS

Considering the economic balance among costs for supplying a fixed seasonal demand provides an interesting variation of the previous problems. This quantity may be made over a considerable time period with a few machines (low capital investment) and a relatively long inventory period, or a large number of machines may be used with a short inventory period. Thus, an economic balance exists mainly between the fixed charges on multiple machines and the inventory holding costs. In such an analysis, it is assumed that the machines start up and operate continuously until the desired production is completed and that then the withdrawal occurs at a uniform rate[7] or all in one batch. The analysis consists essentially of a single lot per year operation, the machine costs (fixed and operating) being a function of M, the number of machines, and the holding charges also being a function of the number of machines, since the total time for manufacture is less with more machines.

Problem

A simple example might be considered as follows. A manufacturer of winter clothing is required to make 48,000 items for the year to be withdrawn during the

---

[7]This fixed demand denotes that the problem is "deterministic" with no variability in the numbers being used. In a following section an elementary problem dealing with variability in demand will be considered.

last three months. The best estimates available indicate that the annual charges on the machines vary as $4000M$ dollars, where M is the number of machines. The product is worth \$10 per item and the holding charges are 12 per cent of its value. Each machine has the same capacity and one machine can manufacture 4000 items per month on a uniform production schedule. What is the optimum number of machines to use on this operation?

Solution

The diagram for this economic analysis is shown in Fig. 13-10. For the first 9 months, production at the rate of 4000 items per month yields 36,000 items stocked, after which the demand is 16,000 items per month for 3 months. The

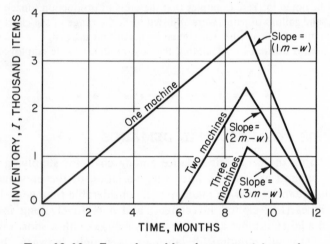

Fig. 13-10. Example problem for seasonal demands.

inventory, therefore, decreases at the rate of $4000 - 16,000$, or 12,000 items per month. If two machines are used, production will not start until the sixth month and, although the maximum inventory also will be reached at nine months, the depletion of inventory is slower. This rate is $(2 \times 4000) - 16,000 = 8000$ items per month. The general relation for the rate of depletion of inventory is $(mM - w)$ where M is the number of machines, m is the make rate, and w is the withdrawal rate during the season; w is equal to $48,000/3 = 16,000$ items per month.

The same reasoning applies for additional machines with smaller inventories. When four machines are considered, the demand rate for the last three months exactly equals the make rate and no inventory is produced. If a larger number of machines are used, the problem becomes a one cycle operation with interim time on one or more machines.

The economic analysis for the problem diagrammed in Fig. 13-10 follows the general analysis of this chapter noting that the total inventory period is a function of the number of machines. In this case, the base of each triangle is $1/M$. The height of each triangle is minus $(3)(mM - w)$ which, therefore, will be a positive number:

$$\text{Total costs} = \text{machine costs} + \text{inventory charges}$$

$$C_T = 4000M + \frac{(0.12)(10)(-3)(mM - w)}{(2)(M)(1)}$$

Since m is 4000 and w is 16,000, this relation can be expressed in terms of one variable, M:

$$C_T = 4000M - 7200 + 28,800/M$$

If this equation is differentiated and the result equated to zero, the optimum number of machines to use can be determined:

$$dC_T/dM = 4000 - 28,800M^{-2} = 0$$

$$M_{opt} = 2.68 \text{ machines}$$

Thus, either two or three machines provides a satisfactory real world situation.

* * * (*Optional*) * * *

A SPECIAL INVENTORY PROBLEM

An interesting variation of the inventory problem is one where shelf life or storage causes a loss in value with age because of obsolescence or deterioration. Typical examples are found in the food processing industries where the processing rate m cannot keep pace with the arrival rate, a, during a season or campaign when the crop is maturing. An economic balance exists between (1) the annual capital investment costs (depreciation) for a process plant large enough to keep the raw material inventory low and (2) the cost of losses. The analysis requires special consideration to determine the inventory losses as follows. As in the section on Analysis for Seasonal Demands, only one lot or batch is considered.

If
S_t = the total quantity available per season, tons
a = the average arrival rate during the season, tons per day
m = the processing rate during the season, tons per day
and
f = the per cent deterioration of inventory per day of storage,

then
S_t/a = the total length of the campaign or arrival days
S_t/m = the total length of process operations, days
$a - m$ = the rate of inventory build-up, tons per day
and $(a - m)S_t/a$ = the maximum inventory, tons

Thus, the inventory model is very similar to Fig. 13-4, and the inventory-days are the area of the triangle of height, $(a - m)S_t/a$, and base, S_t/m, since S_t/m is the total time for which there is some inventory. Thus, the inventory-days are:

$$(S_t/a)(a - m)(S_t/m)/2, \text{ ton-days} \tag{13-17}$$

Therefore, the loss is f times this quantity, or loss equals:

$$\frac{fS_t^2 (a - m)}{2am}, \text{ tons} \tag{13-18}$$

Thus, for a given situation where f, S_t, and a are known, the loss (in terms of raw material) may be evaluated in terms of a production rate, m; since capital costs can also be expressed in terms of plant size, m, an economic balance for minimum cost may be constructed as in Chapter 10 (page 251). The economic analysis may also be made in terms of optimum investment and earning rate as discussed in Chapter 11 (page 302).

AN AGING LOSS EXAMPLE

Problem

Determine the optimum plant size for the following operation involving the processing of a special food product. An annual production of 80,000 tons is expected per season with the investment costs for processing varying as follows:

Process rate, tons per day	200	400	600	800	1000
Capital investment, million dollars	4.0	7.0	9.5	10.9	12.0

Depreciation may be taken as straight line, based on a 10 year life. Essentially, 100 per cent conversion is attained except that a raw material inventory spoilage factor of 0.3 per cent per storage day occurs. The average season lasts 80 days. The product is worth $30 per ton. All other pertinent costs each year are considered essentially constant at any size plant and may be omitted from the analysis for purposes of simplication, since the objective is to study the effect of the spoilage.

Solution

Based on the data available, a tabulation of results can be made up as in Table 13-2 to determine the optimum plant size. If the relation between make rate and investment can be approximated by an algebraic equation, an analytical solution by differential calculus can be obtained. However, this particular problem will be solved by an iterative procedure for each investment level. The average arrival rate, a, equals $80,000/80 = 1000$ tons per day.

The optimum size plant is one that processes about 400 tons per day, since the least annual relative costs are incurred for handling a fixed quantity of input material capable of being converted to a revenue-producing output.

TABLE 13-2. Results for Solution to Aging Loss Example

Process rate, m, tons per day	200	400	600	800	1000
Investment, thousand dollars	4000	7000	9500	10,900	12,000
Annual depreciation[a], thousand dollars	400	700	950	1090	1200
Loss from spoilage[b], thousand dollars	1152	432	192	72	0
Total pertinent costs, thousand dollars	1552	1132	1142	1162	1200

[a] Computed as $1/10$ the investment.

[b] Loss computed from Eq. (13-18) with $f = 0.003$; for the loss is multiplied by \$30, which is the value of a ton at 100 per cent conversion.

REFINEMENTS OF THE ECONOMIC PRODUCTION LOT MODEL

The economic production lot problem is somewhat different from that for the optimum order size. In the optimum order size problem the lead time or make time has no real effect on costs, but in the production lot problem the lead time should be considered insofar as it affects the in-process inventory.

There may be three inventories to consider in the production problem. There is a "work-in-process" inventory, a raw-material inventory associated with the production of a batch, and a finished product inventory. Inventory of finished products has been considered in all of the preceding discussions and has been illustrated as Q on the figures. However, there is a quantity of raw material equal to Q_B/R units that must be available to make a batch of Q_B units, where R is the fractional conversion of a unit of raw material that yields one net full unit of product. (The inclusion of R allows for losses, spoilage, etc.). Thus, some average quantity of raw material at some price or value p_r must be carried in inventory during the lead time or make time, t_m, unless it is provided just as fast as needed without a carrying charge.

In-Process Inventory

The problem for the quantity of in-process inventory involved (at a varying price since work in process increases in value as it more nearly approaches a finished product) must be considered. For the *strictly batch processes* such as illustrated in Fig. 13-1 and 13-2, the quantity involved is, simply, Q_B/R for a lead time per cycle (or batch) of t_m time units. The new carrying charge that must be added to the finished-product holding charge in Eq. (13-11) is:

$$C_h' = (Q_B/R)p_p f_h(t_m/t_t), \text{ dollars per year} \qquad (13-19)$$

All terms except p_p and f_h have been defined previously. The value of p_p, the work in process during the make period, is often taken as the material cost plus one half the labor charge during the make period. The f_h term is the fraction of p_p that is chargeable as carrying cost. The term t_m/t_t is the fraction of the time for each cycle that there is an in-process inventory. It is actually the ratio of Nt_m/Nt_t, that is, the fraction of the total year for which the inventory must be considered. Note that $t_m/t_t = Nt_m$, since Nt_t equals unity in any repetitive process. Thus, for a constant annual withdrawal rate Q_A in the strictly batch process with constant make time t_m, the in-process inventory is independent of batch size because $Q_B/t_t = NQ_B = Q_A =$ constant.

Different considerations for lead time in the economic production lot problems occur for various problems as illustrated in Figs. 13-3 to 13-5. For Fig. 13-3 the work-in-process inventory is similar to that for the batch except that quantity of material actually in process during the make time is a fixed quantity which is designated as Q_p. This fixed quantity may be thought of as a batch of goods of constant size at a production station for an inflow of raw material with a production of m units per unit time of finished product. The make time, or lead time, for the goods in process is t_m, and the carrying charge for work in process corresponding to Eq. (13-19) is:

$$C_h' = Q_p p_p f_h(t_m/t_t) \qquad (13\text{-}20a)$$

There is an increasing product inventory during the make time, t_m, for which the holding charges are of the same form as in Eq. (13-11) (for a decreasing inventory in the economic order problem). However, if there is an aging time component, the diagram of Fig. 13-5 would apply with the total work-in-process charges being:

$$C_h' = \underset{\text{In process}}{Q_p p_p f_h(t_m/t_t)} + \underset{\text{buildup of semifinished}}{(1/2)mt_m p_p f_h(t_m/t_t)} + \underset{\text{aging}}{mt_m p_p f_h(t_a/t_t)} \qquad (13\text{-}20b)$$

where mt_m is the buildup of a batch to be aged. It is equal to Q_B and is the quantity in a finished batch that is aged for a period of time, t_a. In a strictly batch operation the batch sizes Q_B and Q_p are equal. It is assumed that time for charging and discharging at the processing station are included in the times as given. The specific values of p_p may be permitted to vary in the individual terms if refining the economic analysis is desirable.

Raw Material Inventory

There is still one further inventory consideration in connection with operations of the pattern found in Fig. 13-3. This inventory consideration is the quantity of raw material stocked in *batch processing*. If it is assumed that the amount is stocked each time for one batch of product is just sufficient, the quantity of raw material inventory is $(1/2)(Q_B/R)$, where Q_B/R is the quantity of raw material necessary to make a batch of size mt_m units of product in the make period of t_m with R being the conversion factor. (However, since the raw material is used at a uniform rate, the holding charges are based on one half of the quantity for the make period.) The raw-material holding charges thus become for materials worth p_r per unit:

$$C_h'' = (1/2)(Q_B/R)p_r f_h(t_m/t_t), \text{dollars per year} \qquad (13\text{-}21)$$

where P_r is the value of a unit of raw material, f_h is the holding charge (interest), and the other terms are as defined in the text.

For a *semicontinuous process* of the type illustrated in Fig. 13-4, there are three inventory charges to consider which are similar (but not identical) to those for the *batch processes* of Fig. 13-2. The separate values are:

(1) *Product Inventory charges* are obtained from Eqs. (13-8) and (13-9) for an inventory that increases from 0 to Q_I; at the same time, there is withdrawal:

$$I_a p f_h = (1/2)(m - w)t_m p f_h$$

(2) *Work-in-process inventory charges* from Eq. (13-20a) are:

$$I' p_p f_h = Q_p p_p f_h(t_m/t_t)$$

(3) *Raw-material inventory charges,* where the raw stock is delivered in one batch size (mt_m/R) sufficient for one run and where its inventory decreases from (mt_m/R) to zero, are:

$$I'' p_r f_h = (1/2)(mt_m/R)p_r f_h(t_m/t_t)$$

In summary, the total holding charges are:

$$C_h = 0.5t_m[(m - w)p f_h + 2Q_p p_p f_h/t + (m/R)p_r f_h t_m/t_t] \qquad (13\text{-}22)$$
$$\text{Product} \qquad\qquad \text{in process} \qquad \text{raw material}$$

(NOTE: f_h may or may not be the same for different classes of inventory)

From the relative values of the individual product values for p, p_p, and p_r and the quantities involved, the relative dollar importance of the various items making up carrying charges can be estimated by visual inspection of Eq. (13-22). Those items of negligible importance can then be omitted from further consideration in the economic analysis.

Note that Eqs. (13-19) to (13-22) are often written in different forms in the literature, and it is useful to remember that for $Nt_t = 1$ year, $N = Q_A/Q_B$; $Q_A = Q_B/t_t$, and $Q_A t_m = Q_B(t_m/t_t)$.

It should also be emphasized that it is assumed in all the above economic equations that the respective unit prices employed are unaffected by the batch size. Otherwise, each individual unit value becomes another variable in the equation. Since the general analysis assumes constant annual demand, the minimum annual costs and minimum unit cost occurs at the same optimum batch or lot size.

A special kind of combination semicontinuous and batch process can be demonstrated in Fig. 13-5, where the time from a to d was utilized to make a batch of "green" product either semicontinuously or as a batch which was then cured or aged during the period from d to c. The aging of whiskey or the drying and curing of finished painted goods may be in this category. Here, an additional holding cost for a batch of size, Q_B, must be considered for the time period from d to c for each batch; this holding cost is added to the other work-in-process charges. Where the time period from d to c is merely a clean up period for the equipment, there is no charge for hold up costs; this is merely equipment down time and the inventory charges would be the same as for those of the other situations presented.

The loss on storage reduces the potential revenues by the amount of the loss, but in the computation of profit the cost of shelf, or age, loss must not be computed twice; that is, the raw material expense is established and its cost does not change regardless of loss—only the amount of revenue is changed. However, concerning conversion of feed stock, the units of product per unit of feed must be determined under such conditions as would exist if no losses had occurred. This problem will be recognized as a raw-material-inventory type with two special considerations. First, the raw material arrives at some special rate, and second, a storage loss must be taken into account along with actual conversion.

* * * * * * * *

THE STORES INVENTORY PROBLEM

A practical application of the cyclic model is the stores inventory problem where a combination of several types of economic analysis can be useful in selecting an optimum policy.

The general stores problem is given by Fig. 13-2 or 13-11, which is essentially the economic order size problem. However, these diagrams assume no "stock outs," that is, at the exact time the inventory drops to zero another lot of stock is available. Since in the real world it is quite possible for stock outs to occur, a "reserve" ("protective," or "safety") stock is kept on inventory. This reserve stock is shown at the bottom of Fig. 13-12.

Therefore, for the most economical policy, analysis of the stores inventory problem consists of two parts. The first is the economic order size for the active stock. The lead time for the order size then determines the reorder point for the prevention of a stock out at the normal usage or withdrawal rate. The second part of the analysis consists of selecting the quantity of protective stock to hold. Management may arbitrarily fix the stock at some fraction of the lot size, or more elaborate methods may be employed.

The economic analysis of stock outs consists of a balance of the costs of carrying the excess stocks against the cost or loss resulting from a stock out. This analysis can be based on statistical procedures.

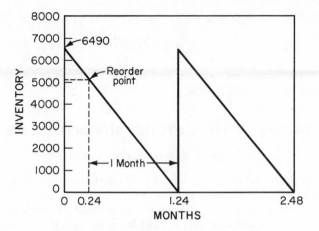

Fig. 13-11. Stores inventory example with no re-serve stock.

In general the withdrawals from an inventory will not be at a uniform rate. Thus, if the withdrawals over an extended period of time (say one year) follow some random pattern for day by day values, then a statistical analysis can be utilized to meet the demand variability. This can be done by a suitable replenishment schedule or by the use of a reserve (or safety) stock. In this manner, the possibility of a stock out can be minimized at some level of probability. Some preliminary treatment of the procedure is presented in the next section, and additional discussion of uncertainty in elementary economic models is given in Chapter 15.

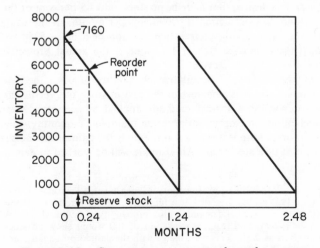

Fig. 13-12. Stores inventory example with reserve stock.

One method[8] for selecting the protective stock, I_p, is:

$$I_p = k \sqrt{E(D)} \tag{13-23}$$

where k is a dimensional constant and $E(D)$ is the expected or mean value of demands for the cycle period.

UNCERTAINTY WITH A STORES INVENTORY EXAMPLE

* * * (Optional) * * *

A more sophisticated method of selecting the reserve stock is to base the amount on stochastic or probability considerations where data are available. For example, Fig. 13-11 provides a solution to an economic lot problem (to be discussed below), where the average withdrawal rate is 5250 items per month. In the standard procedure for solving the economic lot size problem, this withdrawal rate is assumed to be a fixed quantity with no variations. Thus no stock outs occur, because withdrawals are considered to occur exactly as predicted. However, in real world situations these withdrawal demands may fluctuate so that stock outs occur whenever the demand rate exceeds the average rate expected for the entire cycle time. If a distribution of these expected variable withdrawal rates is available from past records and a probability analysis is made, a cumulative probability curve for them such as shown in Fig. 13-13 may be drawn. This curve states that at a given probability all demand rates will be less than a certain amount. Thus, if management adopts a policy of an 85 per cent probability basis, then in the long run the chart says that 85 times out of 100 the stores required will be at a rate of 6200 items per month or less. Therefore, the protective stock will be the difference in the policy rate and the actual rate, corrected for the elapsed time of a cycle. For example, if it is desired that there be no stock outs 85 per cent of the time, the rate for providing this protection is 6200 items per month. If the actual average rate is 5250 items per month, there is a protection of 6200 − 5250 = 950 items per month; if the time of a cycle is 1.1 months, the actual protective stock is (1.1)(950) = 1045 items for a cycle.

Instead of just selecting a cumulative probability of 0.85, the proper probability for minimum cost as a balance of the cost of stock out versus carrying cost may be determined by statistical methods applied to economic balance. The principle is as follows: the optimum program maintains a quantity where the cost of stocking the unit and not needing it just equals the gain of having the unit and avoiding the cost of a stock out. An example will be worked to demonstrate the

---

[8]Whitin, T. M., "The Theory of Inventory Management," Princeton University Press, Princeton, N. J., 1953. The value of k is a function of many variables. However, for one type of distribution (Poisson) of demands with a probability of 0.99 that there would be no stock outs in the cycle period (e.g., 99 weeks out of 100 would show no stock outs), the value of k was given as 2.33. This will change with the desired probability limit. It is not intended to develop this relation further in this text; it is given here for reference only.

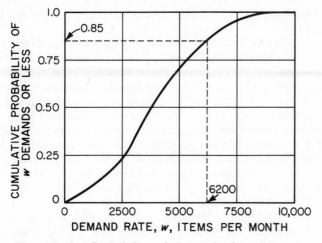

Fig. 13-13. Probability of demands for stores inventory example.

procedure; the results are given in Figs. 13-11 and 13-12. A more comprehensive discussion of the basic principles dealing with uncertainty in economic analysis is given in Chapter 15.

Problem

Small gears must be stocked for assembly into a finished machine. The set up cost for a batch is $200. A gear is worth $2, and carrying charges are 30 per cent. Annual demand is 63,000 gears. The lead time is one month, and the cost of a stock out is equivalent to $2.90 per gear. (a) Determine the economic lot size and draw a graphical model, neglecting consideration of stock outs. (b) Determine the economic protective stock, assuming that variable demands follow the probability relation given in Fig. 13-13.

Solution

Part a. This is an economic order size problem with uniform withdrawals and therefore fits the model of Fig. 13-2, although the lead time is not necessarily equal to the cycle time. From Eq. (13-12) the annual costs are:

$$C_T = \frac{63,000}{Q_B}(200) + \frac{Q_B}{2}(2)(0.30), \text{ dollars per year}$$

$$\frac{dC_T}{dQ_B} = -12.6 \times 10^6 Q_B^{-2} + 0.30 = 0$$

$$Q_B = 6490 \text{ gears}$$

Thus, the total cycle time, $t_t = (6490/63,000)(12) = 1.24$ months, and the model is as in Fig. 13-11[9].

Part b. To select the economic amount of protective stock, management must obtain the long-run probability for optimal policy. This is because a stock out may be quite costly to the company. For a fixed time cycle of 1.24 months as determined in part a, the distribution of actual *demands* for gears is the same as the monthly distribution of *withdrawal rates*, since 1.24 times the rate, w, is equal to the actual number of gears. The distribution of rates is illustrated in Fig. 13-13.

The economic analysis follows principles employing probability that are not developed at this time. The present discussion accepts these principles without proof, utilizing the incremental-concept technique discussed in chapter 9. However, these stochastic principles are amplified in Chapter 15. Thus, for the best policy the incremental cost for stocking each additional gear (the carrying charge, C) must just equal the marginal revenue of having the gear available when needed. The marginal revenue of having the gear available is the probable saving that results from avoiding a stockout. This marginal revenue is the value of a stock out, S, times the probability, $P'(w)$, that demands will be w_{opt} or *more*, where w_{opt} is the optimum demand for minimum cost of stock outs. In other words, the one additional unit cannot be reached until at *least* w units have been passed on a cumulative probability plot of s demands or more. The mathematical expression solved in terms of probability of w_{opt} or less is:

$$C = SP'(w) = S[1 - P(w)] \qquad (13\text{-}24)$$

and for the problem where C and S are known:

$$P(w)_{opt} = (S - C)/S = [2.90 - (0.3 \times 2)]/2.90 = 0.794 \qquad (13\text{-}25)$$

This probability, $P(w)$, is the probability of a w demand rate or *less* which permits reading the optimum demand rate directly at 5800 from Fig. 13-13. Note that $P'(w) = 1 - P(w)$.

This probability on Fig. 13-13 corresponds to 5800 gears per month or less to be demanded 79 out of 100 cycles as an optimum to balance stock-out costs versus

---

[9]If the analysis were to stop at this point as is the case for the usual inventory analysis, the implication is that the withdrawal rate is constant at $63,000/12 = 5250$ gears per month, or 6490 per cycle of 1.24 months. There is implied further for the real case that any withdrawal rate in excess of the above which results in stock outs is of no economic consequence. However, in the real world, this is not acceptable, and stochastic analysis can provide a better policy when probability data are available or can be estimated. In other words, the assumed rate of 5250 gears per month is a desired value or long-run average of a large number of withdrawal rates, which may vary above and below that figure. Furthermore, it corresponds to some cumulative probability (for the distribution of these various rates) that is acceptable to management as an inventory policy. From Fig. 13-13 the cumulative probability at this rate is 0.75, or 75 out of 100; this figures times the demand will be less than 5250 per month. Where other data such as the cost of a stock out are available and where management wishes to use a policy based on economic analysis including this cost, the procedure followed in part b of the problem may be applied.

carrying charges. Since the economic order size cycle time is 1.24 months, the total inventory at the start of a cycle must be $(1.24)(5800) = 7160$ units. Thus, the protective stock is $7160 - 6490 = 670$ units, and the graphical model appears as in Fig. 13-12.

* * * * * * * *

A few comments on this model may be of interest. The cost of a stock out, S, is per unit gear, although it may really mean the cost of shutdown averaged over all the gears involved in an assembly operation. As such it represents the best estimate for a stock out and in itself may be an averaged value of different costs for a large number of stock outs.

The analysis allowed only for the variability in demands or calls for one gear each. In other cases, the number of gears per call or demand may follow some kind of distribution. If a variability of lead time is considered, reserve stock can be added to allow for this variation. For example, delivery times may show such a random distribution that management sets a policy of accepting one month or less for 95 times out of 100. Thus, stock outs occur 5 times out of 100 because of delivery variance alone unless protective stock is available. An over-all protective stock, allowing for all of these individual variances, can be incorporated into the model for management control.

If management adopts a policy of reordering only when the inventory drops to the reorder amount, the time between orders may vary, since the demands in any individual cycle may vary from the average of 6490 gears. In the long run, however, the average time between reorders will approach a constant value.

An alternative policy is to reorder according to a uniform time schedule and to use a different amount for the order. The purpose here is to raise the stock quantity back to the established level required at the anticipated delivery date. The variations in the order size will average out to a constant batch size.

Where a protective stock based on one or more allowances for variabilities is considered, as in Fig. 13-12, an over-all model results. Such a model becomes an established management policy that can be explained and interpreted readily.

EXTENSIONS OF THE CYCLIC MODEL

There are many variations of the cyclic model, and the analysis of these variations follows the principles demonstrated in the previous sections. Although these variations go beyond the actual intent of this book, as a matter of interest certain extensions of the model are introduced.

In general, the operations discussed have assumed a going plant where the annual fixed costs were already established and constant. Accordingly, these constant fixed costs were not included in the basic Eq. (13-1) because they do not affect the results for the optimum. (As previously noted, the constant fixed costs would disappear in the differentiation). However, if any annual fixed costs are affected by the batch size (such as in the design of new equipment or modification of equipment for different cycles), then an annual fixed charge, C_F, should be included in Eq. (13-1). This may then be carried through the economic analysis (as dC_F/dQ_B, for example) and the numerical value included when the design cost, C_F, is known in terms of batch size, Q_B.

Where design of equipment or variations of production scheduling affect the in-process quantity, Q_p, of Eq. (13-20a), the variations must then be considered as a function of batch size.

Another extension which is subject to many variations deals with operating-procedure variations and the objectives of the economic analysis. For example, in the batch production operations with a fixed lead time, there may be considerable interim or dead time which can be employed to increase the annual production; if there is a demand for the additional production, more profit will result because unit costs will decrease. There are two variations of this approach: (1) the case where the same make time as the optimum at a fixed annual production is employed for all cycles, and (2) one in which the optimum batch size with a fixed lead time is based on determining the operation for minimum cost with no restriction on the annual production.

When the lead time or make time is not constant but varies with the size of the batch, the problem becomes slightly more complicated; however, analysis follows basic Eq. (13-1), where cycle costs are expressed in appropriate units. (For example, if hourly costs are known, the costs per cycle can be expressed in terms of the make time and the hourly costs). The analysis may include all of the variations for operations with a fixed lead time that were given in the previous paragraph. This will be amplified in the next chapter.

One further consideration of the variable lead time depends upon whether the process is strictly batch as in Figs. 13-1 and 13-2 or whether goods are produced in a semicontinuous manner as in Figs. 13-3 and 13-4. If the operation is a semicontinuous one, the production rate may often be considered linear with $Q_B = mt_m$; this has been generally used throughout this chapter. In the real world, however, the production may vary as some power of the lead time because the rate slows down with elapsed time. When the relation between Q_B and t_m is known, it can be substituted in the equations; the mathematical analysis can then be made as covered in Chapter 14.

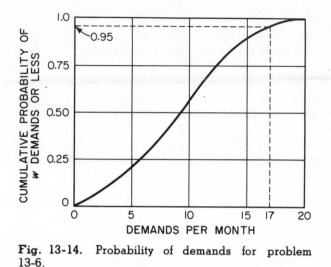

Fig. 13-14. Probability of demands for problem 13-6.

RECAPITULATION OF THE CYCLIC PROBLEM

A summary of the principal concepts and relations is given in this section for study and reference purposes. In general, the mathematical model is proposed to evaluate minimum annual costs for a constant annual demand or production. Thus, minimum costs per unit of production also apply.

The basic relation is for annual costs and repetitive cyclic operations:

$$C_T = NC_c + C_h + C_o, \text{ dollars per year} \qquad (13\text{-}1)$$

where C_c are the costs per cycle, C_h are the total charges for holding materials and products in process or storage, and C_0 are other costs of production including fixed charges and other expense not included in the first two terms. The number of cycles is:

$$N = Q_A/Q_B = H/t_t, \text{ cycles per year} \qquad (13\text{-}10)$$

Lot size or batch size, Q_B, varies with the type of operation; in a strictly batch or lot process the size of a lot is Q_B, whereas in a semicontinuous process, $Q_B = \int_0^{t_m} m \, dt$, where m is a constant or a known function of machine size and make time. The total cycle time, t_t, equals the sum of the make time, the down time, and the interim time. From Eq. (13-10) the batch size required is $Q_B = Q_A t_t/H$, which establishes a second relationship between batch size and total cycle time. This relation together with the technical relation between batch size and make rate provide for elimination of either total cycle time or batch size in the final equation.

Cycle costs, C_c, include all repeating costs for one cycle including get-ready costs, set-up costs, and operating costs for the cycle.

The holding costs, C_h, are based on the fractional value of each product unit that is charged for hold-up either as raw material, semifinished stock, or finished product. The average inventory, I_a, must be computed and is a function of both the quantity produced during a cycle and the time distribution for the cycle. In general, the average product inventory from consideration of Eqs. (13-3), (13-6), and (13-7) is:

$$I_a = \frac{\int_0^{t_t} Q \, dt}{t}, \text{ items-years per year} \tag{13-7a}$$

where this particular Q is units of net inventory at any time during the total cycle. From the preceding equation and technical considerations, the working equation for a strictly batch process is:

$$I_a = Q_B/2 \tag{13-5}$$

and for a semicontinuous process with constant m and w it is:

$$I_a = (m - w)t_m/2 \tag{13-8}$$

(Where raw material inventories are to be considered, the appropriate raw material and semifinished product inventories may be included.)

Thus, the holding costs may be formulated as:

$$C_h = I_a p f_h, \text{ dollars per year} \tag{13-26}$$

In all of these relations for average inventory, the total cycle time is based on the relation that $H = Nt_t = 1$ year. If a shortened year (less than 365 days) is used in the analysis, then all average inventories should allow for that fraction of a year in which the operation is not being conducted. This correction is based on the stipulation that Nt_t is less than one year.

The annual withdrawal rate controls the slope of the withdrawal line for demonstrating cyclic operations on charts. Accordingly, the setting of the withdrawal time from $Nt_t = H$ (that is, the selection of a calendar year or a shortened fictitious year for H) together with the withdrawal rate controls the distribution of the producing times and the interim time left over.

The last term C_o for operating or other costs in the basic relation of Eq. (13-1) must be considered where any of the costs vary with the make time employed in the operation. These variations must be known through the technical relations. For example where the annual fixed charges, C_F, are so many dollars per square feet of machine size M, then one component of the other charges is MC_F in annual dollars, where M is the machine size in square feet. The machine size also probably influences the make rate, thus affecting the cycle costs and the holding charges. Where the make time, t_m, and machine capacity or size, M, are related by a known mathematical function, one less variable must be considered in the economic analysis. This is discussed in the next chapter.

In general, simplifying assumptions are employed in the analysis so that a criterion equation as given by Eq. (13-1) may be manipulated to optimize any one of the variables such as make time, batch size, or number of cycles. Usually, the down time is considered constant, and the interim time is taken as zero; this establishes a continual repetitive operation, where $t_t = t_m + t_d$, and eliminates one variable from the analysis, since otherwise both t_m and t_t would be independent variables.

Where annual production costs, other costs, and operating costs per cycle are constant (leaving only a balance of startup costs and holdup costs), the problem reduces to the economic lot or batch size one where:

$$Q_{B_{opt}} = \sqrt{\frac{2Q_A C_c}{p f_h}} \text{ , items} \tag{13-12}$$

The corresponding economic production lot size for a semicontinuous process with a constant make rate and constant withdrawal rate is:

$$Q_{B_{opt}} = \sqrt{\frac{2Q_A C_c}{(1 - w/m)p f_h}} \text{ , items} \tag{13-15b}$$

Where the fixed costs are included as part of the other costs, C_o, and expressed as a function of Q_B, then Eq. (13-12) for the strictly batch operation becomes:

$$Q_{B_{opt}} = \sqrt{\frac{2Q_A C_c}{p f_h + 2dC_o/dQ_B}} \text{ , items} \tag{13-27}$$

where dC_0/dQ_B is the first derivative of the other costs with respect to the batch size. A similar correction is applied to Eq. (13-15b) for the economic production lot size in the semicontinuous operation.

The results given by the preceding equations (13-12 and 13-15b) presume that interim time is either positive or zero. If the optimum operation results in negative interim time as shown by a test of the following equation, then overlapping production schedules with or without multiple production units must be considered:

$$t_t = t_m + t_d + t_i \qquad (13\text{-}28)$$

Thus, after the make time, t_m, is computed and for a known down time t_d, the interim time is:

$$t_i = H/N - (t_m + t_d) \qquad (13\text{-}29)$$

If t_i is negative, multiple producing operations must be used. In this situation, the *average* number of producing operations occuring simultaneously is $(t_m + t_d)/t_t$. An example of this type of overlapping operation is given in Chapter 14.

SUMMARY

It is evident that management may meet a large variety of cyclic problems for which an operating policy may be required. The real world problems are usually analyzed by mathematical models employing simplifying assumptions.

There are three basic types of cyclic models (a) the economic order size for stock available all in one batch followed by uniform usage, (b) the economic production lot problem where the stock is made at a uniform .rate during the lead time with uniform withdrawal either during or upon completion of a batch and (c) nonuniform production and withdrawal rate models.

Economic analysis consists in general of a balance of increasing and decreasing costs, in which the number of cycles, the decreasing make rate with increased cycle time, fixed costs at different capacities, and carrying charges on inventory, semifinished stock or in-process inventory, and raw material supplies for operations where such materials may constitute a significant economic contribution to the analysis.

The cyclic model may incorporate probability techniques to provide a sound basis for protective stocks in the long-run situation.

Most of these variations can be incorporated into a model to provide a solution for optimum policy.

PROBLEMS

13-1. A stock of parts must be purchased periodically at a reorder cost of $3.00 per order for a product worth $1.20 per unit; there is a carrying cost factor of 20 per cent. What number of orders should be placed each year, if the annual demand is (a) 100 parts, (b) 1000 parts, and (c) 10,000 parts? Calculate answers, and check with chart.

13-2. A small plant manufactures furniture items to meet an annual demand of 12,000 pieces. Product value is $12, with inventory carrying charges of 10 per cent. Set up costs for each lot are $50. Calculate the economic lot size and the number of lots per year if all items of a lot are assembled and finished at one time.

13-3. A distributor of radio tubes buys 768,000 each year. Procurement costs for purchasing, financing, freight, etc., are constant at $240 for each lot. Carrying charges are $0.01 per tube. What is the optimum lot size? Check calculated value using chart.

13-4. The distributor for a television set handles 25,000 items a year. Incidental expenses for ordering, freight, unloading, etc., average about $500 for each order. The sets average $100 each, and carrying charges are about 18 per cent. What is the economic order size?

13-5. A soap-making process produces 10,000 pounds per day and has sales of 6000 pounds per day for 300 days per year. Set up costs are $200, soap is worth $0.15 per pound, and carrying costs are about 16 per cent. What is the economic lot size and number of batches per year? Use both the calculated value and the chart value.

13-6. Develop the diagrammatic model (graphical) for the following stores inventory problem and state the management policy regarding stock outs. Annual demands are 140 per year, product value is $700, and carrying charges are 15 per cent. Each make-ready charge for a new production batch costs $2800. Lead time is one tenth year, and the cost of a stock out is $2000. Assume Fig. 13-14 applies for the distribution of demands in any one month.

13-7. Derive Eqs. (13-13b) and (13-16b) from the basic Eq. (13-9), showing all the algebraic steps.

13-8. Develop the complete analysis as in Fig. 13-8 for the following two programs, assuming demand rates over a range up to 30,000 units per year based on Eq. (13-16b). Two alternative procedures A and B on different equipment are available to management, and a policy must be selected for an annual demand rate of (a) 6000 annual units and (b) 20,000 annual units. Which procedures should be used? Data available are:

Procedure	A	B
Constant cycle costs, C_c	$30.00	$65.00
Product rate, m, units per year	60,000	200,000
Operating costs, V_i, dollars per unit	0.10	0.05

Product value is $18.00 and the value of f_h is 0.08.

HINT: Make a table of calculations for the assumed values of withdrawal rates w at each of the two production rates, and determine the minimum unit cost by computation and plotting as in Fig. 13-8.

13-9. A food processing plant operates on an annual amount of 100,000 tons of feed worth $35.00 per ton for the season, with a loss of 0.2 per cent per storage day and a crop maturing rate of 1000 tons per day. All investment costs including depreciation may be assumed to be 10 per cent per year; other costs excluding feed cost are constant at $1 million per year. (a) What is the optimum plant size for minimum annual costs if the plant investment costs are $22,000$m^{0.8}$, where m is production rate in tons of feed per day? (b) If the product sells for $100 per ton with a profits tax of 52 per cent, what are the maximum profit and the earning rate when the conversion is 0.9 ton of product per ton of feed if no losses occur? (c) Do the answers in (a) and (b) occur at the same investment. Plot the data and comment.

13-10. For the manufacture of a special glass, the cost of a furnace in dollars is $75Q_B^{0.8}$, where Q_B is the size of a batch in pounds. The furnaces have an estimated life of five years. The operating costs for each batch are $1200. (a) What is the optimum batch size in terms of any constant annual production, Q_A? (b) If the desired production is 40,000 pounds per year, what is the minimum for these costs per pound? (c) If the annual production is doubled, what is the optimum batch size?

13-11. Discuss the possible overlap in a process to make a liquid plastic for a production installation which costs $2400Q_B^{0.8}$ dollars, where Q_B is the size of a batch in gallons, with a 12 year life. Start up costs per batch are $600 and operating costs vary as $12(Q_B/1000)^{1.8}$ dollars per hour; equipment production time is fixed at 1 day, and aging time for a batch outside of production time is 3 additional days. Neglect inventory charges and any extra storage facility charges because of overlap. Annual production is 200,000 gallons.

CYCLIC MODELS FOR OPERATIONS

The preceding chapter presented a reasonably comprehensive coverage of cyclic procedures for the purpose of determining the optimum operation. An extension of this subject leads to an analysis of the experimental data utilized in the manufacturing operation. Thus, the present chapter is concerned with the procedures employed in setting up models for real data. In particular, processes are analyzed where production decreases with time. This decrease in production rate may stem from the nature of the operation; for exan.,.le, an operation will slow up in time as the result of the dulling of machine tools. In some cases, a natural law functioning within a biological process may retard the production rate. Other technical causes include such factors as the clogging of water pipes with deposits. Therefore, it becomes expedient to stop the operation at some optimum time and *restore* the facility to its initial production rate. An optimum balance exists between the *cost of restoration* (cleanout or the regeneration of active components such as catalysts) and the *value of lost production at lower rates for extended times*. Although the batch size gets larger with time, the over-all average rate may go through a maximum.

Maximum economic return is not the only consideration. A secondary problem is that another operation may actually be optimum for yielding the *maximum production*. Thus, if maximum production is of more interest than maximum economic return, the operating procedure itself will be different. Although both of these objectives are discussed in this chapter, the basic principles presented herein are similar to those of the preceding chapter.

CYCLIC PRODUCTION OPERATION

The instantaneous production rate, m, in items per unit of time varies as some function of time, t.

$$m = f(t), \text{ items per hour} \tag{14-1}$$

377

NOMENCLATURE

a = a constant	m = make rate, items per unit of time
b = a constant	M = machine capacity factor
C_c = operating and down time costs per cycle	N = number of cycles per period of time
C_d = costs per unit of time for down time	p = product value
	Q = quantity of production items
C_F = annual fixed charge per unit of capacity factor	Q_f = a quantity of raw stock
	Q_A = annual production
C_h = annual holding charges for inventory	Q_B = quantity of production in a lot or batch
C_O = operating costs per unit of time during make time	Q_I = maximum inventory attained
	R = over-all average production rate
C_T = total annual costs	t = time, various units
f_h = fraction of product value chargeable for holding charges	t_d = down time, various units
	t_i = interim time, various units
H = total time, in hours or in fraction of a year	t_m = make time, various units
	w = average withdrawal or demand rate
I_a = average inventory	
K = a constant	

---

where $f(t)$ is some relationship involving machine capacity and any other pertinent operating conditions that may affect the rate:

The total production in a lot or batch during a make time, t_m, from time zero would be Q_B:

$$Q_B = \int_0^{t_m} m \, dt, \text{ total items} \tag{14-2}$$

The technical relations are given in Fig. 14-1. The amount produced per cycle, Q, increases with make times, t, t', or t'', as illustrated, the down time being shown (distances b to c, b' to c', b'' to c'') for the respective make times. The over-all average production rate, R, is the accumulated production, Q, divided by the total of make time plus down time, t_d, or:

$$R = Q/(t_m + t_d), \text{ items per unit of time} \tag{14-3}$$

The over-all average rates are the slopes of the lines ac, ac', and ac'', which demonstrate a maximum slope, $a'c'$. Thus, operations to point b' will give the maximum over-all rate. However, this may not be the optimum operation for minimum costs because minimum costs depend upon the relative costs for the producing time and down time; in addition,

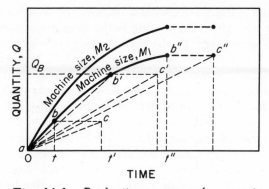

Fig. 14-1. Production curve for semi-continuous process with down time.

when machine capacity is a variable, minimum costs depend upon consideration of the inventory charges plus fixed charges for equipment of a different size M, as indicated on the diagram.

When for the visualization of the problem the production relations and the withdrawal conditions are combined on a chart, the result is as in Fig. 14-2. Case 1 demonstrates the rates and inventory for the repetitive operation with some interim time. In Case 2 there is no interim time and a continual repetitive operation is employed with the same total cycle time as Case 1. In both cases the down time, bc, on the machine is necessary for cleanup and preparation for a new batch.

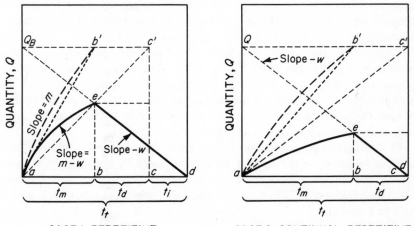

CASE I REPETITIVE   CASE 2 CONTINUAL REPETITIVE

Fig. 14-2. Semicontinuous process with decreasing make rate (given cycle time, t_t).

The diagrams in Fig. 14-2 illustrate the three production rates in cyclic problems: (a) the instantaneous rate is the tangent or slope, m, of the curved line ab' at any time; (b) the average design rate is the constant slope of the dotted straight line ab', which is equal to Q_B/t_m; and (c) the over-all average production (or make rate) is the slope of the straight line ac'. The absolute value of this rate must be equal to, or greater than, the absolute value of required withdrawal or demand rate $w = Q_B/t_i$; the slope of the line ed is minus w. (If the make rate does not exceed w, then the demand will be greater than production). The relations hold for both diagrams in Fig. 14-2. However, in Case 2 the production cycles follow each other continually (that is, there is no interim time); thus, if a longer make period is used, the required production takes place in the same total cycle time, even though a smaller inventory results because of smaller machine size or different operation.

The over-all average make rate in Case 1 is *higher* than the withdrawal rate; this is *always* true if there is *interim time* in the cycle. In Case 2 the two rates are *exactly equal;* this is *always* true for *continual repetitive operation.* It will be observed that the number of units made in any cycle is based on the average design rate during the production period, regardless of machine down time. All of these rates are critical, and one must have an understanding of their relationships to analyze the variations in the cyclic problem.

Where the instantaneous production rate, m, is constant, then the total production in time, t, is simply mt units of product (this was considered in the previous chapter). However, if m varies as some known function of time, an integrated value of $\int_0^t m\,dt$ must be employed to determine the production in a batch. If the variation of m with time can be expressed algebraically, the integration can be performed readily.

The average product inventory on which interest and storage expense must be charged is the sum of two parts. The first is during the make period (a to b in Fig. 14-2) at a nonuniform rate with depletion at the withdrawal rate, w. The second part is the remaining time period (b to d in Fig. 14-2), where no production takes place and withdrawal is occurring at the same uniform rate as previously.

Following the principles in the preceding chapter, one finds that the average inventory is the sum of the areas divided by the total cycle time. This is the difference between the accumulated production and accumulated withdrawals, weighted according to the respective times each unit is held in inventory.

$$I_a = \frac{N \int_0^{t_i} (Q - wt)\,dt}{Nt_t}, \text{ items-year per year} \qquad (14\text{-}4)$$

NOTE: for conditions of *constant rates* for m and w this relation was given by Eq. (13-7) which reduces to Eq. (13-8):

$$I_a = Q_l/2 = (m - w)t_m/2 \tag{13-8}$$

This type of operation (constant make rate) does not show a maximum production rate, since the over-all average production rate continues to increase.

CYCLIC PRODUCTION VARIABLES

The cyclic production problem generally has three independent variables which are (a) the average demand rate, w, (b) the process or make time, t_m, for the cycle which together with the machine size or capacity determines the batch quantity since none of these variables are independent[1], and (c) the total cycle time, t_t, which is the sum of process time, t_m, down time, t_d, and interim time, t_i, all of which may vary. Where the latter two are constant or defined in terms of other variables, then only the annual production rate and process time are variables of interest.

Economic analysis of this type of problem means that first the operation must be evaluated in terms of a plot such as Fig. 14-2 to envision the technical relations. The analysis must show that the annual demand amount, Q_A, will be produced. (The terms Q_A and w are synonymous if w is an annual rate).

The interrelation between process time, machine capacity, and batch size is the next consideration. Where machine size or capacity is fixed or constant, the possible variation of fixed costs in the analysis is eliminated. Similarly, the analysis for operating costs and inventory charges is then limited to the effect of process time on batch size for one machine capacity only.

The total cycle time distribution must be considered specially, since it is usually the sum of three possible parts: the make time, t_m, down time, t_d, and interim time, t_i. The down time or get-ready time is generally regarded as a constant, but the interim time may vary from zero to some finite value. If interim time is not specified, one must deal with two time variables, the total cycle time and the process or make time. If the interim time is equal to zero as in Case 2 of Fig. 14-2, the total cycle time is t_m plus

---

[1] In this analysis the machine capacity is often selected or designed to just meet the capacity required. If complete freedom of choice with any degree of excess capacity is allowed then this item might be considered as an independent variable but this circumstance would control the interim time as demonstrated in Fig. 14-2.

t_d. Where the total cycle time is fixed as in Fig. 14-2, the batch quantity is also fixed for a given down time. Under these conditions, only one machine capacity and its corresponding make time will meet the conditions required for Case 2. However, if the total cycle time is not fixed, then machine size or its equivalent, batch size, is a variable and the interim time will vary. As the interim time approaches zero for the machines of Fig. 14-2 at constant total cycle time, then the machine requirements to do the job also become smaller. Note, however, that larger machine capacities are needed for a given batch size if the make time is reduced, that is, the total cycle time is decreased.

The variations of the interim time and machine size are illustrated in Fig. 14-3 for comparison. In Case 1 the make time is decreased but the

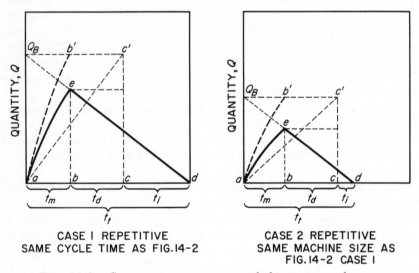

CASE 1 REPETITIVE
SAME CYCLE TIME AS FIG.14-2

CASE 2 REPETITIVE
SAME MACHINE SIZE AS
FIG.14-2 CASE 1

Fig. 14-3. Semicontinuous process with decreasing make rate.

batch size Q_B remains the same as in Fig. 14-2; therefore, the number of cycles are also constant, which can only be true with the total cycle time remaining constant. Thus, the interim time must increase for the same total hours of operation for all cycles. To make this change in operations the machine size would have to increase over that for Case 1 in Fig. 14-2 in order to produce the same batch size in less time.

In Case 2 the same machine size is used as for Case 1 of Fig. 14-2 but for a smaller batch size. In this situation a larger number of batches are required for the same annual production, which reduces the total available cycle time. This will then affect the interim time accordingly.

MATHEMATICAL RELATIONS

The mathematical relation, which expresses these points for the number of cycles during a total elapsed time period H such as one year, is as follows:

$$N = Q_A/Q_B = \frac{Q_A}{\int_0^t m \, dt} = H/t_t \qquad (14\text{-}5)$$

where t_t is the total cycle time and includes both down time and interim time if both are present in the cycle. Interim time always occurs if the total process time plus total down time is less than 8760 hours per year when an annual basis is studied. The annual production, Q_A, is normally considered to apply for a calendar year of $365 \times 24 = 8760$ hours. Therefore, it is the same as an average withdrawal rate, w, based on the total cycle time and is equal to 8760 w when w is items per hour. For the repetitive operation cycle that includes interim time the sum of make time plus any necessary down time will be less than the total cycle time. Where the individual parts of the total cycle period are not fixed or known, it is necessary to include the time function in a proper manner.

The total cycle time general equation for make time, down time, and interim time is:

$$t_t = t_m + t_d + t_i, \text{ total cycle time} \qquad (14\text{-}6)$$

In general, the quantity, Q, made in any one cycle must equal the demand for the cycle and:

$$Q_B = wt_t = w(t_m + t_d + t_i) \qquad (14\text{-}7)$$

Thus, from Eq. (14-2) also:

$$Q_B = \int_0^{t_m} m \, dt = w(t_m + t_d + t_i) \qquad (14\text{-}8)$$

this relation relates the machine size (as included in the make rate function m) and the cycle time components. When any of these variables are constant, the problem becomes more simple.

In the analysis the annual requirement, Q_A, is the critical factor, since it determines the slope of the withdrawal lines employed in the illustrations. If a shortened year (less than the calendar year of 365 days) is employed as a basis of operations, all equations will hold; the shortened time period is considered as H and as a fictitious year for all production computations. All calendar year cost factors and the withdrawal rate must be corrected to apply for this period; the total costs apply strictly for the fictitious short year. However, if the withdrawal is spread evenly over a full calendar year, the real effect is merely to increase the interim or dead time in the cycle over whatever is already present. The shortened year for operations simply fixes the maximum available process time for the given operation. The extra interim time is often considered to be available for scheduling a second operation, which would have its own fictitious year.

Care must be exercised in the selection of the process or make time. For example, if the down time represents a part of the operation necessary for obtaining the product, there is no withdrawal during the operations until the down time is completed. The operation is treated as the strictly batch one, with the size of a batch, Q_B, and the maximum inventory Q_I being synonymous. This was discussed in connection with Fig. 13-5. Although the operation does not necessarily have to be continual, the effective make time is the process time plus the down time.

However, where the product made in a semicontinous process is immediately available but where the equipment is out of production during the down time after a lot or batch is completed, the analysis is as illustrated by Fig. 14-2. In this operation, the make time for production is only the time in process without the dead time, but the machine cannot be utilized for a second batch until the down time is completed. In this situation, the average design rate for production is greater than the over-all average rate. The over-all average rate must equal or exceed the required average demand rate.

The analysis of the cyclic production problem can take many forms because of the variations in the type of problem encountered. Some involve only maximum production rates without regard to economics; see the next section. The maximum production problem is then followed by a consideration of optimization for best economy where product inventory is taken into account as well as other pertinent costs.

PRODUCTION RATE MODELS

The instantaneous production rate, m, is usually constant or decreases with time in a cyclic operation. (There are some possible exceptions where m increases with time in certain selected operations.) Where m is constant, the integration of Eq. (14-2) is quite simple and gives:

$$Q_B = mt_m, \text{ total items per cycle} \qquad (14\text{-}9)$$

and there is no maximum production rate.

For other types of make rates, the relationship $f(t)$ in Eq. (14-2) must be known from theoretical considerations or approximated by empirical relations. This section will demonstrate certain relations and how they may be utilized for economic analysis.

The most simple type of relation is for the case where m decreases linearly with time. If available data for m on a given machine are plotted against time on an arithmetic plot and a straight line results, then:

$$m = at + b, \text{ items per unit time} \qquad (14-9)$$

where a = the slope of the line and may be negative
and $\quad b$ = the value of m when t is zero

Under this condition, the amount made per cycle with a make time, t_m, is:

$$Q_B = \int_0^{t_m} (at + b)\ dt = 0.5at_m^2 + bt_m, \text{ total items} \qquad (14-10)$$

The constants a and b include the machine capacity factor which may or may not be evaluated depending upon other knowledge concerning the operation.

A common relationship found for certain technical processes is a plot of the reciprocal of the square of the make rate against time which gives a straight line on arithmetic coordinates. Thus:

$$\frac{1}{m^2} = at + b \qquad (14-11)$$

From this, then, the function for m to use in Eq. (14-2) would be:

$$m = (at + b)^{-0.5}, \text{ items per unit time} \qquad (14-12)$$

The batch quantity, Q_B, thus is:

$$Q_B = \int_0^{t_m} (at + b)^{-0.5}\ dt = 2[(at_m + b)^{0.5} - b^{0.5}]/a, \text{ total items} \qquad (14-13)$$

When the make rate, m, is plotted on semilog coordinates against time, a straight line often results and the relation is:

$$\log_{10} m = at + \log_{10} b \qquad (14-14)$$

in this case, a is the slope of the plot, and b is the value of m at t equal to zero. Equation (14-14) may be written as:

$$m = be^{kt}, \text{ items per unit time} \tag{14-15}$$

where e = the transcendental number 2.718
and k = 2.303a to convert the $\log_{10}$ data to the base e

Based on this relation, the batch size would be:

$$Q_B = b \int_0^{t_m} e^{kt}\, dt = b(e^{kt_m} - 1)/k, \text{ total items} \tag{14-16}$$

Still another common analysis is to plot log m against the log of time to give a plot as shown in Fig. 14-4:

$$\log m = a \log t + \log b \tag{14-17}$$

where a is the slope of the plot and b is the value of m when t equals unity. This relationship can be rewritten as:

$$m = bt^a, \text{ items per unit time} \tag{14-18}$$

which may be integrated to give the batch quantity, Q_B.

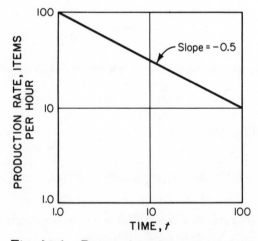

Fig. 14-4. Empirical analysis of production rate.

$$Q_B = b \int_0^{t_m} t^a \, dt = b t_m^{a+1}/(a + 1), \text{ total items} \qquad (14\text{-}19)$$

For the example shown, the slope is minus 0.5, and the value of b is 100 items per hour. Thus, the real world equation would be:

$$Q_B = 200 t_m^{0.5}, \text{ items per cycle}$$

A variation of the cyclic production problem occurs when the production rate, m, is given in terms of efficiency or conversion of the raw stock to product. For example, the instantaneous rate, m, may vary as follows:

$$m = y Q_f, \text{ units or product per unit of time}$$

where y = instantaneous conversion, units of product per unit of raw
stock
and Q_f = units of raw stock per unit time

When y and Q_f are constant, m is also constant. However, if y varies with time, the average production rate varies with the size of the batch in much the same manner as illustrated in Fig. 14-1. Typical process operations of this nature are exemplified by the change of a catalyst efficiency in chemical operations. For this case, with constant, Q_f, the batch size becomes:

$$Q_B = Q_f \int_0^{t_m} y \, dt, \text{ units of product} \qquad (14\text{-}20)$$

Substitution of some known relationship between efficiency, y, and time in this equation permits the use of other equations where Q is required. The instantaneous value of y may vary in a manner illustrated for m in the preceding discussion. The evaluation of the integral yields a cumulative over-all conversion for the cycle.

Although other models apply for different processes and operations, the principles are the same. Usually, the data can be approximated by one of the forms presented in this section, a more sophisticated analysis not being warranted except in very special cases.

MAXIMUM PRODUCTION RATES

When economics are neglected as illustrated in Fig. 14-1, the purely technical problem of maximum production rate requires only an analysis of the over-all average rate equation with no costs involved. The indicated procedure is to define the over-all rate, R, in terms of the batch quantity and total cycle time, t_t, with no interim time. The differential of this equation is then equated to zero, and the value for optimum make time determined.

For example, from Eq. (14-19) an over-all rate for no interim time would be:

$$R = Q_B/t_t = \frac{bt^{a+1}}{(a + 1)(t + t_d)} \tag{14-21}$$

If this is differentiated and the result set equal to zero ($dR/dt = 0$), the optimum make time is obtained:

$$t_{opt} = -\frac{(a + 1)t_d}{a} \tag{14-22}$$

The optimum make time gives the maximum over-all average production rate.

This same relation would apply for every production operation showing a variation of make rate as given by Eq. (14-18). It will be observed that for $a = -0.5$ if t_{opt} is substituted into Eq. (14-18) the instantaneous make rate will be exactly equal to the value of R from Eq. (14-21). In other words, the optimum make time occurs when the slope of the make rate curve of Fig. 14-1 is exactly equal to the slope of the line that defines the over-all average rate. This will be true if the function for Q is increasing continuously with a decreasing slope.

It can also be shown that the maximum rate based on no interim time for operations following Eq. (14-13) is:

$$t_{opt} = t_d + (2/a)(abt_d)^{0.5} \tag{14-23}$$

A similar procedure (of differentiating R with respect to time) is followed for all cases, but for complex relations t_{opt} is best obtained by trial and error solution of the initial rate equation. Such procedures can be programmed readily on automatic computers.

THE OPTIMUM ECONOMIC RATE

The process time for maximum rate as determined by the previous procedures may not be the best operating time for minimum costs. An economic balance may exist in the operation of a cyclic process whenever certain costs increase with increase in make time per cycle and other costs decrease with increase of make time. This is demonstrated in Fig. 14-5 for the idealized case. As shown, the annual costs for a given annual production go through a minimum at some optimum make time. This is true because as the size of a batch increases with increased cycle time then fewer cleanout and set-up costs are required each year. However, when larger batch sizes are made the average operating costs per unit of product may be proportionately greater since the make rate decreases with time and therefore relative operating costs per unit increase with larger make

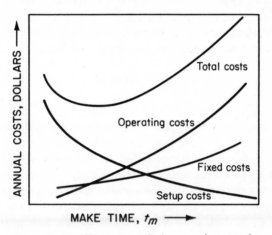

Fig. 14-5. Economic balance for cyclic process.

times. Also for restraint conditions of constant annual production the fixed costs may increase with increased make time as shown. For example, lower production rates may require proportionately higher fixed charges per item. The form of this variation is specific for each process because of technical and economic reasons.

If it is assumed that there are no costs for interim time except the fixed costs, then from Eq. (13-1) the basic equation for annual costs may be written as follows:

$$C_T = N(t_m C_O + t_d C_d) + C_h + M C_F, \text{ annual dollars} \quad (14\text{-}24)$$

where N = the number of cycles per year equal to Q_A/Q_B

$\quad t_m$ = the make time per cycle, various units

$\quad C_O$ = the hourly operating costs during the make time

$\quad C_d$ = the cleanout and setup cost per unit of time during down time

t_d = the down time

C_h = the total holding charges for inventory and storage

M = some capacity factor

and C_F = the annual fixed charges per unit of capacity factor; these might also vary depending upon the magnitude of M

The number of cycles may be replaced by its equal Q_A/Q_B; Q_B is as defined in Eq. (14-8). When the operation is fixed as continual repetitive with the interim time as zero or fixed as a constant value, then there is only one machine capacity that will give the required batch size for any given make time, t_m. This relationship is defined by the particular form of Eq. (14-2) that applies for the technical operation concerned. For example, the illustration in Fig. 14-4 gave a relation for the batch size as:

$$Q_B = 200 t_m^{0.5}, \text{ items per cycle}$$

As was discussed in connection with Fig. 14-2, there is one machine size that will give this lot size in a given make time, but other combinations of production time and machine capacity can be used to make the same lot size. These combinations are included in the constant 200. Thus, the relationship is known if the constant b in the general equation is equal to KM, where K is a constant and M is a machine-capacity factor. However, when for experimental data b has a real value of 100 and the machine size is known, K is evaluated, since $K = 100/M$. Thus, having such information available and knowing $m = KMt^{-0.5}$, one can compute the required machine size from Eq. (14-8) as follows:

$$Q_B = w(t_m + t_d + t_i) = (K/0.5) M t_m^{0.5} \tag{14-25}$$

Considering all these factors and conditions, one may now write Eq. (14-24) in a form suitable to serve as a mathematical model for analysis. Thus, where $N = Q_A/Q_B = \dfrac{Q_A}{\displaystyle\int_0^{t_m} m \, dt}$ then from Eq. (14-5) one obtains:

$$C_T = \frac{Q_A(t_m C_0 + t_d C_d)}{\displaystyle\int_0^{t_m} m \, dt} + C_h + M C_F, \text{ annual dollars} \tag{14-26}$$

This equation may then be manipulated to determine the optimum make time, t_m, for minimum annual cost. However, the equation is subject to the following restrictions: the interim time must be non-negative (that is, zero or greater) and any additional charges during any interim time must be included in the fixed costs. The restriction of non-negative interim time is stated by Eq. (14-6).

The relations for the holding charges developed in Chapter 13 are by the use of Eq. (14-4) restated as:

$$C_h = pf_h I_a = (pf_h / t_t) \int_0^{t_t} (Q - wt) \, dt \qquad (14\text{-}27)$$

where I_a is the area under the plots shown in Figs. 14-2 and 14-3. The integration is best performed in two parts. The first part includes the area up to the end of the make time, and the other part is the triangle for the down time plus interim time (if any).

In many cases of analysis of optimum rate at minimum cost, the holding charges are omitted entirely. The problem may be simplified still further by fixing the capacity factor. This applies where the equipment is already installed or where some desigr capacity has been preselected. The make rate function, m, then is known in terms of time only, and the fixed charges of Eq. (14-26) are constant. Thus, this term has no influence on the optimum value when the equation is differentiated. It should be noted that if the machine capacity is fixed, the interim time cannot be fixed also for an optimum analysis. This is true because any make time with a fixed machine capacity will set the batch size. This in turn determines the number of annual cycles and the residual interim time after make time and down time are known. The total cycle time for a known batch size is given from the relation in Eq. (14-25) where $t_t = Q_B / w$.

An alternative form of Eq. (14-26) is sometimes written where Q_A / Q_B is replaced by $N = H / t_t$ to give:

$$C_T = \frac{H(t_m C_0 + t_d C_d)}{(t_m + t_d + t_i)} + C_h + MC_F \qquad (14\text{-}28)$$

The variable t_i is a function of t_m; t_i may be replaced in terms of t_m from Eq. (14-5) as follows:

$$HQ_B / Q_A = t_t = t_m + t_d + t_i \qquad (14\text{-}29)$$

$$t_i = HQ_B / Q_A - (t_m + t_d) \qquad (14\text{-}30)$$

Substitution of Eq. (14-30) into Eq. (14-28) yields Eq. (14-26) when Q_B is replaced by $\int m \, dt$.

AN OPTIMUM RATE EXAMPLE

Problem

The optimum rate analysis for minimum cost will be illustrated by a problem using the known make relation given in Fig. 14-4. There is a down time of 6 hours and zero interim time. Thus, the make time and its related machine capacity operate on some cycle like the one in Fig. 14-2, the exact cycle being determined by the optimum make time. The startup and down time costs are $120 per cycle, and operating charges are $15 per hour during production. The product is worth $4 per pound. The product has 8 per cent charged against it for storage and interest; however, these charges will be neglected in this problem. The constant b as determined from Fig. 14-4 is based on a machine size of 2000 units; that is, $b = 100 = 2000\,K$, and $K = 0.05$ in the make relation, Eq. (14-18), where $m = 0.05\,Mt^{-0.5}$. Each unit of capacity requires an investment of $2400 with a 13 year life. What is the optimum production cycle for minimum cost for a fixed annual production of 100,000 units withdrawn at a uniform rate throughout the year? As previously indicated, holding charges are neglected to simplify the presentation.

Solution

The annual costs will go through a minimum when Eq. (14-26) applies, which requires evaluation of the integral in the first term. This may be accomplished readily at constant M, since the machine size is not an independent variable but is a function of Q_B through Eq. (14-25). The relation among machine size and cycle times is:

$$M = \frac{w(t_m + t_d + t_i)}{2Kt_m^{0.5}} = \frac{(100,000/8760)(t_m + 6)}{2(0.05)t_m^{0.5}} = 114(t_m^{0.5} + 6t_m^{-0.5}) \tag{14-31}$$

If the interim time and down time are fixed, the machine capacity required is a function only of the make time for a known withdrawal rate (fixed annual production). In fact, the make time is also determined under these conditions because the quantity at point e of Fig. 14-2 is uniquely specified, since it is equal to $w(t_d + t_i)$ which is constant. Thus, the quantity Q_B must be made at b' by some machine size in any make time t_m that is selected. The corresponding machine size will be given by Eq. (14-31). The make time may be selected from economic considerations which will then determine the complete cycle of operations and the machine capacity.

With zero interim time and for $t_d = 6$ hours, the economic relation of Eq. (14-26) may be written as follows (when holding charges are neglected and $C_F = 2400/13 = \$184$ per unit of capacity):

$$C_T = (H/t_t)(15t_m + 120) + \$184\,M, \text{ annual dollars} \tag{14-32}$$

Since M is given by Eq. (14-31), the final working equation becomes for $H = 8760$ hours:

$$C_T = \frac{8760}{t_m + 6}(15t_m + 120) + 20{,}970(t_m^{0.5} + 6t_m^{-0.5}) \qquad (14\text{-}33)$$

This expression can be differentiated to obtain the optimum, t_m; however, the mathematical analysis is tedious, and substitution of assumed values of t_m for a trial and error solution is probably the best procedure. Such a solution can be programmed readily on an automatic computer. Results are given in the following tabulation.

Make time, t_m, hours	Annual cost, C_T, dollars
4	262,500
8	253,970
9	253,770
10	253,940

The optimum make time is shown to be about 9 hours.

It should be noted here that not all real equations set up as Eq. (14-32) will yield an optimum cycle time. Whether or not a minimum exists depends upon the relative magnitudes of the individual cost items.

Where the interim time is a variable in addition to t_m in Eq. (14-32), it becomes necessary to replace t_i by its equal from Eq. (14-30). This action leads to an equation of the form of Eq. (14-26) with two variables, t_m and M; this equation requires partial differentiation to solve for the optimum make time and the corresponding optimum lot size. This case will not be considered here.

THE OVERLAPPING REPETITIVE CYCLE

An interesting variation of the cyclic production problem is the situation where the optimum batch size, Q_B, is such that the computed time for one complete cycle $t_t = (Q_B/Q_A)H$ is less than the allowable process time. As has been discussed, the relationships between cycle times are as follows:

$$\text{make time} + \text{down time} + \text{interim time} = \text{total cycle time}$$

$$t_m \quad + \quad t_d \quad + \quad t_i \quad = \quad t_t \qquad (14\text{-}6)$$

Thus, if $t_m + t_d$ for optimum operation exceeds t_t, then t_i must be negative, and the only possible operating schedule is an overlapping one where additional production is started on a new cycle before the make time on the previous cycle is completed. In fact, more than two batches may be in process simultaneously during some part of the production cycle. Where all of the available time ($t_m + t_d$) is required for the machine, then multiple machine installations are required. However, if part of the make time is an aging or finishing time not requiring manufacturing or fabricating facilities, the succeeding cycles may overlap the preceding ones until there is no free time on the production part of the facility.

For the semicontinuous process, the make rate must always be equal to, or in excess of, the withdrawal rate; otherwise, multiple machines (overlap) are necessary. Therefore, the make rate, m, is the sum of the make rates of the individual machines. If the make rate is variable and is adjusted to just equal the withdrawal rate so that there is no inventory, then the operation will be continuous rather than cyclic.

The overlapping repetitive cycle is illustrated in Fig. 14-6 for the following situation. A required production of parts is 200,000 items a year. These parts are made in a series of operations requiring 10 hours and then

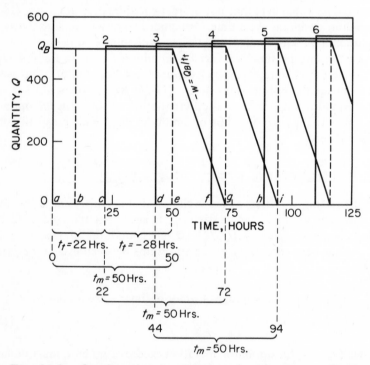

Fig. 14-6. Overlapping repetitive cycle operation. Single production facility.

are subjected to finishing operations which require 40 additional hours for a total production time cycle of 50 hours. The operations are strictly batch, that is, no items are available until the entire batch is ready. Although there is only one production line, the finishing operations can be made in overlapping cycles. An economic analysis demonstrated that the optimum batch size is 500 units. On this basis, a total of 200,000/500 = 400 batches are necessary each year. Thus, for the 8760 hours per year available (365 days) the effective cycle time available is 8760/400 = 21.9 hours. Accordingly, a batch must be completed approximately every 22 hours to provide the needed withdrawal rate $Q_A/8760 = Q_B/t_t$. This can be accomplished only by overlapping the finishing operations as shown in Fig. 14-6.

From Eq. (14-6) the interim or dead time in a cyclic operation is:

$$t_i = t_t - (t_m + t_d) \qquad (14\text{-}34)$$

where t_d represents the aging time during which the production facility is free to start a new batch. Since the allowable cycle time, t_t, is only 22 hours for a required effective make time of 50 hours, then the interim time must be negative, which means there must be overlap. The amount of the overlap is computed from Eq. (14-34) as:

$$t_i = 22 - 50 = -28 \text{ hours}$$

The overlap is shown in Fig. 14-6, where the time, ab = 10 hours, is used for production of the first batch and where time, be = 40 hours, is employed for finishing. The time for finishing is dead time for the fabricating facility operation. However, since overlap is needed in this schedule of operations, a new batch must be started 28 hours before batch 1 is finished. This is shown at point c; the dead time on the production operation has been decreased from 40 to 12 hours.

If the steady operation is considered to start at point e, the stock will be withdrawn during the period eg at a rate of Q_B/t_t. When the stock is depleted at point g, the finished stock from batch 2 is ready for withdrawal until point i is reached. Note that if a third batch is not started at point d at least 28 hours before point g, there will be no finished stock available when the product supply is depleted at point i.

For steady operation, then, two finishing cycles are always running and part of the time, three, which is apparent from the diagram. Numerically, on the average, there are $(t_m + t_t)/t_t = 50/22 = 2.27$ batches in some stage of production; that is, during $0.27 \times 22 = 6$ hours, three finishing cycles are in operation.

If the results for an overlap situation require such a short cycle time that the production time must be overlapped, then multiple production machines are necessary.

The average inventory for the overlap situation of Fig. 14-6 becomes $Q_B/2$ for the product; however, the in-process inventory will be nQ_B where n is the average number of batches which exist for the entire time period of one cycle. This is constant if production is continued all year.

SUMMARY

The models in this chapter have considered the general problem of cyclic production problems of which there are two types. In the first, the objective is to attain the maximum over-all average production where the make rate decreases with time. The second type of problem deals with the selection of the optimum make time for operations with a decreasing make rate in order to produce at minimum cost.

The generalized procedure for converting observed production data into suitable models has been presented by example. However each new problem must be considered as a separate one utilizing the principles illustrated.

As a matter of record, it may be of value to note that the general make equation is:

$$m = f(t) \tag{14-1}$$

When $f(t)$ is of the form (as an example):

$$\frac{1}{m^2} = at + b \tag{14-11}$$

Then the optimum make time for maximum production rate is:

$$t_{opt} = t_d + (2/a)(abt_d)^{0.5} \tag{14-23}$$

Also, when for a given installed machine capacity only operating costs per unit of time, C_O, and down time costs per unit of time, C_d, are considered, the optimum make time for minimum cost is:

$$t_{opt} = t_d C_d / C_O + 2(ab C_O t_d C_d)^{0.5}/(a C_O) \tag{14-35}$$

When $f(t)$ is of the form (as an example):

$$m = bt^a \tag{14-18}$$

then the optimum make rate *for maximum production* is:

$$t_{opt} = - \frac{(a + 1)t_d}{a} \tag{14-22}$$

In addition, when inventory is neglected and machine capacity is fixed, then the optimum rate *for minimum cost* is:

$$t_{opt} = - \frac{(a + 1)t_d}{a} (C_d/C_o) \tag{14-36}$$

PROBLEMS

14-1. A cyclic operation with a down time of 2 hours shows an instantaneous production rate of $m = 100t^{-0.5}$ items per hour where t is time in hours. (a) Draw a plot of accumulated production Q versus time. (b) What is the optimum cycle time and maximum average production rate for this process?

14-2. For the following instantaneous make rate data determine the total cycle time for maximum over-all average production rate and calculate the maximum rate when the turnaround time is six hours for cleanup and startup of equipment.

m, items per hour	2510	2270	2100	1740	1010
time, hours	0	10	20	40	100

Assuming none of the product is available until a batch is completed, draw the cycle diagram and compute the average in-process inventory.

14-3. A heat-exchanger arrangement is subject to fouling and requires periodic shutdowns and cleanouts in order to maintain a desired production rate. A calendar year of 8760 hours is available to produce 3×10^9 pounds per year of throughput. The instantaneous reciprocal throughput, m^2, is linear with time on stream as follows:

$$\frac{1}{m^2} = 4 \times 10^{-14}t + 2 \times 10^{-12}$$

where m is pounds per hour. What is the optimum production cycle for minimum cost per year if the hourly labor cost for down time is $16 per hour and operating costs are $4 per hour? There are no charges for any interim time, and fixed costs can be omitted, since the equipment is already installed. The down time for cleanout and startup is 8 hours.

14-4. A process shows a make rate $m = Q_f y$, items per hour. Q_f is feed rate in pounds per hour, y_t being efficiency of production. Experimentally, it is found that $y = 0.96^t$ where t is in hours. Turnaround or down time for the process is 8 hours. What is the best cycle time and batch size to obtain a maximum over-all production rate where Q_f is equal to 1000 pounds per hour?

14-5. The instantaneous make rate for a semicontinuous cyclic process is $m = K/(at + b)^c$, items per unit time, where K, a, b, and c are constants and t is process time. The down time is constant as t_d. (a) What is the general equation for the optimum time for maximum production rate with no interim time? (b) If $c = 0.5$, what is the optimum time?

14-6. A process shows an instantaneous make rate of $m = 375/t^{0.23}$, items per hour, with a down time of eight hours. Operating costs are \$10 per hour when processing; startup costs are \$50 for each cycle. Fixed charges can be neglected. What is the process time for (a) maximum average over-all production rate, and (b) for minimum annual costs at a constant annual production of 200,000 items.

14-7. Discuss the possible overlap in a process that is to make a liquid plastic for a production installation which costs $24 Q_B^{0.8}$ dollars, where Q_B is the size of a batch in gallons with a 12 year life. Startup costs per batch are \$600, and operating costs vary as $12 (Q_B/1000)^{1.8}$ dollars per hour, with an equipment production time fixed at 1 day and an aging time for a batch outside of production time of 3 additional days. Neglect inventory charges and any extra storage facility charges because of overlap. Annual production is 200,000 gallons.

ECONOMIC ANALYSIS AND UNCERTAINTY [1]

15

Many problems in management economy involve variability in the experimental data. Since economic analysis generally employs an average value for data, there is some uncertainty about the exact meaning of these values. In the long run, however, over a wide range of values the average value for an item of interest is a central one about which all real values approximate. In uncertainty analysis, this long range value, or mean value, is called the "expected value." The manner in which the real values distribute themselves around an average value can be studied mathematically by "probability" or "stochastic" analysis. Then, the results can be combined with economic analysis in order to determine and set management policies and decisions.

By the utilization of probability concepts, management can control or estimate the risk of the available data being in error and thereby determine the influence of variability on the results of an analysis. For example, the amount of sales for a product on any given day may vary from 25 to 150 items, but the average value over a period of days may be 100 items. The use of 100 items as a firm number for computations can lead to serious errors. However, if consideration is given to the variability encountered, such errors can be minimized. This chapter explains some of the theory of uncertainty analysis, models being employed to demonstrate how the theory is applied to real world problems. Only very simple situations are discussed here with the assumption that the probabilities are known. However, an abridged elementary presentation of probability analysis is given in the next section.

In general, the uncertainty effects are included in the theoretical economic analysis by making use of the values for critical elements in the analysis as determined by probability. Thus, if a possible profit of $100 is expected for sure (no uncertainty), the profit is $100 \times P$, where P, the

---

[1] Prepared in collaboration with Dr. F. P. May.

NOMENCLATURE

A	= a calling rate, number per unit time		$P'(n)$	= the cumulative probability of n or more
c	= a cost per item		$P(S)$	= the cumulative probability of S sales or less
C	= cost dollars			
e	= the transendental number 2.718		$P'(S)$	= the cumulative probability of S sales or more
$E(S)$	= the expected (or average) value of S		$P(x)$	= the cumulative probability of x or less events occurring
$E(t)$	= the expected value for time between calls		$P'(x)$	= the cumulative probability of x or more events occurring
$E(T)$	= the expected value of T		P_0	= probability of zero state
$E(x)$	= the expected (or average) value of x		P_x	= probability of state x
			p	= a price per item
$E(Z)$	= the expected value of profit		r	= number of service facilities or channels
f	= frequency of an occurrence			
f_u	= a use factor		S	= a designation of items, sales dollars, pounds, number of items, etc.; or a service rate, completions per unit time
$g(S)$	= an expression meaning function of S			
$g(t)$	= an expression meaning function of t			
			t	= a time interval, various units
m	= a constant; the mean value of a group of items fitting a Poisson distribution; or a profit margin per item		T	= wait time in queue plus service time
			T_q	= queue time only per call for service
n	= a number of items		x	= a variable; a specific item; or state of a service system
$P(A)$	= a probability of event A occurring			
			Z	= profit dollars
$P(n)$	= the cumulative probability of n or less			

probability, is 1.00. However, if there is uncertainty (that is, either $100 *or* no dollars will be made as profit for the given set of circumstances) and if it can be estimated that the probability of obtaining the $100 profit is 0.75, then the expected profit is $(100)(0.75)$ or $75. This means that in the long run the average profit for all cases will be $75. The principle can be applied to either over-all economic balance equations or incremental analysis procedures.

In many situations the applications require the utilization of tables or graphs for making the calculations; such tables or graphs are needed when the equations are too complex for formal mathematical analysis. The application of stochastic principles in management economy calculations may include the use of random number procedures to generate data for calculations when the probability data are known. This is the Monte Carlo method and may be applied to over-all economic balance calculations.

A number of these applications will be discussed and demonstrated by simple examples.

PROBABILITY ELEMENTS

In an elementary sense probability may be considered as the fraction of the total number of relevant events which correspond to a particular event A; that is,

$$P(A) = \frac{\text{number of ways } A \text{ can occur}}{\text{total number of events considered}} \qquad (15\text{-}1)$$

where $P(A)$ is read "the probability that A will occur." If a company sells 20 items, two of which are defective, and if the likelihood that a customer will get any one item is the same as that of getting any other item, then the probability that a customer will get a defective item is $\frac{2}{20} = 0.10$.

In applying probability principles to economic analysis, management must have certain data available in a useful form. Such a form is illustrated in Figs. 15-1a and 15-1b for the cumulative probability. Graphs provide a sounder basis for the analysis than intuition or guesses. The data and procedure for obtaining such a graph are given in Table 15-1.

For example, suppose analysis of operations indicates that to maintain the optimum inventory the inventory should be such that the probability is 0.75 that the stock will be adequate for sales on any given day. From Fig. 15-1a, it is seen that the probability of sales in dollars of $200 or less is 0.75. Accordingly, a stock equivalent to $200 should be on hand at the start of each day to satisfy the requirements of the analysis. This means that for a large proportion of days, or 75 per cent of them, sales will be $200 or less and for the other 25 per cent of the days sales demand will exceed $200 worth of merchandise. The selection of a probability of 0.75 is based on some criterion selected by management for a best policy; the $200 inventory represents that probability applied to some known data available to management.

Although it is helpful, it is not necessary to have a complete background in statistical analysis to utilize probability charts in management economy decisions. An understanding of the simple cumulative probability chart in Fig. 15-1a is desirable since this chart is the most generally useful one for different types of computations in economy studies. The chart in

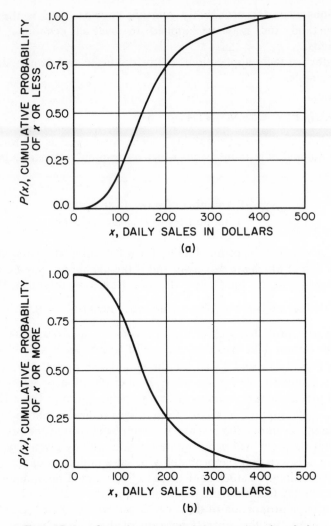

Fig. 15-1. Cumulative probability curves for daily sales. (a and b).

Fig. 15-1a presents the accumulated probability of all occurrences; that is, it shows the probability of x occurrences or less and is denoted as $P(x)$. Figure 15-1b gives a frequently employed variation of this chart. It is the accumulated probability of P' of x occurrences *or more* (that is, *at least x* occurrences). The two diagrams may be related as $P'(x) = 1 - P(x)$. An expanded discussion of the development of these charts is given in the next section.

PROBABILITY CHARTS

The cumulative probability chart in Fig. 15-1a is obtained from real data (past operating experience). For example, these data may be daily sales for 50 days with values of $40, $200, $160, $200, $75, $90, $125, etc., occurring on any one day. A frequency count is made for each range of sales dollars per day as shown in Table 15-1. A frequency distribution may then be plotted as in Fig. 15-2. This is a histogram in which the horizontal axis is divided into intervals (on the abscissa or x axis). Each interval corresponds to the group, and on each, a rectangle is constructed whose area is proportional to the frequency within the group limits. With uniform intervals, as used here, the heights of the rectangles are propor-

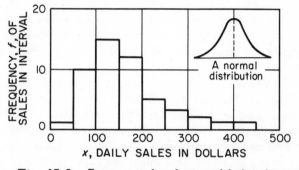

Fig. 15-2. Frequency distribution of daily sales.

tional to areas and also to the frequencies. If the number of observations was much larger and the intervals were much narrower, it is to be expected that the irregular step-form of Fig. 15-2 would become smoother with smaller steps and would approach a smooth curve as a limit such as occurs in Fig. 15-3. This smooth curve (frequency distribution curve) may be thought of as defining the true underlying distribution. Thus, if the

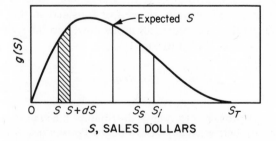

Fig. 15-3. Idealized probability distribution function for sales.

histogram or frequency-distribution curve is expressed in terms of *fractional* proportions of the total frequencies instead of as *actual* frequencies, these fractional proportions of the total frequencies will be related to the probability as defined above, and the total area under the curve will total unity. The graphs will then represent probability distributions, and if a mathematical expression can be obtained for the smooth curve, this equation is a *probability distribution function* defined mathematically as $g(x)$.

The probability distribution curve does not give the probability of a specific value of a variable where the variable is considered to vary continuously over its whole range. However, the probability that a randomly chosen value will lie between any two values of the variable can be obtained. Thus, the area under the probability distribution curve, between two specified values, S_1 and S_2, of a variable (such as daily sales) gives the probability that an observation will have a value *between* S_1 and S_2. This probability, when multiplied by the total number of observations made, gives the number of observations which are expected to have a value between S_1 and S_2.

In economic analysis, however, it is usually of more importance to have the probability that the variable will be greater than or less than some specified value. This is obtained from the cumulative probability distribution curve or function. The cumulative probability distribution is the cumulative area under the probability distribution curve (Fig. 15-3) obtained by summing up all the area under the curve up to a given value of the variable. The cumulative probability for all values of the variable must terminate at unity.

In many situations, data can be considered to be distributed approximately as a normal distribution or as a Poisson distribution, although there are numerous other mathematical relations that describe the distribution of real world data. The normal or Gaussian distribution is symmetrical about a central value, which is demonstrated by the insert in Fig. 15-2. The Poisson (a distribution function of a discrete variable) may take a variety of configurations. If the numerical value of the mean exceeds unity, the curve is usually skewed or lopsided so that it rises more rapidly on the left side and decreases more slowly on the right side of some most frequent value as shown in Fig. 15-2.

* * * (*Optional*) * * *

In the practical problem of fitting a distribution function to actual data for economic analysis, there are many cases where a Poisson distribution can be employed as a good approximation. When counting *units* which are approximately randomly distributed in an interval of measuring space and when certain desired information is the probability of a given number of units being in an interval, the Poisson expression for the distribution of the units is proper; this applies for many situations. The Poisson distribution function is:

$$P_x = \frac{m^x e^{-m}}{x!}, \quad x = 0, 1, 2, \ldots, \text{(that is, integers)} \qquad (15\text{-}2)$$

where P_x = the probability of x events in a stated period of time, unit of length, unit of area, or the like

x = the number of events

m = the expected number of events in unit interval, or the average value of x for all observations

e = the transcendental number, $2.718\ldots$

and $x!$ = the factorial x, with $0!$ equal to one

When the actual distribution function is known from considerations other than the observed data being considered, then the actual distribution function obviously should be used. With only one set of observed frequencies the simplest form of distribution function which will fit the data satisfactorily should be employed. The fitting of a distribution function to a set of empirical data can be a complex problem beyond the scope of this presentation. However, when a simple distribution function is assumed, the goodness of fit can be tested readily by means of a chi-square test as outlined in any standard text on statistics. It should be noted here that the shape of the distribution function does not always appear as in Fig. 15-3. As a matter of fact, if the value of m in Eq. 15-2 has a numerical value of unity or smaller, the plot starts at a finite value on the ordinate for the zero value of the variable, and the value of $g(x)$ continuously decreases as x becomes larger. (Such a plot is demonstrated in Fig. 15-8).

Where the distribution data are assumed to fit a Poisson distribution function, the probabilities may be calculated by the use of Eq. (15-2). Cumulative probabilities for all values of x or less may also be calculated by summing up the probabilities for all values of x up to the same specified value. This is expressed mathematically as:

$$P(x) = \sum_{i=0}^{n} \frac{m^{x_i} e^{-m}}{x_i!} \qquad (15\text{-}3)$$

where $P(x)$ = the cumulative probability of x or less; $P(x)$ must equal 1.00 when all values of probability are considered

and x_i = an individual value of x

Calculated frequencies for any given total number of observations can be obtained by multiplying the calculated probabilities by the total number of observations. These calculated frequencies can be plotted in the same manner as the observed frequencies are plotted in Fig. 15-2.

The mathematical relations for a frequency distribution, $g(x)$, may take a variety of forms, some of which are quite complex. This expression may be read as some function g of a variable x. However, determining the analytical expression is not necessary when a cumulative probability distribution curve, $P(x)$, such as is given in Fig. 15-4, is available from estimates or other known data (as for example, sales of a similar item or sales from a previous season). Note in Fig. 15-4 that S (dollars) or n (items) may be used interchangeably.

Figure 15-3 is a continuous probability distribution curve, based on data such as in Table 15-1; the assumption is made that this smooth curve is a good approximation to a true underlying continuous curve for the universe of all possible values from which the grouped data (such as in Table 15-1) were taken as a sample. Any other group of values taken from the curve then would have the distribution of the assumed universe. The values along the S axis are fractions where S_T may be considered as unity for the universe. Therefore, the value of S (sales dollars) is the fraction of maximum possible sales dollars which corresponds to a particular value of $g(S)$. Its real dollar value is the fraction multiplied by the maximum possible sales dollars corresponding to S_T. For any given case, of the type illustrated in Table 15-1, the maximum dollars for S_T is the size of the largest observation.

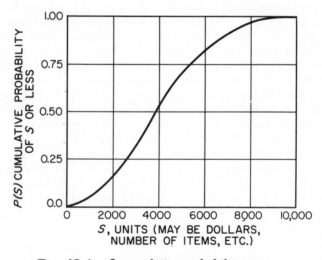

Fig. 15-4. A cumulative probability curve.

When a smooth probability distribution function such as is shown in Fig. 15-3 can be established for actual data, this plot becomes a useful tool for economic analysis. Since it is a mathematical model that approximates the real world data, it can be utilized in further mathematical analysis. In connection with this analysis it should be pointed out that the average or mean value of a group of values under consideration is the same as the "expected" value as determined by statistical procedures.

On Fig. 15-3 the probability that any sales dollars S will be between specified limits is equal to the area under the curve between the specified limits. Note that any one specific value of S has a probability of zero since S may be any value from zero to infinity. For this reason in real world problems it is necessary to consider the probability of S being between the values of S and $(S + dS)$ where dS is a differential value of S. It follows that the probability of a given value which lies in the interval S to $(S + dS)$ is equal to the area $g(S) \, dS$. Thus, the probability that S in any interval, S_1 to S_2, is given by the integral $\int_{S_1}^{S_2} g(S) \, dS$. The cumulative probability that a randomly chosen S will be equal to S_2 or less is obtained when the lower limit of integration is set equal to zero. The cumulative probability curve $P(x)$ is obtained directly from the distribution curve $g(x)$ by measuring areas under the curve from zero up to values of S and plotting these areas against S. The total area under the distribution curve (Fig. 15-3) will be 1.00. (It might be noted at this point that a physical significance of the numerical

TABLE 15-1. Probability Calculations for the Value of Daily Sales During a Study of 50 Days

Daily Sales x	No. Days of x Sales, Frequency, f	Product[a], $f \times x$	Probability of Sales in an Interval[b] $P(x) = f / \sum f$	Cumulative Probability Observed[c]
0–50	1	25	0.02	0.02
51–100	10	750	0.20	0.22
101–150	15	1875	0.30	0.52
151–200	12	2100	0.24	0.76
201–250	5	1125	0.10	0.86
251–300	3	825	0.06	0.92
301–350	2	650	0.04	0.96
351–400	1	375	0.02	0.98
401–450	1	425	0.02	1.00
451–500	0	0	0	—
	50	8150	1.00	—

[a] The product of the frequency of column 2 and the midpoint of the interval of column 1. In extensive statistical analysis these calculations are greatly simplified by coding techniques and calculating machines. The mean value for daily sales is:

$$\sum (f \times x) / \sum f = 8150/50 = 163$$

[b] The interval limits are those for the rectangles of the histogram as plotted in Fig. 15-2 and shown in column 1. The probability of any random occurrence falling within the limits of an interval is the ratio of the area of a rectangle to the total area for all rectangles. Whenever uniform intervals are employed, as in this case, the ratio of areas is also in proportion to $f / \sum f$.

[c] The cumulative probability distribution is the plot (Fig. 15-1a) of these numbers versus x and represents the fraction of total events corresponding to the x at the upper limit of the interval; all of these numbers are smaller than x.

value of $g(S)$ at any S is that it is the slope of the corresponding cumulative probability distribution curve, $P(S)$, at the same value of S.)

The term expected value which has been used previously needs to be clearly understood before proceeding further. The expected value of a random variable is obtained by finding the average value of the variable over all possible values of the variable. The procedure for computing the expected or average value depends upon the information available. For example, if the known data are in the form of Fig. 15-1a, the mathematical expected value is:

$$E(S) = \frac{\int_0^{S_T} S g(S)\ dS}{\int_0^{S_T} g(S)\ dS} \tag{15-4}$$

When the probability function $g(S)$ involves all ranges of S and the denominator totals unity, only the numerator need be evaluated. For discrete increments of sales dollars as determined from a probability curve, Eq. (15-4) may be evaluated as follows:

$$E(S) \cong \frac{\sum_{i=0}^{i=n} [(S_i + S_{i+1})/2] \Delta P_i}{\sum_{i=0}^{i=n} \Delta P_i} \tag{15-4a}$$

where S_i and S_{i+1} refer to the limits of discrete intervals along the abscissas of Fig. 15-1a and $\Delta P_i = P(S_{i+1}) - P(S_i)$ where $P(S_i)$ and $P(S_{i+1})$ are the corresponding values of cumulative probability with n being an upper limit. The expected or average is the value of the abscissa of the centroid of area from 0 to S_T of a plot such as Fig. 15-3.

Where the data are not plotted but are in the form of a table such as Table 15-1, the expected or average value is computed as shown in footnote a of the Table to give $E(S) = \$163$.

When the individual items x are known to follow a Poisson distribution then the expected or average value of x will be:

$$E(x) = \sum_{i=0}^{\infty} \frac{x_i m^{x_i} e^{-m}}{x_i!} = m \tag{15-5}$$

* * * * * * * * *

APPLICATION TO ECONOMIC ANALYSIS

With a mathematical expression of the probability distribution function or with a graph of the cumulative probability curve available, it is possible to incorporate the variability into the solution of a variety of economic problems. For example, during a fixed period it may be desirable to estimate the optimum merchandise that should be stocked for a perishable item, where reordering during the season is impractical and items left unsold are essentially a total loss. The profit equation is simply $Z = S - C$, where Z is profit dollars, S is total sales dollars, and C are costs in dollars. Where the sales demand is considered as an unvarying quantity (no uncertainty) each season, then only an amount of stock to meet this quantity of sales is required, and the profit will be a fixed number every season. In the real world situation, the total sales dollars will fluctuate for different seasons, and some losses will result where the stock exceeds the demand. The profit for a same given amount stocked every season will vary from season to season but will have a long-run average value which is determinable from a cumulative probability curve such as Fig. 15-4 if one is available[2]. This average profit will be the result of two types of calculations: (a) For those seasons where the demands were less than the amounts stocked, there will be an individual profit on the items sold and a loss on those unsold—the net result may be a profit or a loss for that particular season at a given stocking policy amount; (b) for those seasons where demands equal or exceed the fixed amount stocked, the profit will be a constant value equal to the sales dollars minus the costs for the amount stocked. (None of the excess demand could be met because the stocking policy had been established and there was no product available.) The average profit then must be based on the relative quantities under (a) and (b) for any one specific stocking policy. By considering a number of stocking policies, one finds that the profits for each may be computed and a curve plotted as in Fig. 15-5 from which the optimum policy may be selected. An abbreviated set of data will demonstrate the calculations; after this, it will be shown that the mathematical model permits a solution for the best management procedure.

Consider a policy where a stock at a cost of $1728 is ordered every year. This exact stock has a sales value of $2400, but actual sales are expected to vary from year to year according to the numbers in Table 15-2 for the

---

[2] Such a plot is obtained from available data based on past history and the frequency distribution for the data of the type illustrated in Table 15-1. This type of plot could be the result of any kind of a probability distribution function that might approximate the data. It is not limited to normal, Poisson and binominal distributions which are the more common ones used.

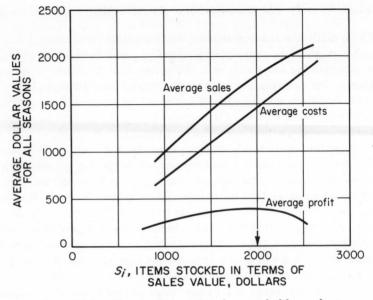

Fig. 15-5. Best stocking policy for perishable product.

past 10 years. (If past history can be used to forecast the future, then these data may be employed for development of useful models.) Excess stock at the end of the season is a total loss. It is desired to compute the long range average or expected profit for this specific stocking policy. The calculations are shown in the table, and consist of the determination of the average profit (or loss) for (a) those seasons where sales are less than \$2400 and (b) those seasons where, although sales demands exceed \$2400, only \$2400 can be sold because this amount of sales consumes the stock. Table 15-2 shows that an average profit of \$315 is the expected value for a stocking policy equivalent to \$2400 worth of sales. Similarly, with the same seasonal distribution of sales, another stocking policy could be assumed by management and the average profit determined in the same manner. If the individual seasonal demands are not known but a cumulative probability curve such as Fig. 15-1b is available, the expected profit at each stocking policy can be computed by means of Eq. (15-4a). The steps for each stocking policy would be as follows: (a) The average or expected value of sales from zero up to the value of each stocking policy as a limit would be obtained by Eq. (15-4a) using a stepwise evaluation with the cumulative probability curve that applies. The profit for this stocking policy would then be determined by subtracting the cost at the specific stocking policy from the expected sales. The results would be average profit for all sales demands less than the stocking policy. (b) The

TABLE 15-2. An Abbreviated Table of Calculations for a Fixed Stocking Policy of $2400 Worth of Sales Using 10 Different Seasons (All Values in Dollars)

Season No.	Seasonal Sales Demands[a]	Actual Sales Possible	Costs[b]	Profits
1	2000[c]	2000	1728	272
2	1500[c]	1500	1728	−228
3	2400[c]	2400	1728	672
4	2600	2400	1728	672
5	4200	2400	1728	672
6	1728[c]	1728	1728	0
7	2200[c]	2200	1728	472
8	2400[c]	2400	1728	672
9	1000[c]	1000	1728	−728
10	3000	2400	1728	672
Totals		20,428	17,280	3148
Average Values		2042.8	1728	314.8[d]

[a] Size of demands are based on previous experience or as estimated by special methods.

[b] All individual entries in this column are $1728, since this is the cost associated with the fixed stocking policy of $2400 of sales as agreed upon arbitrarily.

[c] Seasons with sales of $2400 or less.

[d] Note that this average is also a composite of the individual values weighted according to their frequency of occurrence, or in other words weighted according to probability. That is there are 7 seasons out of ten with sales of $2400 or less, which show an average profit of $1122/7 = \$160.3$. There are 3 seasons with sales above $2400 for an average of $(672 \times 3)/3 = \$672$. Weighting the averages according to their 70 and 30 per cent probabilities of occurrence gives a net over-all average of

$$\text{Profit} \qquad \begin{array}{l} 160 \times 0.7 = 112.2 \\ 672 \times 0.3 = \underline{202.6} \end{array}$$
$$\text{Net over-all profits} \qquad \$314.8$$

average profit for all sales demands equal to or greater than the stated stock policy is a constant and equal to the profit for sales equivalent to the stocking policy. The relative number of these latter sales is equal to the probability of the stock policy sales or more. The relative number of the average profit for sales less than the stock policy in (a) above is the probability of the stock sales policy or less and is equal to $P(S)$ as read from the cumulative probability curve. The over-all average profit at each stock policy is the weighted average of the separate averages in (a) and (b) above. The weighting is done on the relative probabilities of each component.

Another alternative procedure to use when the cumulative probability curve is available is to generate a table of demands by the Monte Carlo method using random numbers and then make an analysis such as in Table 15-2. This will be discussed later in the chapter.

A problem that will serve as an exercise in covering these methods is given at the end of the chapter with certain aids to the solution.

The results for various stocking policies are listed in Table 15-3, and it is noted that the profits go through a maximum. These data could be plotted on a chart such as shown in Fig. 15-5 to determine the optimum stocking policy which shows a maximum profit at $2000 in sales.

TABLE 15-3. Variation of Profit for Different Stocking Policies with Same Seasonal Demands as in Table 15-2.

Stocking Policy in Terms of Sales Dollars	Average Sales	Average Costs[a]	Average Profits
1000	1000	720	280
1500	1450	1080	370
2000	1823	1440	383
2400	2043	1728	315

[a] Cost of any one item sold is $1728/2400 = 0.72$ of sales value, which is also the ratio of total costs to total sales applying for this problem.

Alternatively, a more sophisticated analysis of the perishable stock problem can be made using the basic relation of Eq. (15-4) as shown in the following section.

* * * (*Optional*) * * *

MATHEMATICAL ANALYSIS OF OPTIMUM

It will be noted in Table 15-2 that the average profit at any one specific policy is a composite of the variable net profits and losses for sales demands less than the stock policy plus the constant profits for all demands equal to or in excess of the stock policy. Thus, each seasonal demand entry provides profit according to its frequency of occurrence, and the average profit is the weighted value of all these which may be expressed for any stocking policy equivalent to S_i in sales dollars as follows, noting that the profit per sales dollar (Z/S) is considered constant:

$$E(Z) = \frac{\int_0^{S_T} Zg(S)\, dS}{\int_0^{S_T} g(S)\, dS} = \int_0^{S_i} (S - C_i)g(S)\, dS$$

$$+ \int_{S_i}^{S_T} (S_i - C_i)g(S)\, dS \qquad (15\text{-}6)$$

where $E(Z)$ = the expected or average profit[3]

        S = sales dollars; S_T is maximum demand and S_i is sales dollars equal to value of items stocked

        C = cost dollars; C_i is all costs for stocking and selling S_i dollars worth of sales

  $g(S)\,dS$ = the probability that S will be in the interval dS and is used for weighing each component of profit; the denominator term $\int_0^{S_T} g(S)\,dS$ is numerically equal to one, and was omitted when the equation was split into two parts

and Z = profit dollars

Equation (15-6) is the basic relation for the profit and may be solved stepwise as demonstrated in Tables 15-2 and 15-3 when data for the distribution of demands are known. However, these tedious calculations can be avoided by determining the optimum conditions under which Eq. (15-6) maximizes. This maximum expected profit occurs when the first derivative of Eq. (15-6) with respect to S_i is set equal to zero and the solution completed. The mathematics is complex because one limit of integration contains a parameter of the equation which differs from the variable of integration. The resulting manipulations yield at the optimum over the range zero to S_i:

$$\int_0^{S_i} g(S)\,dS = P(S_i)_{\text{opt}} = (S_i - C_i)/S_i = Z_i/S_i \qquad (15\text{-}7)$$

Equation (15-7) states that the maximum profit in a stocking program occurs when the cumulative probability of S demands or less (such as given in Fig. 15-4) equals the ratio of profit margin to sales dollars. Here the maximizing of the profit follows the procedure for a regular economic balance with stochastic considerations added. The long run economic balance is formulated to compare the profits returned (by having stock available) with the loss resulting if there is an overstocking. In the preceding problem the value of $Z_i/S_i = (2400 - 1728)/2400 = 0.28$. If the probability curve that applies in this case were available, the probability of 0.28 for S or less would occur at a stocking policy of 2000 items, thus confirming Fig. 15-5.

*   *   *   *   *   *   *   *

---

[3] This must be written in two parts because the relation between S and Z is not continuous over the range of 0 to S_T as was noted above. The relation $g(S)dS$ in both Eq. (15-6), and Eq. (15-4) as well, represent areas of differential width dS under a specific distribution curve such as Fig. 15-3 whose total area is unity. Thus, over a finite range of sales dollars from 0 to any S the expected profit over that range is given by the sum of all the differential expected profits (for $S + dS$) over the range, and this is the average or expected value (weighted for their probability of occurrence) up to any limit of S selected. Note that the denominator $\int_0^{S_T} g(S)dS$ is equal to unity, and thus only the numerator must be considered.

INCREMENTAL ANALYSIS WITH UNCERTAINTY

It has been demonstrated (Chapters 7 and 9) that maximum over-all profit results when the sales dollars for one additional unit just equals the cost for that unit. When uncertainty enters, however, the situation is different because allowance must be made for the uncertainty. Thus, for the general case at the optimum policy the *expected* or average sales value for each additional unit of sales will equal the average cost for the sales unit. Where the average cost is a constant, its value is not subject to fluctuation and the incremental equation may be written:

$$[P'(S)] S = C \qquad (15\text{-}8)$$

where $P'(S)$ = the probability of an amount of sales S or more

S = total sales dollars
C = total cost dollars

This expression states that at the optimum, the incremental gain from one additional sale (that is, the probability of that sale times its sales value) must equal the incremental cost for the unit of sales. It is implied that the lowest probability of making an additional sale is the cumulative probability value for that sales unit or more, that is, the concept of incremental analysis provides that one additional sale shall be made. To make this sale an amount S at least must be sold before the critical value of sales corresponding to $P'(S)$ is attained.

The probability of not selling the next unit above the critical one is its cumulative probability $P(S)$ or less. The two cumulative probabilities are related by $P'(S) = 1 - P(S)$ as noted previously and an alternative equation for Eq. (15-8) may be written as[4]:

$$[1 - P(S)] S = C \qquad (15\text{-}9)$$

Equation (15-8) can be manipulated algebraically to yield:

$$P'(S) = C/S \qquad (15\text{-}10)$$

Equation (15-9) can be manipulated to give:

$$S - P(S)S = C$$
$$P(S) = (S - C)/S = Z/S \qquad (15\text{-}7)$$

---

[4]These statements are analogous to the "an event *will* occur or *will not* occur" situation which is described by a binomial distribution, and therefore multiplying the magnitude of an event by its probability as in Eq. (15-8) gives the *expected value* to be used.

where the optimum probability result is the same as that obtained by differentiating the profit equation as shown in the earlier derivation of Eq. (15-7).

The selection of either Eq. (15-7) or (15-10) is entirely a matter of choice, depending upon the form in which the data are available. If the data are on a plot like Fig. 15-1a, Eq. (15-7) is directly applicable; for data in the form of Fig. 15-1b Eq. (15-10) applies. However, either probability can be obtained readily from the other.

It also should be noted that the evaluation in Eq. (15-7) may take several other alternative forms. For example, the entire analysis starting with Fig. 15-3 could be made on a basis of unit values rather than dollars of sales. In this type of analysis, both the numerator and denominator of Eq. (15-7) would be divided by n units to give an expression in terms of sales dollars, costs, and margins per unit rather than in total dollars. This has some advantages where unit costs and unit sales values are constant regardless of the number of units involved. Thus, Eq. (15-7) may be written for the sought probability in terms of units when unit costs and sales price are given. When both unit costs and sales price are independent of the units of production or when averaged values are used, the equation becomes:

$$P(S)_{\text{opt}} = \frac{Z/n}{S/n} = \frac{m}{m + c} = \frac{p - c}{p} \qquad (15\text{-}7a)$$

where m = profit margin per item (difference in sales price and cost)
c = cost per item
and p = the sales price per unit

A variation of the development of Eq. (15-7a) can be based on the incremental concept. Thus, if the profit or margin (instead of revenue or sales) per unit is considered as an incremental gain, then the corresponding incremental loss is the cost per unit for unsold items, and the following relation is valid:

$$[P'(n)]m = [1 - P'(n)]c \qquad (15\text{-}8a)$$

where the terms have the same meaning as previously discussed. The term $[1 - P'(n)]$ is the probability of not selling n units or more at a unit cost c. Thus, it is the probability, $P(n)$, of selling n units or less, and Eq. (15-8a) may be rewritten:

$$[1 - P(n)]m = [P(n)]c \qquad (15\text{-}8b)$$

These equations differ from Eq. (15-8) because the incremental gain is in terms of the margin or profit rather than in terms of total sales or revenues. However, manipulating Eq. (15-8b) algebraically to solve for $P(n)$, gives exactly the same result obtained from Eq. (15-7a).

TWO UNCERTAINTY EXAMPLES

Management has made an analysis of potential sales for a perishable product which shows a cumulative probability curve as in Fig. 15-4 where S is sales dollars. The product is a complete loss due to spoilage if it is not sold. Accordingly, a management policy has been set up which requires that only the optimum quantity based on a stochastic analysis should be ordered for sale, even though a maximum of $10,000 worth is possible.

Problem 1

What is the optimum quantity for purchase based on Eq. (15-7)? Assume that the cost of each unit is $1.20 and that it sells for $1.60 with a profit of $0.40 per unit. For maximum expected profit, how much should be stocked?

Solution

Based on incremental analysis where the probability of one more sale times the value of that sale must equal the profit for one more sale, the analysis has shown in Eq. (15-7a) that:

$$P(S) = (1.60 - 1.20)/1.60 = 0.25$$

The result states that at a probability of 0.25 for S sales dollars or less, the maximum expected profit in the long run would be obtained. This probability corresponds to a perishable stock equivalent to $2300 worth of sales from Fig. 15-4 for an average profit of $2300(0.40/1.60) = 575. However, this profit is the average that would be obtained from a large number of seasonal sales. Some seasons (25 per cent of them) would yield less than $2300 of sales, with losses for unsold items if more were stocked. In some seasons (that is, 75 per cent of the time) more sales would have been possible with profits greater than $575. Considering all of these factors the maximum average profit for all seasons will result when only $2300 worth of perishable product is stocked.

A second type of problem is often encountered which is helpful in understanding relations between "expected" values. Consider a situation where management has an opportunity to buy a stock at a good price now for sale during a later season where the probabilities of sales follow Fig. 15-4. During the height of the season any restocking will necessitate paying a premium price to the manufacturer. This is an economic balance problem where the constant costs of storage of a stock at a fixed level are balanced against the variation in premium costs due to the variation in demands. The balance is demonstrated in Fig. 15-6. The preseasonal stocking costs increase linearly with the amount stocked. However, because of the variation in demands, during some seasons there may be insufficient preseason stock to meet the demand, thus making it necessary to purchase additional stock at a premium cost. Similarly an excess of stock may be available for some seasons; however, it is not necessarily a loss, since it can be sold at a future time with the holdover cost being added to the carrying cost. (This latter complication can be considered as part of the long run average carrying cost to simplify the analysis.) The premium charges for the excess stock decreases with an increased preseason inventory, and the sum of the two costs goes through a minimum as demonstrated. Note that each abscissa point in Fig. 15-6 represents one stocking policy and that for a distribution of demands there are some that are less and some that are more than the amount stocked. The expected or average number greater than those stocked for each stocking policy must be determined from distribution data in order to compute the premium costs for the expected excess number.

For any given distribution of demands, there will be an expected or average value from the probability distribution, which will be the long run number of sales per season. Where the sales price per unit does not vary with sales volume, the long run expected revenue per season will be constant. Thus, the policy that gives the minimum cost for the season in the long run will also give the maximum profit.

In general, the analysis is based on Eq. (15-7) where the only pertinent costs to consider are the carrying charges for preseason stock and the premium charges for items purchased during the season. The purchase price is a constant that applies to all items alike and merely raises the position of the curves (but not their shape) in Fig. 15-6. It is not a loss if the items are not sold since this is not the perishable case. In the economic analysis, a stocking policy that saves paying the premium price may be considered as a yield or revenue equivalent to the premium. Thus, the premium saving may be conceived as sales revenue per item and so used for p in Eq. (15-7a) where c is the carrying charge per item. This is an incremental analysis where the long-run average premium saved allowing for varying demands) must equal the average stocking cost at the optimum preseason stocking policy.

Problem 2

Suppose it costs management $1.20 to stock an item at present which would entail a premium cost of $3.60 per item if purchased at a future date (because of anticipated wage and material increases) during the height of a season where the total sales distribution is expected to follow Fig. 15-4 up to 10,000 items. Calculate (a) the economic optimum quantity of stock required at present, and (b) compare the optimum stocking policy of (a) for expected costs with other possible policies if the regular purchase price is $11.40 at the present time.

Solution

Part a. Utilizing a form of Eqs. (15-8) and (15-9) for the optimum probable incremental unit of sales or more yields:

$$[1 - P(n)]3.60 = 1.20$$

$$P(n) = (3.60 - 1.20)/3.60 = 0.67$$

The probability $P(n)$ of n sales or *less* is used for convenience, since the available data in Fig. 15-4 are on this basis. From Fig. 15-4 a cumulative probability of 0.67 corresponds to 4800 items. This means that the long run optimum policy will occur if 4800 items are stocked and if other purchases are made at the premium price when any one season demand exceeds 4800 items.

Part b. It will now be demonstrated through detailed step by step calculations that the solution in part *a* is the best policy.

The 0.67 cumulative probability in part *a* means that 67 seasons or sales programs out of 100 will sell 4800 items or less (or conversely, $100 - 67 = 33$ seasons out of 100 should have a demand for more than 4800 items). Thus, the solution involves the cost for the optimum which will be the cost of stocking 4800 items (regardless of how many are sold) plus the cost at a premium price for the expected items that will be sold in excess of 4800. Similarly, for any program the total cost will be the cost for those stocked plus the cost for additional expected sales. The latter will vary with each amount stocked as will be demonstrated in the calculations shown in Table 15-4a and 15-4b.

The results in Table 15-5 show that an optimum policy of 4800 items stocked at $1.20 each plus expected sales of 566 items at a premium of $3.69 will cost $7798 above the purchase price. In order to demonstrate that this is an optimum consider some arbitrary management policy of say stocking 3000 items plus expected sales above 3000 obtained by paying a premium cost for the excess sales. These calculations are shown in Table 15-4b where 1465 items above the 3000 would be expected to be sold for a total cost of $(3000 \times 1.20) + (1465 \times 3.60) = \$8874$ above the purchase price.

Finally, consider a policy of stocking 7000 items which yields an additional 115 items expected above 7000 at a cost of $(7000 \times 1.20) + (115 \times 3.60) = \$8814$ above the purchase price. Thus, the theoretical optimum procedure of 4800 items is the best policy based on a stocking cost versus a premium cost for purchase when needed. A summary of the results is given in Table 15-5 together with the costs for stocking none or all of the items; the data are plotted in Fig. 15-6.

TABLE 15-4a. Calculations for Uncertainty Example Using Optimum Policy of 4800 Items[a]

Sales Range above 4800 Items	Probability for Range $g(n)\Delta n$	Incremental Additional Items with Probability in Column 2[b]	Probable Additional Items[c]
4801–5000	$0.71 - 0.67 = 0.04$	$\dfrac{200 + 0}{2} = 100$	4
5001–7000	$0.90 - 0.71 = 0.19$	$\dfrac{2200 + 200}{2} = 1200$	227
7001–9000	$0.99 - 0.90 = 0.09$	$\dfrac{4200 + 2200}{2} = 3200$	288
9001–10,000	$1.00 - 0.99 = 0.01$	$\dfrac{5200 + 4200}{2} = 4700$	$\dfrac{47}{566}$

[a] In principle the expected value of n items *above* the stocked number n_i is the x value for the centroid of area as stated by Eq. (15-4), for the data above n_i. This expected value is thus weighted by the relative number of items and their distribution above n_i. The calculations may be performed in several ways of which Table 15-4a is but one example. The values of $g(n)\Delta n$ in Table 15-4a are obtained from the curve in Fig. 15-4. The values of n in column 3 of Table 15-4a are referred to a zero datum at the amount stocked of 4800.

[b] Midpoint of range above 4800 referred to a 4800 datum.

[c] Column 2 times column 3 to give incremental additional sales. Since the probabilities in column 2 are based on the total of all items, the entries in column 4 are automatically corrected for only the proportion of those items above the 4800 stocked.

TABLE 15-4b. Calculations for Uncertainty Example Using An Arbitrary Policy of 3000 Items in Inventory[a]

Sales Range above 3000 Items	Probability for Range	Incremental Additional Items with Probability in Column 2	Probable Additional Items
3001–5000	$0.71 - 0.33 = 0.38$	$\dfrac{2000 + 0}{2} = 1000$	380
5001–7000	$0.90 - 0.71 = 0.19$	$\dfrac{4000 + 2000}{2} = 3000$	570
7001–9000	$0.99 - 0.90 = 0.09$	$\dfrac{6000 + 4000}{2} = 5000$	450
9001–10,000	$1.00 - 0.99 = 0.01$	$\dfrac{7000 + 6000}{2} = 6500$	65

[a] The calculation is made in same manner as for Table 15-4a except 3000 items will be inventoried.

TABLE 15-5. Summary of Results for Uncertainty Example

Stocked Items	Probable Additional Items Sold	Average Excess Costs[a] Dollars
none	3930	14,184
3000	1465	8874
4800	566	7798
7000	115	8814
10,000	0	12,000

[a] Costs for stocking items in column 1 at $1.20 each plus premium costs of $3.60 each for additional items probably sold.

The objective of this table and Fig. 15-6 is to demonstrate a number of points concerning economic analyses.

(1) In the long run (that is, a large number of sales programs) the expected or average number of sales will be 3930. This is the result from Eq. (15-4) starting at zero inventory. Any specific single program may have sales above and below this number, varying from zero to 10,000.

(2) The numbers in the second column represent, in the long run, the number of additional items expected to be sold when sales are in excess of the amount inventoried in column 1.

(3) On the average, the excess costs (sum of carrying charge and premium price) will go through a minimum when 4800 items are inventoried (at a cost of $1.20 each and 566 are purchased as needed

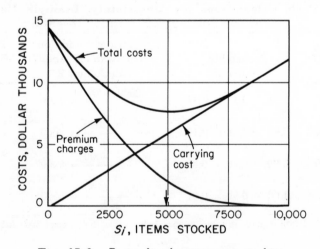

Fig. 15-6. Best policy for preseason stocking.

at a premium cost of $3.60 each) compared to any other policy of inventory and purchase as needed as shown by the results in the third column. In 67 per cent of the programs, the sales will be 4800 items or less but 33 per cent of the programs will require purchases at a premium to meet sales demands in excess of 4800 items.

(4) Since the long run average is 3930 items, the unit costs will go through a minimum at the same optimum stocking policy as for minimum total excess costs. Then if the unit excess costs are added to the purchase price of $11.40 to obtain the total unit costs the economic analysis is unaffected, since the minimum total unit cost will also occur at the same optimum value.

COMMENTS ON UNCERTAINTY ANALYSIS

There are numerous variations of Eqs. (15-7) and (15-9) in their applications. These expressions permit management decisions which are more reliable than guesses when the cumulative probability curve is available. It will be noted that Fig. 15-3 is for a distribution of sales less than some fixed maximum value, whereas Fig. 15-2 is a distribution of sales over a period of 50 days. If a sample period of 100 days had been used a similar distribution would be expected, but all quantities would be expected to be about double.

A variation of Eq. (15-10) is available since C and S can be on a unit basis; if an expression in terms of profit or margin per unit is desired, Eq. (15-10) can be written in terms of cumulative probability of number of units n or more, $P'(n)$, as:

$$P'(n)_{opt} = \frac{c}{m + c} = c/p \qquad (15\text{-}10a)$$

Note that:

$$P(n)_{opt} = 1 - P'(n)_{opt} = 1 - \frac{c}{m + c} = \frac{m}{m + c} = \frac{(p - c)}{p} \qquad (15\text{-}7a)$$

It is interesting to speculate on the various average values available that may be used in the solution of uncertainty problems. For example, from the data in Fig. 15-4:

(a) The median value of S is 3800 (half of number of items is above and half is below the median)—the 50 per cent point on the cumulative curve.

(b) The weighted average, expected value, or mean S is 3930. This differs from the value in (a) and (c) unless the distribution curve is symmetrical about the mean value.

(c) The mode (most frequent) S is approximately 3600 as the curve is drawn.

(d) The average of the range of values is 10,000/2 or 5000.

From these figures, it is obvious that numerous possible values can be selected as a basis for stocking decisions; these values will give different final results, even though all may be justified on some basis or other. The use of probability techniques employing the "expected value" *reduces the chances and magnitude for error in a management policy* in the long run. However, where the available probability data do not reflect the true real world situation, the stochastic approach will also result in errors. For example, where changes in population, buying habits, or quality of product invalidate the available distribution of sales data, corrections must be made in the cumulative probability plots if they are to be applicable and useful.

THE MONTE CARLO METHOD

The Monte Carlo method is a procedure by which random sampling methods are used to solve mathematical problems which have solutions otherwise very tedious or impractical to obtain. This method often involves the artificial generation of data representative of expected data which simulate the real world. A valuable procedure in economic analysis for generating simulated experience is the utilizing of random numbers as decimal values of the cumulative probability for the variable of interest. A value of the variable of interest is thus obtained, and a sequence of such values will have the same statistical characteristics as the distribution function employed to generate them.

For example, the curve in Fig. 15-4 may be considered to represent the actual distribution of sales units for a company. With such a curve known and with a table of random numbers available, a set of sales demands can be listed which will be distributed in the same manner as the original data from which the probability curve was constructed. For example, the data in column 3 of Table 15-6 represent a distribution of demands generated by the Monte Carlo method in the preseason stocking problem that was studied previously. When a table of random numbers (such as given in the Appendix D) are employed, the first two digits are considered as probability and by the use of Fig. 15-4 the sales units or less, n, are read directly. The random number tables are read in any manner desired (by rows, columns, skipping around, etc.) provided no subjective selection or rejection of certain numbers is made. The greater the number of trials the more closely the values obtained will correspond to the true distribution.

TABLE 15-6. Calculation by Monte Carlo Method of Expected Value of Sales in Example Problem With Zero Storage

Trial No.	Random Number	n Units[a]	Average[b] n	Storage Charges	Premium Charges[c], Dollars	Total Charges, Dollars	Average[b] Total Charges, Dollars
1	50	3850	3850	0	13,860	13,860	13,860
2	35	3100	3475	0	11,160	11,160	12,510
3	06	900	2617	0	3240	3240	9421
4	90	7000	3712	0	25,200	25,200	12,365
5	42	3400	3650	0	12,240	12,240	13,140
6	07	1050	3217	0	3780	3780	11,580
7	76	5500	3543	0	19,800	19,800	12,754
8	73	5100	3738	0	18,360	18,360	13,455
9	89	6900	4089	0	24,840	24,840	14,720
10	42	3400	4020	0	12,240	12,240	14,472
—	—	—	—	—	—	—	—
—	—	—	—	—	—	—	—
29	22	2300	3945	0	8280	8280	14,201
30	61	4400	3960	0	15,840	15,840	14,256
31	50	2400	3910	0	8640	8640	14,075
32	71	5200	3950	0	18,720	18,720	14,220
33	62	4500	3967	0	16,200	16,200	14,280

[a] As read from Fig. 15-4 at probability corresponding to random numbers in column 2.
[b] Running average for each added row.
[c] Since none are stocked, all items are obtained at a premium of $3.60 each.

Continuing the application to the stocking problem, one can use Table 15-6 to show the results for estimating the expected value of sales when no items are stocked. Note the final long run average after 33 trials is 3967 (but it fluctuated above and below this number), whereas in Table 15-5 this number was given as 3930 as computed from Eq. (15-4a). Similarly, the premium charges can be computed for any specific policy of stocking such as 3000 items, as shown in Table 15-7, and the combination of stocking charges and premium charges can be determined for each demand and then averaged. Thus, for all times that less than 3000 items are sold, there is no premium charge, and only the base storage cost of $3600 is considered. Here again the results are in good agreement with those in Table 15-4b where 1465 extra items were computed (compared to 1480 in Table 15-7) at an excess cost of $8874 (compared to $8929 in Table 15-7). Similar tabulations for various other conditions of a stocking policy can be made from the same set of Monte Carlo data with the results as shown in Table 15-8.

TABLE 15-7. Calculation by Monte Carlo Method of Expected Value of Sales in Example Problem When 3000 Items Are Stored

Trial No.	Random Number	n Units	Extra Needed	Storage Charges, Dollars	Premium Charges, Dollars	Total Charges[a] Dollars	Average Total Charges, Dollars
1	50	3850	850	3600	3060	6660	6660
2	35	3100	100	3600	360	3960	5310
3	06	900	—	3600	—	3600	4740
4	90	7000	4000	3600	14,400	18,000	8050
5	42	3400	400	3600	1440	5040	7452
6	07	1050	—	—	3600	3600	6810
7	76	5500	2500	3600	9000	12,600	7637
8	73	5100	2100	3600	7560	11,160	8078
9	89	6900	3900	3600	14,040	17,640	9140
10	42	3400	400	3600	1440	5040	8730
—	—	—	—	—	—	—	—
—	—	—	—	—	—	—	—
—	—	—	—	—	—	—	—
29	22	2300	—	3600	—	3600	9031
30	61	4400	1400	3600	5040	8640	9018
31	50	2400	—	3600	—	3600	8843
32	71	5200	2200	3600	7920	11,520	8927
33	62	4500	1500 1480[b]	3600	5400	9000	8929

[a] Cost for stocking 300 items at $1.20 each plus premium costs of $3.60 for the extra items needed.

[b] Average of extra items needed.

TABLE 15-8. Comparison of Expected Values in Example Problem By Mathematical Analysis and by Monte Carlo

Items Stocked	Expected Value Calculations		Monte Carlo Calculations[a]	
	Additional Items	Excess Costs[b]	Additional Items	Excess Costs[b]
none	3930	$14,184	3967	$14,280
3000	1465	8874	1480	8929
4800	566	7798	594	7898
7000	115	8814	97	8749

[a] Based on 33 trials.

[b] Sum of storage costs for items stocked plus premium costs for additional items required.

Thus, the Monte Carlo method has provided a great variety of results from one set of data. The agreement is good for the two methods, and if more than 33 trials by Monte Carlo had been made, a still better result (higher confidence) would be obtained.

This very elementary example of Monte Carlo has been used to demonstrate its application; the straight calculation of expected value as shown in Tables 15-2 and 15-4 is frequently less laborious. However, there are more elegant applications of the Monte Carlo which can be used to solve problems with a considerable saving of time and manpower.

COMMENT ON MONTE CARLO

The reason this procedure is useful is that each value for demands will be produced with about the same relative frequency in this process as in the world represented by the curve and they will appear in a completely random sequence. This is equivalent to marking a set of cards to represent sales in such a way that the relative frequency of numbers on the cards is the same as that given by the cumulative distribution curve. Based on the distribution given in Fig. 15-4, 50 per cent of the number would be less than 3800 and 50 per cent greater, 25 per cent would be less than 2600, and so on. Then random drawing of the cards, with replacement after each drawing, would be a Monte Carlo procedure and equivalent to the use of random numbers as probabilities as described above. In this way, an inexhaustible set of data can be generated which will simulate the results for which the probability curve was valid.

The Monte Carlo results have the same statistical characteristics as the statistical model used; therefore, the results simulate the real world only as well as the statistical model does. This is an incentive, where possible, to fit empirical data to a mathematical model arrived at by logical and technical considerations. For example, if a mean or average value for a set of observations can be assumed, it is possible to construct a calculated cumulative probability curve for demands by means of Eq. (15-3) if they are considered to follow a Poisson distribution as is illustrated in Table 15-1 and Fig. 15-1a. By means of random numbers, it is then possible to generate a series of demands for subsequent cost calculations. All of this information has been developed merely by knowing an average or mean value and by the assumption of a Poisson distribution.

Monte Carlo procedures are more useful than indicated by this simple example. In more complex problems they often provide solutions which cannot be carried through by available mathematics; in other instances, the person making the analysis does not have the mathematical background required. The methods are perhaps most useful where a sequence of more than one set of probabilities is involved.

For example, consider that the probability of occurrence for a series of events *A* is known and is followed by a series of events *B* whose probability distribution is also known. Then certain alternative policies may be proposed for solving the associated problems when both events *A* and *B* occur. Each alternative policy may have a different cost associated with its specific solution to the problem. From Monte Carlo procedures a large number of examples may be generated (without further experimental work) and the costs for the alternatives then compared to determine the best program.

It should be emphasized that best results with the Monte Carlo are obtained when a large number of observations are used. The 33 trials in Table 15-6 were probably much too few for a rigorous analysis, and the use of 200 to 300 trials would be a better procedure, using computers to save labor and time.

The value of the Monte Carlo method in providing an unlimited supply of simulated experience should be apparent for those cases where experimental data are expensive or considerable time is required to obtain them. The method often has a considerable economic advantage in the solution of complex problems, the techniques for which can be found in more advanced texts.

THE QUEUEING PROBLEM

One of the areas of management economy studies employing operations research techniques is in the application of probability to situations where some operation calls for a service and has to wait if the service function is busy. For example, in a large plant, where special tools are needed periodically, the tools may be held in a central location for withdrawal when needed. Each call for a tool requires that the attendant provide a service which takes time and accordingly a queue or line may form waiting for service. Other examples are checkout counters in grocery stores, ticket offices for railroad or airline service, docks for ships, toll booths on highways, and many others. The analysis has many applications and carries various names such as queueing, waiting lines, assignment of machines or personnel, repair of machines, accumulation, congestion, servicing, or classification.

Essentially, the economic analysis consists of an economic balancing of the cost for increasing service facilities against the saving in the cost of waiting time resulting from the increased service rate. The results of such an analysis are useful in helping management decide whether or not to increase the service rate or to reduce the calls for service.

The analysis requires the determination of the average or expected mean time that an operation is halted because of waiting time and the service time for any one call. Thus, the product of the number of calls in a given elapsed period times the waiting time for any one call gives the total time the operation is halted. From the cost of the operation per unit time the cost for the total down time can be computed.

When a servicing problem involves probability because the calls for service are considered to occur at random, then an expected or average condition for the queue that forms must be evaluated. In the analysis, the average queue length is desired, which means that all queue lengths occurring must be weighted by their respective probabilities to give the expected long-run average. In other words, the queue that forms varies in length because at certain times the number of calls for service exceed the capacity of the service facility to keep itself clear. From a distribution of these queue lengths, it is then possible to estimate the average value by the probability methods discussed earlier in the chapter.

The expected waiting time (queue time plus service time) is a function of the number of service facilities (channels) involved, the average arrival (or calling) rate and the average service (or completion) rate. Some of the underlying theory is quite complex, but the final economic equations are not difficult to apply. However, an elementary presentation of theory will be given here for the simple case of the queue length not being biased by the time of day for which the data apply. Another basic simplifying assumption is that the number of calls is made from a universe of infinite number of possibilities. (If a finite source of calls is considered, the number of possible calls remaining is continuously decreasing, which is a more difficult problem.)

The real value of queueing analysis in most management policy decisions is the determination of what action should be taken for the problem being considered. Thus, under the hypothetical situation where the calling rates and completion rates are constant and equal, all service channels are kept busy one hundred per cent of the time. This condition is approached in many process industries by automatic devices (automation) where the calling or completion rates are controlled to prevent accumulations. The same technique is applied by management where arrivals or calls for service are encouraged by special sales during dull periods and discouraged when calls exceed the capacity for service. Alternatively, management can take steps to decrease or increase the service facility operation whenever an economic analysis so indicates.

Where the arrival or call rate exceeds the service completion rate, the queue will grow to infinity, and all channels will be busy all of the time. The queue will in reality adjust itself if it involves people who will not wait, but if the callers are machines, management must take some action to prevent the loss of waiting (that is, idle) machine capacity. An infinite queue prevails, whether or not the calling rates and service times are constant or varying, provided the average calls per unit time exceed the average completion rate. If the average completion rate exceeds the average calling rate a queue may or may not form. Management's problem is to adjust the service facility or needs for servicing for the best economic results. For the case where the calling rates and completion rates are constant with the latter being larger, then the service facility will be idle part of the time and it can be closed down part time.

QUEUEING PROBLEM DIAGRAM

In Fig. 15-7 a simple queueing problem is diagrammed for a short period of 20 hours with service at a varying completion rate and the assumption that only one call is made at any time (no simultaneous call at any one instant). The diagram is for a textile mill operation where an operator is required to change the weaving machinery at periodic intervals because production lots are completed. The next lot arrangement must then be set up.

The calls are identified in consecutive order, and the shaded blocks indicate a call being serviced. The open blocks show a call (identified by

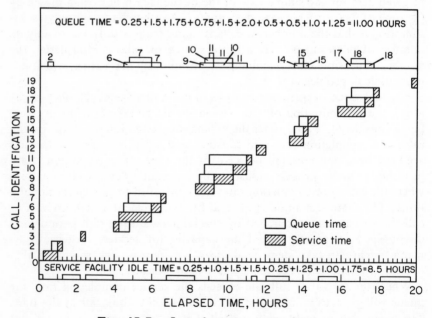

Fig. 15-7. A single station queue diagram.

number) waiting, and open blocks in a vertical pile indicate queues of varying lengths which are illustrated in the upper portion of the diagram. Starting at zero time, one finds that the first calls occurs at 0.25 hour and is serviced in 0.75 hour. In the meantime, another call for service occurs at 0.75 hour but has to wait until 1 hour for service; thus, a queue of one is formed. The third call occurs after 2.25 hours and is serviced in 0.25 hour with no queue or waiting period, and so forth for subsequent calls. At 8.5 hours a queue of one is formed; a new call for service occurs at 8.75 hours, which adds to the queue; this call is followed by another arrival at 9 hours to form a queue of 3. From such a diagram set up using real world data, it is possible to make the economic analysis. For this purpose, some base unit time period such as an hour, a day, a year, etc., is selected.

The first items of interest are the average service rate and average arrival rate. These are computed readily from the chart data. For example, in an elapsed time of 20 hours, the service time is 20 hours minus 8.5 hours of service-facility idle time, or 11.5 service hours for 19 calls serviced. Thus, the average service rate, S, is $19/11.5 = 1.65$ arrivals serviced per hour. There are 19 calls or arrivals in 20 hours, and therefore the arrival rate, A is $19/20 = 0.95$ arrivals or calls for service per hour.

A most important item of information is the time lost by the arrivals or calls waiting in the queue. The mathematical relations as a function of the calling rate and service rate are developed and defined in the next sections but can be demonstrated from Fig. 15-7. For example, the total lost time spent in the queue is shown at the top of the diagram to be 11.0 hours in 20 hours, or $11.0/20 = 0.55$ total queue hours per hour of time. The numerical value can be greater than unity.

A desired item for calculations is the total hours of queue time plus service time for all calls, since this is the amount of time the caller is nonproductive. As shown for the diagram for a 20 hour period, this total time is 11.0 hours of queue time plus 11.5 hours of service time or $(11.0 + 11.5)/20 = 22.5/20 = 1.13$ wait-hours per hour of elapsed time.

As will be discussed in the next sections, the long-run average values of calling rates and service rates can be utilized to obtain the desired information for economic analysis. For example, the queue time plus service time is given by the expression, $A/(S - A)$, or from the data here this would be $0.95/(1.65 - 0.95) = 1.36$ wait-hours per hour of elapsed time. The values of 1.36 and 1.13 wait-hours per hour are considered to be in agreement, since the sample of data based on Fig. 15-7 is a relatively small one. With an extended chart a better distribution of the necessary service times and idle times could be obtained. It will be shown that where the calls follow a Poisson distribution it is only necessary to count the arrivals over a total period and determine the average calling rate, A, in calls per unit time. Similarly, where the service times are not constant, it is only necessary to establish the average service rate, S, of completions per unit time in order to proceed to the economic analysis.

QUEUEING PROBLEM THEORY[5]

There are three main considerations in queueing which involve theoretical concepts and certain known information:

(a) the average arrival rate, A, is based on a Poisson distribution and the average service rate, S, is based on an exponential distribution.
(b) the probability distribution of the queue length and its expected value, and
(c) the conversion of the results in (a) and (b) to expected waiting time in the system.

In the following discussion using analytical procedures, there must be data available to ensure that their distribution does fit the requirements of the expressions assumed to apply for the real world problem.

The equations which will be developed are based on a few simplifying assumptions. First, the "state" of the system is defined as the number of calls in the system equal to the one being serviced plus the ones waiting in line to be serviced.

Second, the average state as established by probability analysis is not a function of any particular time interval; that is, the average queue length in the long run is equally correct for any one selected time period. Third, the probability for any other state can be computed as a fraction of the probability of the zero state, P_0, for no calls waiting in line or being serviced. Finally, P_0 can be a function of the average calling rate, A, and average service rate, S.

It is not necessary to know the theoretical background in order to use the following equations, but some explanation is in order to assist the reader who is interested.

The Average Arrival and Service Rates

It is necessary to know the time between calls or arrivals for calculation of the total waiting time. If A is the average number of calls per unit time, then by reasoning, the expected time between calls is $1/A$. From probability theory, it can be demonstrated that where calls are distributed randomly and fit a Poisson distribution the expected or average value of experimentally measured times between calls can be established to be

---

[5] R. Vaswani, lecture notes (adapted), North Carolina State College.

equal to $1/A$ by the methods developed in the early part of the chapter. Thus, the average time between calls, $E(t)$, for a real problem can be obtained experimentally by measuring the average number of calls per unit time and taking its reciprocal.

$$E(t) = 1/A = \frac{\text{total time}}{\text{total calls}} \qquad (15\text{-}11)$$

The time units may be in seconds, days, or other units.

The same reasoning applies for computing the average service time when the average servicing rate, S, is the number of calls divided by the units of time actually used for servicing (not the total elapsed time for the period).

The Expected Value of the Queue Length

The total waiting time for any call before it receives service depends upon the number in line waiting to be serviced and the service time. It is necessary, therefore, to know the average or expected number of calls waiting in the system. This is obtained from Eq. (15-12) when the individual probabilities of there being 0, 1, 2, 3 or n calls in a given time interval can be determined. The number of calls waiting plus the number being serviced (one in service for a single service facility) is called the state of the system and on the average is considered to be independent of the period of hour or day. Thus, at state 4 there are 3 calls waiting while one is being serviced, and the expression for the probability of this situation is P_4; the expression for any state is P_x.

By definition, the expected or long-run average number of items in the system $E(x)$ is equal to the sum of the products of each state times its probability, from zero to infinite states, or for the general case:

$$E(x) = \sum_{0}^{\infty} (xP_x) \qquad (15\text{-}12)$$

where x = the state, and
P_x = the probability of state x

In order to solve Eq. (15-12) it is necessary to evaluate the probabilities for each individual state starting with zero (that is no calls in the system). When P_0 is known, the probability at any other state x for calling rate of A per hour and a service rate of S per hour is given by the following equations:

$$P_x = (A/S)^x P_0 / (r! r^{x-r}) \qquad (x > r) \qquad (15\text{-}13a)$$

$$P_x = (A/S)^x P_0 / x! \qquad (x < r) \qquad (15\text{-}13b)$$

Thus, it is only necessary to compute P_0 and recognize that the sum of all state probabilities must equal unity.

The general equation for the determination of the reciprocal of the zero state probability is as follows for multiple channels (store clerks, counters, lathes, chemical reactors, etc.) where r is the number of channels or service facilities:

$$1/P_0 = \sum_{x=0}^{r-1} \frac{(A/S)^x}{x!} + \frac{(A/S)^r}{r! \left(1 - \dfrac{A}{rS}\right)} \qquad (15\text{-}14)$$

For the special case of a single channel ($r = 1$), the direct relation for P_0 is computed from Eq. (15-14) as:

$$P_0 = 1 - A/S \qquad (15\text{-}15)$$

Thus, the expected value for the state in a single channel from Eq. (15-12) becomes for a converging series:

$$E(x) = (A/S)/(1 - A/S) = \frac{A}{S - A} \qquad (15\text{-}16)$$

The units of $E(x)$ are, for example, total wait-hours per unit of time (in this case per hour). A wait-hour should be considered like a man-hour unit of labor.

Since P_0 is the probability of being idle, a use factor f_u may be defined as $f_u = 1 - P_0$, the fraction of the time the system is in use.

Calculation of the Expected Waiting Time

The expected waiting time, $E(T)_m$, for *each call for service* is the sum of the average waiting time in the queue plus the average time for service. The general relation involving probability of all states and the calling and service rates is given by the following useful equation for multiple channels:

$$E(T)_m = \frac{(A/S)^r}{(rS)r!} \times \frac{P_0}{\left(1 - \dfrac{A}{rS}\right)^2} + \frac{1}{S} \qquad (15\text{-}17)$$

The value of $E(T)m$ is in wait-hours per service call.

The first term on the left is the queue time or the time that any one call waits in line for service and $1/S$ is the average service time. The terms A and S are respectively the average number of calls per hour and the average number of service completions per hour; r is number of channels and P_0 is the probability of the system being in state zero (no other calls in the system).

For the special case of a single channel system where $P_0 = (1 - A/S)$ the preceding equation reduces to:

$$E(T) = \frac{1}{S - A}, \text{ wait-hours per service call} \qquad (15\text{-}18)$$

In any given interval of time selected as a basis (such as 1 hour, 1 day, or 1 week), the total wait-hours in the basic time interval would be the product of the wait time per call multiplied by the calling rate for the base time selected. Thus, if the calling rate is A calls per hour, then the total wait-hours per hour of elapsed time is $AE(T)$, or for a single channel:

$$AE(T) = \frac{A}{S - A}, \text{ total wait-hours per hour of elapsed time} \qquad (15\text{-}16)$$

It will be observed that this is exactly the same equation as for the average value of the state $E(x)$ given previously. This is as it should be when the rates and base time are in similar units, since the average length of the system divided by the average call rate should be the average time any one call stays in the system (that is, $E(x)/A = E(T)$ for both single and multiple channels). Thus, the general expression for multiple channels is:

$$E(x)_m = AE(T)_m, \text{ total wait-hours per hour of elapsed time} \qquad (15\text{-}16a)$$

These equations provide the working relations, as summarized in Table 15-9, for handling observed data. For simple servicing problems, it is only necessary to know the calling rate, A, and the service rate, S, which may be taken by observation or obtained by estimation as an approximation to the real world situation when no data are available and some answer must be given. This will be demonstrated in the following sections.

SIMPLE APPLICATIONS OF QUEUEING EQUATIONS

The application of the equations may be demonstrated by considering the data available from Fig. 15-7 and the relations in Table 15-9. Thus, the arrival rate, A was found to be 0.95 calls per hour. The expected

TABLE 15-9. Summary of Equations for Queueing Problems For Single Channel System with an Infinite Universe of Calls[a]

Description	Expression	Equation No.
Expected queue time plus service time per call	$E(T) = \dfrac{1}{S - A}$, wait-hours per service call	(15-18)
Expected *total* queue time plus service time for all calls in a given time period; this is also the expected state of the system	$E(x) = AE(T) = \dfrac{A}{S - A}$, total wait-hours per unit of time	(15-16)
Expected queue time only per call	$E(T_q) = E(T) - 1/S = \dfrac{A}{S}\left(\dfrac{1}{S - A}\right)$ queue-hours per service call	—
Expected *total* queue time only for all calls in a given time period	$AE(T_q) = \dfrac{A^2}{S}\left(\dfrac{1}{S - A}\right)$, total queue-hours per unit of time	—
Probability of zero state	$P_0 = 1 - A/S$	(15-15)
Probability of state x	$P_x = (A/S)^x P_0$	(15-13)

[a] For multiple channels see text; for the finite universe of calls special tables are needed which can be found in more advanced texts.

waiting time, including service for a single channel arrangement, is computed from Eq. (15-18):

$$E(T) = \frac{1}{1.65 - 0.95} = 1.43 \text{ wait-hours per service call}$$

Thus, the total hours of idle time for the callers is $AE(T) = 0.95(1.43) = 1.36$ wait-hours per hour of elapsed time for the operation. This is the expected or average state of the system, which means that on the average there are 1.36 calls in the line or being serviced.

The probability of any state is given by Eq. (15-13), and for a single channel at state 3 (two in queue, one being serviced) the probability is:

$$P_3 = (0.95/1.65)^3(1 - 0.95/1.65) = 0.08$$

For a two channel system with the same arrival and service rates as before, the corresponding waiting time can be determined from Eq. (15-17); this requires computation of P_0 from Eq. (15-14). The determination is as follows:

$$1/P_0 = \frac{(0.95/1.65)^0}{0!} + \frac{(0.95/1.65)^1}{1!} + \frac{(0.95/1.65)^2}{2\left(1 - \dfrac{0.95}{2 \times 1.65}\right)}$$

$$= 1.000 + 0.575 + 0.231 = 1.806$$

$$P_0 = 0.555$$

$$E(T) = \frac{(0.95/1.65)^2}{(2)(1.65)(2)} \times \frac{0.555}{\left(1 - \dfrac{0.95}{2 \times 1.65}\right)^2} + \frac{1}{1.65}$$

$$= (0.05)(0.78) + 0.606 = 0.645 \text{ wait-hours per service call}$$

A QUEUEING PROBLEM EXAMPLE

Problem

A distributing operation operates several delivery trucks which are loaded and dispatched from one service center at an average rate of 2.25 per hour. The trucks arrive at an average rate of 2 per hour, 24 hours per day, 5 days per week. The cost of truck operation is estimated at $3 per hour per truck, and the loading facility is not otherwise productive when idle (that is, there is no special charge for its use when in this service because it is out of service for another operation). It is proposed to install a special loading facility at a cost of $149,000, which will increase the service rate to 2.5 trucks per hour. (a) Is the rate of return obtained worth the cost, if the value of money is considered to be 20 per cent at an estimated life of 10 years? (b) Would it be better to spend $200,000 for a second service facility which has the same loading capacity as the original one?

Solution

Part a. The solution is straightforward with Eq. (15-18) and perhaps could be seen intuitively without the elaborate previous calculations for the single service facility situation; for multiple service centers the solution is not so simple, although Eq. (15-17) is easily applied. Thus, from Eq. (15-18) the cost for the present operation is the average total idle time of each truck for waiting and servicing. From Table 15-9, the necessary calculations are:

$$E(T) = \frac{1}{S - A} = \frac{1}{2.25 - 2.0} = 4 \text{ wait-hours per call}$$

The total wait-hours per hour, therefore, are $4A = 4 \times 2$, or 8 wait-hours per hour of elapsed time.

In one year, there are $24 \times 5 \times 52 = 6240$ hours. The total queue time plus service time for all calls would be:

$$8 \times 6240 = 49,920 \text{ truck hours idle per year}$$

After installation of the special loading facility, the idle time would be:

$$\frac{1}{2.5 - 2.0} \times 2 \times 6240 = 24,960 \text{ truck hours idle per year}$$

or a saving of:

$$(49,920 - 24,960)(3) = \$75,000 \text{ per year}$$

In this example, the service facility is not otherwise productive, and there is no lost other production during the period it is servicing. (This comment is included since in certain queueing problems the value of this lost production would have to be charged against the service operation).

Continuing the problem solution by applying the economic analysis to determine the earning rate, i, for an annual saving, R, of \$75,000 on an investment of \$149,000, one finds that the present worth factor from Eq. 2-4 is:

$$a_{n/i} = P/R = 149,000/75,000 = 1.98$$

At 10 years, from the table in the Appendix C or Fig. 2-2, this result corresponds to an i equal to about 50 per cent, which is above the 20 per cent the company considers necessary and therefore the investment would be made.

Part b.   For two channels with the same arrival and service rates as before, the expected waiting time is computed from Eq. (15-17) for $r = 2$ channels. However, that equation requires calculation of the probability of state zero, P_0, from Eq. (15-14):

$$1/P_0 = \frac{(2/2.25)^0}{0!} + \frac{(2/2.25)^1}{1!} + \frac{(2/2.25)^2}{2!\left(1 - \dfrac{2}{2 \times 2.25}\right)}$$

$$1/P_0 = 1 + 0.89 + 0.71 = 2.60$$

$$P_0 = 0.384$$

Referring to Eq. (15-17) gives:

$$E(T) = \frac{(2/2.25)^2}{(2)(2.25)2!} \times \frac{0.384}{\left(1 - \frac{2}{2 \times 2.25}\right)^2} + \frac{1}{2.25}$$

$$= 0.099 \times 0.69 + 0.44 = 0.51 \text{ wait-hours per call}$$

The total idle time would be $(0.51 \times 2)(6240) = 6365$ hours.

This is a saving of $49,920 - 6365 = 43,555$ truck hours per year at $3 per hour, or $130,665 annually on a $200,000 investment. This gives a present worth factor of $200,000/130,665 = 1.53$, which for 10 years is equivalent to an earning rate of about 65 per cent. Thus, the installation of a new duplicate channel is a better decision than modifications to improve the service rate with one channel.

MONTE CARLO METHOD APPLIED TO QUEUEING
* * * (*Optional*) * * *

This section deals with certain aspects of probability analysis as applied to the queueing problem. It is presented as a matter of interest and may be omitted for classroom work without affecting the presentation of the other sections of the chapter.

The objective of this presentation is to demonstrate the application of certain theoretical principles to the real world situations likely to arise as illustrated by Fig. 15-7. For example, consider a set of observed data in connection with calls for service, where the average rate of calls, A, is given by the ratio of total calls to total time, say 3 per second (that is, $A = 3$). It can be developed mathematically for any selected interval of time between arrivals (zero second, one second, two seconds, etc.) that the distribution of the number x of random arrivals in those intervals follows a Poisson distribution described by a relation similar in form to Eq. (15-2):

$$g(x) = (At)^x e^{-At}/x! \tag{15-19}$$

where $g(x) = $ the probability of x calls for service in one selected time interval, t
and $\quad A = $ average rate of calls for service

Note that there would be a separate and different distribution plot for each size (second, hour, day, etc.,) of time interval, t, that is selected as a basis for the probability distribution analysis.

On the basis of the preceding, it is stated without proof that the probability distribution function, $g(t)$, of time intervals between calls can be expressed as a continuous function:

$$g(t) = Ae^{-At} \tag{15-20}$$

which gives the relative distribution of arrivals on a continuous basis as calculated in Table 15-10 and plotted as the solid line in Fig. 15-8. It should be noted that this distribution density function differs in appearance from that in Fig. 15-3 because the functions describing the frequency distributions are different. The histogram plot shown in the figure is the result of a Monte Carlo treatment which will be developed next. The good agreement for the calculated and Monte Carlo results should be noted.

TABLE 15-10. Comparison of Eq. (15-20) Results With Those Generated by the Monte Carlo Method

Arrival Interval, t, seconds	Calculated $g(t)$ by Eq. (15-20)	Monte Carlo Results				
		Distribution Interval	Distribution f	$f/\Sigma f$	Cumulative Probability	Height of Histogram[a]
0	3.00	0.00–0.09	69	0.23	0.23	2.30
0.1	2.22	0.10–0.19	63	0.21	0.44	2.10
0.2	1.65	0.20–0.29	38	0.13	0.57	1.30
0.3	1.22	0.30–0.39	30	0.10	0.67	1.00
0.4	0.91	0.40–0.49	31	0.10	0.77	1.00
0.5	0.68	0.50–0.59	24	0.08	0.85	0.80
0.6	0.48	0.60–0.69	11	0.04	0.89	0.40
0.7	0.37	0.70–0.79	10	0.03	0.92	0.30
0.8	0.27	0.80–0.89	5	0.02	0.94	0.20
0.9	0.20	0.90–0.99	6	0.02	0.96	0.20
1.0	0.17	1.00–1.09	3	0.01	0.97	0.10
1.1	0.11	1.10–1.19	2	0.01	0.98	0.10
1.2	0.08	1.20–1.29	4	0.01	0.99	0.10
⋮	⋮	⋮	⋮	⋮	⋮	⋮
∞						
			300	1.00	1.00	

[a]Computed from column 5 entries divided by interval width of 0.10 second to obtain the histogram height corresponding to column 2 for the continuous function. (This operation is the reverse of obtaining the area under the probability distribution curve.)

The Monte Carlo technique to generate a set of data was employed by picking 300 four digit numbers from a suitable table of random numbers. The selection of random numbers requires that the average calling rate in calls per second must be numerically equal to 3. Thus, if the range of random numbers (that is, the range for elapsed time considered) is to extend to 100 seconds, the number of calls must be 3 × 100, and 300 random numbers are used. Since it is desired to calculate the time between calls to two decimal places, a four digit random number will be employed. Thus, a number 4688 represents a call that occurred at 46.88 seconds.

These 300 numbers were considered to represent arrival times, and they were then arranged in numerical sequence, as noted in Table 15-11, so that the times

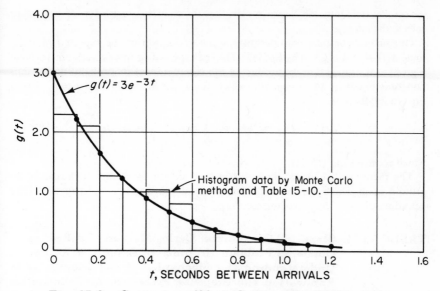

Fig. 15-8. Comparison of Monte Carlo and Eq. (15-20) results.

between arrivals could be determined. The distribution of the times between arrival was then established for the intervals as listed in columns 3, 4, and 5 of Table 15-10.

These times between arrivals, as found in Table 15-11, were sorted and counted and the frequency distribution established as listed in column 4 of Table 15-10. The cumulative probability distribution as shown in column 6 of Table 15-10 was then found.

The cumulative probability in column 6 represents the area under the probability density function plot, and therefore if the values in column 5 are divided by the interval width of 0.10, the height of the histogram for each interval can be determined to develop a probability distribution plot for the times between arrivals. This was done in column 7 of Table 15-10 and the results are plotted in Fig. 15-8 to demonstrate the fit of the calculated curve and the histogram as determined from Monte Carlo data.

Another theoretical point should be noted. Since the area under the curve in Fig. 15-8 is the cumulative probability for values of t or less, $P(t)$, then the area can be found analytically from Eq. (15-20) and Eq. (15-21):

$$P(t) = \int_0^t g(t)\, dt = \int_0^t Ae^{-At}\, dt = 1 - e^{-At} \qquad (15\text{-}21)$$

It will be noted from the configuration of Fig. 15-8 that the biggest part of the area under the curve (that is, the major portion of the total probability) occurs at low values of t. This is as it should be since from intuition it would be expected for the smaller time intervals between calls that the chances are greater for one or no calls in the interval.

One final theoretical consideration is the evaluation of the expected intercall time, $E(t)$, that was given by Eq. (15-11). The expected value of a random variable is given as the integral of the product of the variable times its probability distribution function over the range for which the expected value is desired. From Eq. (15-20) then:

$$E(t) = \int_0^\infty tg(t)\ dt = \int_0^\infty tAe^{-At}\ dt = 1/A \qquad (15\text{-}22)$$

which confirms Eq. (15-11).

The theoretical relations discussed in this section form the stochastic background upon which are based the more useful waiting time and expected state equations given in the preceding sections.

TABLE 15-11. Steps in the Monte Carlo Method Applied to a Queueing Problem (In order to Assist the Reader in Understanding the Mechanics of the Monte Carlo Application the Stepwise Procedure Is Followed in Table 15-11; Note Only the Lowest 10 Random Numbers of the 300 Are Used in Column 3 of This Illustration)

Trial	Random Numbers[a]	Sequenced Numbers (Elapsed time)	Time, t, Between Arrivals
1		01.08	1.08
2		01.59	0.51
3	0108	01.89	0.30
4		02.13	0.24
5	9961	02.56	0.43
6		02.73	0.17
7		03.13	0.40
8	0256	03.28	0.15
9		03.38	0.10
10		03.50	0.12
.	.	.	.
.	.	.	.
.	.	.	.
300	0328	99.61	—

[a] Numbers shown in this column are only those that appear in the lowest 10 of column 3 except for the highest number read which was 9961.

* * * * * * * *

COMMENT ON THE QUEUEING PROBLEM

The presentation here was oversimplified purposely for an infinite universe of calls to demonstrate the principles involved, although it does approximate the real world situation for many problems. The principal difference for the real world case is for the finite number of calls where the expected waiting time equations are specific for each number possible. For this purpose tables are required and are generally available[6].

There are many variations of the queueing problem. These include such variations as parking lots and hotels; here, when service facilities are fully occupied, no queue forms because the customers are turned away. A general equation for these special probabilities is developed as before, with the probabilities for the states being summed only to r channels rather than to infinity for the steady state condition of no change in state during a small time interval. This leads to different expressions for P_0 and P_x. For the zero state, it can be shown that:

$$P_0' = \frac{1}{1 + A/S}, \text{ probability of state zero when system has a maximum of state one} \qquad (15\text{-}23)$$

The resulting general equation for any number r of service facilities is:

$$P_a' = \frac{(A/S)^a/a!}{\displaystyle\sum_{x=0}^{r} (A/S)^x/x!}, \text{ probability of state } a \text{ when system has a maximum of state } r \qquad (15\text{-}24)$$

where r is the number of service facilities. When P_a' is to be the probability of turning away calls at state r because all facilities are taken, then a is equal to r.

For large values of r and x in Eqs. (15-14) and (15-24), the mathematician multiplies the numerator and denominators by $e^{A/S}$ and obtains expressions that are more easily evaluated by tables[7] than by separate calculations:

---

[6] D. C. Palm, "The Assignment of Workers in Servicing Automatic Machines," *J. Ind. Eng.*, **9**, (1):28(1958).

[7] Thus:

$$1/P_0 = e^{A/S}\left[1 - \sum_{x=r}^{\infty} e^{-A/S}(A/S)^x/x! + \frac{e^{-A/S}(A/S)^r}{r!\left(1 - \dfrac{A}{rS}\right)}\right] \qquad (15\text{-}14a)$$

$$P'_r = \frac{e^{-A/S}(A/S)^r/r!}{\sum\limits_{x=0}^{r} e^{-A/S}\dfrac{(A/S)^x}{x!}} \tag{15-24a}$$

These formulas are of a form $a^x e^{-x}/x!$, which can be evaluated by use of Molina's tables[8].

For example, assume a line of 3 chemical reactors which might have available an average of 3.6 feeds or charges per day with an average batch completion rate of 4 per day. What is the probability of all 3 machines being in use with all excess feed batches being diverted to another operation? From Eq. (15-24) where $(A/S) = 3.6/4 = 0.9$:

$$P'_r = \frac{0.9^3/3!}{0.9^0/0! + 0.9^1/1! + 0.9^2/2! + 0.9^3/3!} = \frac{0.122}{2.427} = 0.0502$$

SUMMARY

The element of risk inherent in many management decisions can be considered by the stochastic approach through probability theory. These theories permit a mathematicial evaluation of uncertainty, and such theories may be applied readily in simple situations where the cumulative probability relation is known or can be estimated to achieve a reasonable agreement with the real world situation. In this respect, it may be pointed out that cumulative probability curves such as Fig. 15-1 and 15-4 have a great utility. Furthermore, it is not necessary to have a complete knowledge of probability theory as long as the significance of the cumulative curve is understood. It may also be noted that quite often the cumulative probability curve may be expressed simply by an expression of the form:

$$P(x) = 1 - e^{-ax} \tag{15-25}$$

where $P(x)$ = the cumulative probability of x or less units
and a = a constant

The probability of x or more units then would be simply e^{-ax}.

The vagaries of the future which may affect management economy are minimized to a considerable extent by these applications. However, in certain situations, the analysis is exceedingly difficult, and the applications are limited. Although there are no automatic or formula solutions for this type of problem, judgment and experience employed in conjunction with probability should be profitable in management decision making.

---

[8] E. C. Molina, "Poisson's Exponential Binomial Limit," D. Van Nostrand Company, 1942.

PROBLEMS

15-1. For a certain furniture manufacturer the probability of sales of special two drawer bureaus in a given period is given by Fig. 15-4 based on the history of such offerings. There are no sales after the given period. If each bureau costs $8.16 and sells for $11.54, how many should be produced? Set up the incremental analysis for solution and also check by direct solution in a final equation.

15-2. The assumption is made that the probability curve in Fig. 15-9 applies for sales of a seasonal demand in excess of regular production; it is desired to determine an economic production policy for a possible maximum excess of 10,000 units. The product may be made during the season at a premium cost of $7.20 per unit by overtime and special arrangements, or it may be made in the off season and stored for the demand period at a total storage cost of $6.00. (a) By incremental analysis, determine the optimum number to put in inventory before the demand period. (b) Determine the expected additional demands from the attached figure if the optimum were stocked. Follow the example previously given.

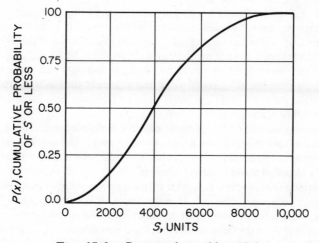

Fig. 15-9. Diagram for problem 15-2.

15-3. A chain store is selling strawberries which deteriorate rapidly and are a total loss if not sold quickly. The probability of x *sales or less* is approximated by the expression $P(x) = 1 - e^{-0.001x}$ for values of x boxes up to 10,000. If the strawberries cost $0.22 per box and sell for $0.38 per box, how many boxes should be purchased for maximum expected profit?

15-4. Using the Monte Carlo method and the probability relation in problem 15-3, generate at least 300 demands and confirm the optimum stocking policy at a 0.42 cumulative probability or less for the perishable strawberry problem.

HINT: Set up a table for random numbers for various different stocking policies.

15-5. The following problem can be worked both by students who have had no statistics and those who have such a background. Data are adapted from Davis[9]. Distribution data for failure of a certain type of lamp bulb are as follows:

Life hrs:	401–600	601–800	801–1000	1001–1200	1201–1400	1401–1600	1601–1800
Freq:	3	31	136	165	67	12	2

(a) Assuming the data follow a normal type distribution, plot the cumulative probability curve for failures by plotting the data on probability paper and reading off the answers. The answer is as follows:

Probability:	0.10	0.25	0.50	0.75	0.90	0.95	0.98
Life (hrs):	810	922	1047	1170	1279	1345	1420

(b) Compare the result in (a) with that which would be obtained for a computed Poisson distribution.

15-6. Calculate the gain that would result in the example problem on queueing in the text if two additional loading facilities were installed at $300,000 each to provide three service centers rather than installing the special loading facility with one service center.

15-7. A certain service facility has no provision for a waiting line (such as a single counter for service where a customer will not wait if he is not served immediately). If the service rate is 2.5 per each 15 minutes and calls are 1.5 per each 15 minutes, at what probability will customers be turned away when in state one? HINT: Since there can be no queue and only two states (zero and one) are possible, the answer will occur at steady state where probabilities for state zero and state one total unity, which can be solved directly for probability of state one.

15-8. A mechanical sorting device can handle 200 items per hour. The installation costs $600 per month for an 8 hour, five day week. The average call rate for the sorter is 150 items per hour, and cost of idle time is $2 per hour. Would it pay to install another sorter? Discuss.

15-9. A parking space for two cars must turn cars away when in state two. If the average occupancy time is 0.75 hour and calls occur at a rate of one per hour, what is the probability of the occurrence of state two (turning customers away)? Add the three probabilities to a sum of unity. Using the formula for P_0' and P_1', solve for P_2', noting that this is a case of two service facilities, or solve by direct formula substitution for P_2' instead of using the summing procedure.

[9] D. J. Davis, *J.A.S.A.*, **6**: 142 (1952).

DECISIONS ON EXPENSE AND CAPITAL OUTLAY 16

From the outset, the purpose of this text has been to demonstrate the type of mathematical relations that management can employ to operate a business firm more profitably. In general, profiitability has been defined in terms of an earning rate on investment or some equivalent of this concept. The models employed for a criterion equation have been set up using real world data so that some practical decision can be made for optimal operation. The types of information required and the manner in which the data are utilized have been demonstrated by example problems. In all of these analyses, certain assumptions (either stated or implied) have been made to simplify the presentation of a major concept.

After the basic concepts are understood, it becomes possible to superimpose more complicated factors on the basic model and consider the effects of these factors on decision policy. This chapter considers certain interesting variations and procedures that are employed in arriving at decisions and policies for the expenditure of funds from the company treasury.

PROCEDURES IN DEFINING EXPENSE POLICIES

In the expenditure of funds for certain annual expenses, decision policies are often established in an arbitrary manner. Among such expenses are those for advertising, for research and development, and for marketing budgets. Policies for these expenditures may be decided on a basis of:

(1) Some percentage of sales or fraction of the unit value of product based on the previous year's sales or a forecast for the next year.
(2) Some percentage of anticipated profit as the amount to be budgeted.
(3) Policies followed by similar businesses.
(4) Some desired minimum achievement and adjustment of the budget to meet this goal.

NOMENCLATURE

C_a	= additional expense	n	= life in years
D	= annual depreciation	n_c	= a time period
D_a	= depreciation on additional investment	P_Y	= expected probability of success
F_R	= annual fixed expense for research	R	= annual return
		S_R	= savings or credits attributable to research
I	= investment	t	= a tax rate, fractional
I_a	= additional investment	V	= a total cumulative expense
i	= expected annual earning rate	V_R	= annual direct charges for research
i_o	= earning rate in other opportunity		
N_c	= time period selected for evaluations	Y	= net profit
		Z	= gross profit

(5) Demand or pressures for an immediate achievement such as a "crash" program based on top managements analysis of a best policy.

For example, advertising expense may vary from one to 15 per cent of sales, depending upon the type of business operation. Companies with large turnover ratios of sales to investment will have a smaller percentage of sales for their advertising budget. Complex manufacturing companies with large investments may have smaller turnover ratios and allow a larger percentage of sales for advertising.

Decision policies for research budgets may be set by similar procedures. In general, they vary up to about 12 per cent of sales, 3 to 4 per cent being average. The percentage varies with the type of operation. For heavy industry with relatively little obsolescence change from year to year, small percentages of sales are budgeted for research. For pharmaceuticals and other operations in rapidly changing fields, a greater proportion of sales dollars will be spent on research.

Sophisticated procedures apply to budget decisions where data can be obtained to provide a rigorous mathematical basis for policy making. Thus, if the concepts of Chapter 7, are utilized, a study of marginal revenues as compared to marginal costs should be fruitful. Studying how marginal unit profits vary with increasing advertising expense per unit of product may show that some optimum outlay is indicated. The difficulty with such analyses consists in obtaining the technical or marketing data that is applicable and in being sure that the analysis is valid for forecasts.

RESEARCH BUDGETING

In research budgeting, numerous procedures have been proposed. In general, these methods involve: (a) the evaluation of credits or returns that should accrue to the research effort, (b) adjusting the credits for a variety of activities with yearly carryovers to some annual basis, and (c) comparison of the credits with the costs for the research. The basic principles here should be those that apply to any economic analysis. It seems reasonable to compute the earning rates for research by a consideration of the profits directly attributable to research on a base of some datum year present worth of the costs (annual expense plus capital) for some specified period. This procedure appears quite simple but in practice raises a number of questions. Among these are what time period of study should be employed and what proportion of the profits are attributable specifically to research rather than to normal production?

Where quantitative indexes have been developed for measuring research contribution, the time period considered is usually 3 to 5 years; that is, the moving total of costs for the most recent five years may be compared with the moving total of credits for the same five years or with some other respective time period. In certain cases, the ratio of the cumulative credits to the cumulative costs has been used as an index of research profitability, although this appears to have no sound theoretical basis. In starting from scratch this ratio should approach and remain at about unity in total research programs. The figure of unity is based on the assumption that one dollar spent for research will return $15 over a life productive span of about 15 years, or about one dollar per year. Where a cumulating period of five years is employed for research evaluations, the remaining 10 years of productive return will supply profits in excess of the research costs. The credits in these analyses are profits or savings after taxes for: (a) improvements in plant operations developed by research, (b) new products from research, and (c) improved products from research. The manner in which the magnitude of (b) and (c) are computed will influence the result, and the procedure can be any arbitrary one. For (b) it may be all profits for sales of new products, whereas for (c) usually some portion of the products' profits are considered. For (a) the saving ordinarily can be determined directly.

Thus, the budgeted amount for research can be established from the total estimated credits expected according to the schedule of the three items in the previous paragraph. In essence, this procedure is a form of capitalized payout time where all direct expense and other capital outlay are amortized over the time period used as a basis for evaluation. Thus, an equation for uniform annual direct and fixed costs $(V_R + F_R)$ with uniform annual credits S_R for research may be written as:

$$N_c(V_R + F_R) = N_c S_R, \text{dollars} \qquad (16\text{-}1)$$

where N_c = the time period used for evaluations

V_R = annual direct expense for research

F_R = annual fixed expense and includes all additions of capital outlay that are considered to be amortized in the N_c time period

S_R = the total net annual credit attributable to research after correction for profits tax

Since most research returns and expenditures will fluctuate annually on a moving total basis of say five years, Eq. (16-1) must be written in a more general form of:

$$\sum_{i=1}^{i=5} (V_R + F_R)_i = \sum_{i=1}^{i=5} (S_R)_i \qquad (16\text{-}2)$$

Thus, at the start of the fifth year where S_R for the past four years is known and can be estimated for the fifth year, the allowable expense for the fifth year would be:

$$(V_R + F_R)_5 = \sum_{i=1}^{i=5} (S_R)_i - \sum_{i=1}^{i=4} (V_R + F_R)_i \qquad (16\text{-}3)$$

The data for these moving summations are best kept straight by tabulations for the individual years. Any base time from 2 to 5 years should be acceptable for this type of analysis. In some variations of Eq. (16-2) only the nondeductible part of $(V_R + F_R)$ is used; that is, the left side is $(1 - t)\sum (V_R + F_R)_i$ where t is the profits tax rate.

A RESEARCH BUDGET EXAMPLE

Problem

The research and operations committee of a company have evaluated past performance and estimated the research credits for the start of year 19X6 as shown in Table 16-1. This company employs a 4 year base for research evaluations, and research expenses have averaged $120,000 per year plus an average of $30,000 every 3 years for new depreciable capital which amortized over 3 years is $10,000 per year. The profits tax rate is 56 per cent. What are the allowable research expenses for 19X6 for a four year study period?

Solution

According to Eq. (16-3), the sum of all credits for the 4 year period is $550,000. Expenditures for research in the past 3 years have been (3 × 10,000) plus (3 ×

TABLE 16-1. Research Credit Schedule, Thousands of Dollars

Type	19X3	19X4	19X5	Estimated 19X6	Totals
Process improvements[a]	20	10	15	15	60
New products[b]	80	100	120	100	400
Improved products[c]	10	30	40	10	90
Totals[d]	110	140	175	125	550

[a] Direct dollars saved corrected for tax deduction of 56 per cent.

[b] Estimated as some fraction of net profit contributions of these items.

[c] Estimated as some fraction of net profit contribution of these items; the proportion would be smaller than for (b), since credit is taken only for the improvement.

[d] This sum is sometimes used as an index of research return to compare with expenditures.

120,000) for a total of $390,000. Thus, an excess of $550,000 − $390,000 = $160,000 would be budgeted for 19X6.

Some procedures require that the ratio of $S_R/(V_R + F_R)$ be of the magnitude of 3 or more instead of unity. This is entirely a relative and arbitrary basis, depending upon how credits are evaluated for research contributions and what portions of plant investment are returnable as depreciation credits. In the example, no production facility depreciation was considered. The only depreciable capital was for research capital outlay that could not be charged as annual expense.

AN ALTERNATIVE RESEARCH BUDGET

Another variation of research budgeting is based on the capital investment involved and the risk earning rate (see Chapter 5, Table 5-4). The risk earning rate is the excess earning rate for the project above the going rate or other opportunity rate for alternative investments. In mathematical form, this is:

$$(1 - t)V = n(i - i_o)I, \text{dollars} \qquad (16-4)$$

where V = total allowable expense that can be supported by a project having an investment, I

n = the commercial life of project

i = expected annual earning rate

i_o = earning value in other opportunity or alternative investment

and t = profits tax rate

This equation states in symbols that the total nondeductible expense for research on a project should be equal to, or be less, than the dollars expected to be returned by the investment which are in excess of the dollars that could be made in alternative investment.

DECISION POLICIES FOR CAPITAL OUTLAY

Where there are a number of projects competing for limited capital funds it is necessary for management to establish some policy as to which activity is to receive approval. Certain procedures have been established in earlier chapters which provide an index of profitability so that the projects may be rated as to their relative economic feasibility. However, this simple establishing of priorities based on earning rate or some other profitability index is not always the only basis for decision. For example, the magnitude of the investment, availability of funds, type of financing regarding source of funds, etc., may all affect a decision.

In decisions involving profitability, management often employs the procedures of Chapter 11 based on an economic analysis where capital recovery is considered with or without simplifying assumptions. The most generally accepted procedures are those policies based on earning rate, because this approach permits comparisons with other opportunities for investing available funds. In general, the zero-time present worth of projects seeking capital is known or must be computed when evaluating earning rate. This provides another base for comparison, since it establishes the magnitude of the real or equivalent investment. Thus, where earning rate by itself may not be a sufficient index of profitability for decision making the capital requirement is usually known. Capital requirement by itself is not a good criterion because a low investment is not necessarily a measure of economic feasibility. Annual costs may be known, or they can be computed from present worths for a specified elapsed time. However, annual costs are a poor criterion of profitability except where the investments are of the same order of magnitude. In reality, differences in annual costs for a given investment are a measure of the value of money. Payout times are most useful as measures of profitability for operations of a similar commercial nature. They also are a measure of the value of money, but the relationship is not clearly evident. The net result of these observations is that earning rate is in general the best measure of profitability.

There are three interest rates then that are involved in managerial decisions. The first is the earning rate that will be returned by the proposal. The evaluation of this rate has been covered in earlier chapters of this text. The second rate is the cost of the capital that may be used in the proposals under consideration. This subject will be discussed in later sections of this chapter. The third interest rate is the minimum acceptable rate that company policy decides should be a criterion as to whether or not, to invest in a proposal. This rate is often the opportunity rate, or the rate the company's funds may earn if management decides to invest them in outside opportunities or some other alternative project rather than to commit them for a specific operation. The minimum acceptable rate varies with the type of project and its chance of success, as evaluated by objective and subjective studies. For example, high risk projects may require expected earning rates in excess of 50 per cent. An average risk, such as the enlargement of operations, may require only a 20 to 50 per cent rate, where low risk ventures such as improvements in plant operations may be feasible with indicated earning rates above 10 per cent. All these rates are after profits tax allowances.

There are a great variety of projects vying for funds. These projects vary from those which are absolutely necessary (for example, a worn out power plant) to those whose profitability is not susceptible to reduction to finite values (for example, waste disposal facilities for esthetic purposes). In between these extremes are projects of varying profitability types such as have been suggested earlier in this text where profitability is measured by their earning rates. There are replacement projects (Chapter 12), new projects with expected earnings (Chapter 11), a variable type of expected profitability at different levels of investment for a given project (Chapter 11), alternative investments in other going concerns, and similar types of opportunities for company funds. The distribution of such projects according to earning rate may be as shown in Fig. 16-1 based on Dean[1].

For any given time interval, usually one year, a tabulation of prospective projects with their earning rates and cumulative capital demands can be prepared. Such a chart may appear as in Fig. 16-2. This curve defines

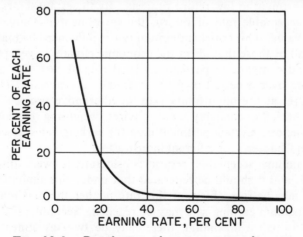

Fig. 16-1. Distribution of projects according to their profitability.

[1] J. Dean, "Managerial Economics," Prentice-Hall, 1951.

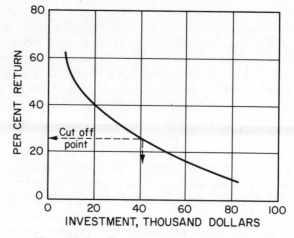

Fig. 16-2. Earning rate-demand curve.

the amount of dollars that can be invested in projects to give some minimum earning rate say 25 per cent. If more than the indicated capital is available and is to be invested, then the company must be satisfied with a lower cutoff earning rate. Conversely, if a company has limited available capital, the cutoff point may be higher with higher yields being obtained.

The demand curve of Fig. 16-2 also serves in a limiting manner to prevent investment in low return projects. By the establishment of some minimum acceptable rate of return, the investment is limited to some maximum value. This critical minimum rate is influenced by many factors including all of those that affect the economic climate in general and the industry and company in particular. In the usual case, the company's average rate over a period of years is used as a criterion. This rate is sometimes set at the opportunity rate for alternative projects or investments outside the company and requires weighting of comparative economic factors. Certain pertinent data for earning rates based on sales are given in Chapter 7 for different industries.

The minimum acceptable return considerations were discussed in Chapter 11, and it should be reiterated that under this limitation certain conditions may be true. First, at this rate neither the maximum dollars nor the maximum earning rate may be reached. Secondly, if the earning rate goes through a maximum, there may be two investment levels at which the minimum acceptable rate can be obtained. Finally, if this minimum rate is considered to apply as a critical marginal rate for new investment, then an investment level different from any of the others may be established.

In connection with the selection of the cutoff rate of Fig. 16-2, it is of interest to consider that total dollars profit at a low rate may be more important than a greater earning rate for a few dollars. In fact, the probability of success in project economic analysis may be greater for the low rate projects. Another factor to consider is that a number of smaller investments at high returns for a limited time period may limit availability of funds. Thus, should a good project with large cash returns at a lower rate become accessible at a later date, there may be insufficient funds to accept the proposal.

The minimum acceptable earning rate as a criterion often is the earning rate for the company based on its stock value. Accordingly, if proposals do not indicate earning rates above this criterion, the company management would not consider investing internal funds (generated by profits and depreciation charges) in such ventures. It would be better to pay any excess available funds out as dividends.

Timing in the life history of a project is also a factor to examine. Thus, the *flush* period of an oil well is at the beginning of its economic life with a *decreasing* production with time. In contradistinction, a gasoline service station has a *low* initial profitability and *increased* growth is expected.

SOURCES OF CAPITAL

There are two sources of capital from which a firm can obtain the funds it needs for new capital investment—*internal funds* generated by the business operation and *external funds* obtained from outside sources. In manufacturing industry, investment funds obtained from the profits (that part not paid out as dividends) and the depreciation charges constitute approximately 75 per cent of the funds. These are the real dollars that an enterprise generates from its operations. The internal funds are made up about 40 per cent from depreciation and 60 per cent from retained profits. The remaining 25 per cent of the investment funds are obtained externally, about 40 per cent of these funds being obtained through the issuing of stocks and bonds and the other 60 per cent being arranged through short term loans and mortgages.

Internal funds are a result of witholding from the stockholders some proportion of the earnings each year. This proportion is of the order of about 50 per cent, but it varies from company to company and also with the economic climate and the future plans of the company. Thus, in periods of inflation, the deposits for depreciation are not sufficient to pay the renewal charges on equipment purchased at lower prices. Accordingly, such internal funds as are necessary are obtained by using more of the earnings. In periods of expansion also, the earnings may be subject to greater inroads to help provide funds. In years of small profits, a company may desire to protect its future by retaining in the surplus a larger proportion of the smaller earnings. All of these considerations influence the decision of management regarding the use of internal funds available.

Ordinarily, external funds are always available to a well managed firm with a history of success. The principal problem in decision making is to determine which are the most economical. The three general types of capital sources—debt, preferred stock, and common stock—are discussed in Chapter 4, but there are many variations of the general types. These variations take the form of special inducements and rights of the lender to encourage investment in the company. The most important inducement is the interest rate returned to the lender, which becomes the cost to the company. These financial costs to the company become part of the operating expense, and they include all charges in connection with handling the funds as well as the interest on them. Since these are legitimate operating costs, they are deductible from gross profit for the profits tax computation, and, accordingly, the apparent interest rate is effectively less in real dollars as will be discussed further.

It is not the intent here to discuss the many aspects of corporation finance but rather to present a simple summation to demonstrate how the economics of external financing varies with the type. In general, the amount of the interest rate a company pays varies with the time into the future for which the funds are committed. Thus, long term debts carry a higher interest rate than short term loans. In addition, the interest rate varies with the risk involved for the lender, being greatest where the security offered by the company is least. The bond debt shows the least yield because since it is secured by the assets of the company, annual interest must be paid; however, the common stock yield is returned only if a profit is made and dividends are declared.

It is interesting to note here that the term risk is ambiguous because it depends upon the viewpoint of the borrower or the lender. The company bears the most risk for financing with bonds because the interest charges and face values at the maturity date for the bond *must* be paid by the company. Thus, the bond obligations are most risky because they could force a company out of business if it did not operate to meet those obilgations. Conversely, common stock is the least risky for the company, since if no profits are made, there is no obligation to pay dividends. However, when profits are made, the company pays dividends at high rates; thus the cost is high for capital obtained by selling stock.

In comparing the cost of capital, the influence of profits tax must also be considered, since interest on debt payments are deductible but dividends paid out are not. Thus, if the profits tax is 52 per cent a comparison of interest rates may appear as in Table 16-2.

TABLE 16-2.   Comparison of Capital Costs (Per Cent)

	Apparent Rate, Omitting Taxes	Actual Rate, Considering Taxes	Comparative Rate
Bonds and loans	4	1.92	4
Preferred stock	6	6	12.5
Common stock	8[a]	8	16.7

[a] Earnings as based on stock value (average figure).

A comparison of rates as a cost should include the influence of taxes, since each real dollar available to the company generated from operations requires $1/(1 - t) = 1/(1 - 0.52) = 2.08$ dollars of profit before taxes. In evaluating relative costs for different capital sources, either the actual rate or the comparative rates may be used for all types, but the basis should be clearly stated.

THE DILUTION EFFECT OF STOCK FINANCING

When a company decides to raise new capital by selling additional stock, then the original stockholders' equity (ownership) is diluted in proportion to the amount of stock issued. Accordingly, the total dollar earnings must increase in proportion if the rate of return is to remain constant. In making these comparisons, the earning rate being considered is based on the stock market price (see item 3 of Table 5-4) and not on the plant investment where the stock value on the balance sheet is carried at its original issue price. Also in calculating the additional earnings required, all mandatory payments from profits (such as financial costs and preferred stock dividends) are deducted before computing the earnings per share of stock. An example calculation follows:

Problem

The Smith Corporation shows a simplified annual report as given in Table 16-3. It has been decided to increase the capital by selling 20,000 shares of new stock which should sell at $25 per share based on current stock prices. What must be the additional earnings to maintain the stock earning rate at a constant value?

Solution

The current stock earning rate is $2.65 per share for 100,000 shares. These shares have the same market value of $25 as do the new ones. Thus, the present stock earning rate is $2.65/25 = 0.106$, or 10.6 per cent. Since there will then be a total of 120,000 shares, the total profit after taxes available must be (0.106) (120,000)(25) = $318,000, or an increase of $318,000 - 265,000 = $53,000,

TABLE 16-3. Simplified Balance Sheet and Income Statement For Smith Corporation (All Figures Not Carrying a Dollar Sign are in Millions of Dollars)

Assets			Liabilities
Current assets	1.2	0.3	Current liabilities
Fixed assets at		0.5	Funded debt at 2 per cent
depreciated value	2.5	1.0	Common stock at $10 for
			100,000 shares
		0.5	5 per cent preferred at
			$100 for 5000 shares
		1.4	Surplus
Totals	3.7	3.7	

Income Statement[a]

	Before Stock Issue	With New Stock Issue
Net profit (before financial costs)	0.300	0.353
Financial costs	0.010	0.010
Net earning available	0.290	0.343
Preferred dividends $500,000 × 0.05	0.025	0.025
Available for common stock	0.265	0.318
Earnings per share	$2.65	$2.65

[a] Profits tax rate of 52 per cent.

which is also the increase in required profit before financial costs, or a total of $353,000.

DEBT FINANCING

It will be observed that if the earning rate for all capital is just maintained, then the objective of the expansion has gained nothing; it would have been better to have financed the expansion of $500,000 by debt financing. Consider that the new $500,000 was financed with 4 per cent bonds and that the new gross profit would be the same as that yielding the $353,000 for stock financing. With debt financing more dividends per share would be available from the greater profits resulting from expansion; this is shown in Table 16-4.

The magnitude of the difference in stock earnings will vary with the amount of the earnings and the relative amount of debt and stock financing. This was mentioned also in Chapter 11.

The comparative cost for the capital in the example problem based on pretax values is the 4 per cent paid for bond financing versus the 10.6 per cent after taxes for the stock earning rate of Table 16-3. This is an equivalent of $10.6/0.48 = 22.1$ per cent on a pretax basis for use of internal earnings. It will be noted that these finance stock earning rates are not the same as the capitalized earning rates of Chapters 5 and 11, which are based on initial or current investments.

TABLE 16-4. Comparison of Stock and Debt Financing

	Stock Financing	Debt Financing
Dollars of:		
Profit before interest	724,583	724,583
Interest	10,000	30,000
Taxable profit	714,583	694,583
Taxes at 52 per cent	371,583	361,183
Net profit	353,000	333,400
Interest	10,000	30,000
Preferred stock dividends	25,000	25,000
Available for common stock	318,000	288,400
Shares of Stock	120,000	100,000
Earnings per share	$2.65	$2.88

CAPITAL FROM INTERNAL FUNDS

When capital for expansion is used from internal funds generated by the firm's operation, the net effect as far as cost of capital is concerned is exactly the same as financing by common stock issues. In Table 16-3 it was shown that a new common stock issue would provide $500,000 to give additional earnings of $53,000 at a rate of 10.6 per cent. Suppose this $500,000 had been taken from internal funds; then the accounting would appear as in Table 16-4 for stock financing, except that the shares of stock would read 100,000 and the additional earnings per share would be $3.18 - 2.65 = $0.53. On the basis of the stock yield rate of 10.6 per cent, the gain in stock value based on earnings would be $0.53/0.106 = $5 per share or a total of $500,000 in value for the stockholders who own the 100,000 shares. Thus, the $500,000 that was taken from the profits and depreciation funds which are owned by the stockholders has been replaced theoretically by an increase in the market value of exactly a like amount. In the real world, the increase may not exactly adjust to the correct amount on the market in which case the cost of capital would vary from the 10.6 per cent figure; that is, the stock earning rate would differ from 10.6 per cent because fluctuations in the stock market price are usually not a true indication of the actual worth of a stock, but rather are an indication of the public demand for the stock.

DEBT FINANCING AND PROFITS TAX EFFECTS

Where part of the investment is financed by debt capital, the interest charges represent a tax credit which shows up as increased net profit. This subject was demonstrated in an example problem in Chapter 11 (page 288) but will be amplified here. Data are given in Table 16-5 to demonstrate the quantitative effect of this tax credit for an operation with various amounts of debt financing at 4 per cent interest assuming a 52 per cent profits tax. The pertinent results are shown graphically in Fig. 16-3.

TABLE 16-5. Influence of Profits Tax and Debt Financing on Profitability

Plan:	A	B	C	D
Thousands of Dollars of:				
Investment	100	100	100	100
Equity capital	100	75	50	25
Debt capital	0	25	50	75
Annual gross profit before				
interest charges	20	20	20	20
Interest charges	0	1	2	3
Taxable earnings	20.0	19.0	18.0	17.0
Tax at 52 per cent	10.4	9.9	9.4	8.8
Net profit	9.6	10.1	10.6	11.2
Earnings after interest	9.6	9.1	8.6	8.2
Per Cent:				
Capitalized earning rate[a]	9.6	9.1	8.6	8.2
Capitalized earning rate[b]	9.6	13.4	21.2	44.8
Capitalized earning rate[c]	9.6	12.1	17.2	32.9

[a] Based on proprietary earning (earnings after interest) and total investment.

[b] Based on net profit and equity capital. Note that net profit exceeds proprietary earning by the amount of financial costs.

[c] Based on proprietary earning and equity capital.

The results in Table 16-5 and Fig. 16-3 demonstrate rather clearly that debt financing returns high yields on the proprietary or equity capital, thus indicating its desirability. However, as has been pointed out previously, debt financing is a high risk from the company viewpoint, since interest charges must always be met. This point will be amplified in the next section.

CAPITAL LEVERAGE

A company's financial structure is often made up of common stock plus other financing with mandatory charges such as debt interest or preferred

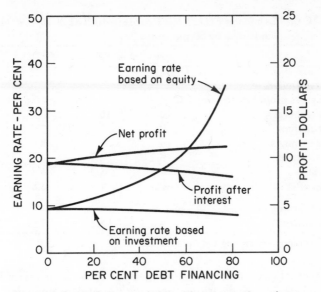

Fig. 16-3. Influence of debt charges and profits tax. (See Table 16-5).

dividends. These fixed demands on the earnings produce a multiplying effect on the residual earnings when the profits fluctuate. This multiplying effect is called capital leverage and becomes more important at higher proportions of mandatory charges which the company is required to pay out. Thus, if there are no mandatory charges and the gross earnings are reduced 50 per cent, then the total earnings and earnings per share of common stock will also drop 50 per cent. However, if bond interest or preferred stock dividends in fixed amount must be paid, then total earnings and earnings per share of common stock will be less than 50 per cent and could drop to zero (or be negative). The explanation, of course, lies in the fact that the fixed obligations do not decrease in proportion as the earnings become less.

Leverage effects are demonstrated in Table 16-6 which uses the same data as Table 16-5, except that the gross profit before interest has been reduced 50 per cent to $10,000. The equity basis or the stock value is assumed to remain constant in all of these computations.

The fifty per cent reduction in gross profit has halved the earnings in Plan *A* but reduced the rate in a higher proportion for the other three plans. The reduction is proportionately greater as the debt financing increases.

By the same reasoning, the earnings per share of common stock will increase at a greater relative rate than the dollar values as the earnings

TABLE 16-6. Leverage Effect for Problem in Table 16-5 with Profits Reduced by 50 per cent

Plan:	A	B	C	D
Thousands of Dollars of:				
Annual gross profit before interest charges	10	10	10	10
Interest charges	0	1	2	3
Taxable earnings	10	9	8	7
Tax at 52 per cent	5.2	4.7	4.2	3.6
Net profit	4.8	5.3	5.8	6.4
Earnings after interest	4.8	4.3	3.8	3.4
Capitalized earning rate[a] (per cent)	4.8	5.7	7.6	13.6

[a] Based on proprietary earning and equity capital.

increase when there are constant obligations to be paid. The factor of leverage affects finance decisions, since the use of debt financing at lower interest rates imparts leverage possibilities to the operation which may be undesired.

NONPRODUCTIVE INVESTMENTS AND TAXES

Nonproductive investments in waste disposal systems, shower rooms, cafeterias, shops, warehouses, and the like, do not earn a profit, but their annual operating costs will reduce taxable income. If the tax credit for the annual depreciation charge exceeds the nondeductible part of the operating expense, the annual return will be increased when a nonproductive investment is made. However the over-all earning rate is reduced, and the payout time will not be reduced except for extraordinary situations. An example of this effect follows:

Problem

A food processing plant is built at a cost of $200,000 and shows a gross profit of $100,000 per year. The plant is located where waste disposal facilities are not mandatory but may be desirable. The additional investment for the waste system is $12,000, with annual operating expense of $200. Equity capital is to be used with a profits tax of 54 per cent and estimated life is 12 years. (a) What are the annual returns for a decision (1) to omit, or (2) to build, the waste system? (b) What are the capitalized earning rates and payout times for the alternative decisions?

Solution

An analysis of the data yields Table 16-7:

TABLE 16-7. Calculations for Problem on Nonproductive Operation

	Omitting Waste Disposal	Including Waste Disposal
Dollars of:		
Investment, I	240,000	252,000
Gross profit, Z	100,000	98,800[a]
Net profit, $Y = (1 - 0.54)Z$	46,000	45,448
Annual depreciation, D	20,000	21,000
Annual return, $R = Y + D$	66,000	66,448
Capitalized earning rate, $(Y/I)100$ (per cent)	19.1	18.0
Capitalized payout time, I/R (years)	3.63	3.79

[a] $1200 added annual costs for expense ($200) and depreciation ($1000).

Part a. The annual return is shown to be larger for the situation with waste disposal because the tax saving of $552 plus $1000 depreciation exceeds the annual expense of $1200. Expressed algebraically, the increase in annual return is the excess of tax credits for depreciation on the nonproductive investment above the effective cost for annual expense for its operation:

$$\text{Increase in annual return} = tD_a - (1 - t)C_a = t(D_a + C_a) - C_a \qquad (16\text{-}5)$$

where t = tax rate
D_a = added annual depreciation
C_a = added annual expense

Since the annual return is increased, it would appear desirable to install the waste system, but a final decision probably would better be based on the effect on earning rate which is reduced.

Part b. The capitalized earning rate is reduced by the additional investment, which must always occur when investment is increased and the profit is decreased by depreciation cost and when operating expense decreases profit still further. The capitalized payout time is increased here as it usually is. The term, tD_a, represents the additional contribution to annual return when there is zero expense for operations where D_a is $(1/n)$th the value of I_a for the additional investment. With n being the life of the main investment, the annual contribution is tI_a/n, and the additional investment is paid out in $I_a/(tI_a/n) = n/t$ years. If this payout of n/t years is shorter than the life of the main investment, then the nonproductive investment can shorten the over-all payout time. This may occur under the special conditions of high tax rate, zero expense, and very short life for the nonproductive investment. Note that for the example used $n/t = 10/0.54 = 17$ years.

THE REAL COST OF ANNUAL EXPENSE

Equation 16-5 has several other implications, since the term $(1 - t)C_a$ evaluates the actual net cost per dollar of expenditure for any annual operating cost and explains why many companies will spend considerable sums of money for consultants, economic surveys, research, and other special studies when tax rates are extremely high. On some government contracts involving renegotiation, the effective t may get very high so that the cost to a company for special services may be very low for each dollar spent, with the possibility that the expenditure may pay off sometime in the future when taxes are lower.

The preceding section illustrates the rather important principle of the use of valuable capital (earning at 19 per cent) for a purpose which is non-productive. The converse is to employ a minimum of the valuable capital for nonproductive purposes. The rent-or-buy problem is of this nature and, as was noted in Chapter 12, is a decision involving the selection of alternatives which utilize the procedures that apply for the data. Buildings may be rented to reduce the capital investment, the rental charges being deducted as an annual overhead cost. Similarly, tank cars, truck fleets, warehouses, and other services may be rented instead of using capital to provide the equivalent effect. The criterion, of course, is the potential earning rate or value of the capital required to conduct the business and provide for the process equipment and its operation. If this capitalized earning rate is in excess of the ratio of annual rental costs (correcting for income taxes) to the capital required for the equivalent service, then the capital should be employed only for productive investment. The point arises that the capital of those providing the rented service must also earn a profit which might just as well be made by the company. The situation is analogous to the use of borrowed money where the potential rate for the company's operations is sufficiently in excess of the going rate for money to make feasible the payment of rent for the use of the borrowed money.

An example follows. A company can rent a building costing $130,000 for an annual charge of $14,000. If the company purchases the building, the annual charges for depreciation, taxes, and maintenance will be $5000; that is a gross saving of $9000 would result which, allowing for profit taxes of 50 per cent, would mean a net gain of $4500 by owning the building at a capitalized earning rate of about 3.5 per cent. If the potential earning rate for the company is only 10 per cent, the capital used for the building could be more profitably spent for additional process facilities where $13,000 net profit could be obtained. If the $130,000 capital is borrowed for either case, the respective annual net profits will be reduced (but not proportionately).

A similar situation exists for the accounting of catalyst charges and process liquids required as part of the initial capital investment for equipment. If the quantity required is kept at a minimum then renewals and make-up costs become part of the annual operating costs rather than an annual depreciation charge. By use of minimum initial quantities, the net profits and taxes are increased because of a smaller investment for a given output of product. The annual return is decreased, but usually a decreased payout time will result.

ALTERNATIVE AND REPLACEMENT DECISIONS

The problem of management having to make decisions on alternative investments and expenses has been treated at different levels of sophistication in Chapter 12. When a policy has been established, any of those procedures can be utilized to provide the criterion for executing the policy in the manner intended.

DEFERRED INVESTMENTS

Management is often faced with the decision of making an investment at the present time or of deferring action until a more favorable time as indicated by economic analysis.

The nature of the decision usually is one of two types; these two types are based on the techniques of Chapters 11 and 12. Where the problem is to decide on an investment of a new or different kind to provide the same service being rendered, a typical replacement analysis is required. However, where the decision is based on a new or expanded service, the profitability must be evaluated and compared with the profitability of other opportunities for investing available funds. Then, the basis for a decision may be any one of those covered in the text.

Decisions on deferred investments bring into importance many economic factors not reducible to dollars and cents. Nevertheless, they must be considered as affecting the decision policy. Such factors as safety, inflation, sales protection, improved customer service, and improved working conditions may favor immediate investments. Other factors such as the anticipation of fast obsolescence, the possibility of more accurate market analyses in the future, and the overcapacity of the new investment from the standpoint of present conditions may operate to make it preferable to delay the investment.

A DEFERRED INVESTMENT EXAMPLE

Problem

The decision required is as follows: Should a company build a full-size plant now for a new product or should management elect to build a half-size one now and an equal-size addition to be available starting the fourth year from now. Annual sales for the first three years are expected to yield $50,000 in net profits for a half-size plant (or $20,000 for the full-size plant because of extra depreciation) and $100,000 per year for the succeeding three years with a full-size operation. A full-size plant is expected to cost $600,000, a half-size one costs $420,000 now; the addition at 3 years costs $300,000 and has a life of 6 years. If the half-size plant is built, there will be a saving of $8000 in deductible (for tax purposes) costs per year for plant operation for the first 3 years. Six years is to be used as the study period, with straight-line depreciation; for this calculation money is considered to be worth a minimum of 12 per cent. Neglect salvage value and profits tax.

Solution

This problem should be recognized as the selection of alternatives based on a comparison of the relative returns with respect to the principal involved at zero time according to the principles expressed in Eq. (12-1). Accordingly, a schedule of calculations is as given in Table 16-8, where the present worth of the expenditures is compared with the present worth of the expected returns.

By the deferment of the additional investment in a full-size plant, a smaller (about $13,600) present worth capital is returned based on a comparison at 12 per cent. Thus, ordinarily the decision would be to build the full-size plant now. This could be overruled if the irreducible factors noted in the previous section can be considered to be worth the $13,600 of present capital, in which case the half-size plant would be built with a later addition.

Note that a variation in the economic analysis can be made using the principles of Chapter 11 to determine the exact earning rates of each separate alternative. This would be a trial and error solution by trying different earning rates until the present worths of the returns were exactly equal to the present worths of the costs. These economic rates of return would then be measures of profitability which could be compared in making a decision.

THE DISCONTINUE DECISION

One of the decisions management is called upon to make is whether or not to discontinue an operation. This in reality is a special kind of comparison of alternatives which can be evaluated by comparing the expected return with some minimum acceptable one.

TABLE 16-8. Schedule of Calculations For Deferment Example (Dollars)

	Half-Size Plant with Later Addition	Full-Size Plant Now
Present worth of costs:		
Initial investment	420,000	600,000
Subsequent investment = $300,000/(1.12)^3$	213,000	0
Total present worth of pertinent costs	633,000	600,000
Present worth of returns:		
Savings of operating costs = 8000×2.402	19,200	
Depreciation of initial investment	287,800	411,000
Depreciation of subsequent investment[a]		
$(300,000/6) \times 2.402/(1.12)^3$	85,500	
Profits for first 3 years:		
(50,000)(2.402)	120,100	
(20,000)(2.402)		48,000
Profits for last 3 years:		
$(100,000 \times 2.402)/(1.12)^3$		171,000
$(80,000 \times 2.402)/(1.12)^3$	136,800[b]	
Total present worth of returns	649,400	630,000
Excess of relative[c] returns over costs at 12 per cent interest	16,400	30,000

[a] One sixth of subsequent investment for 3 years starting at the beginning of 4th year. The present worth at the start of 4th year is then reduced to initial starting date.

[b] Since the depreciation charges are 70,000 + 50,000 = $120,000, or $20,000 greater than for the initial full-size plant. The annual net profit is reduced by $20,000.

[c] The term relative is used because the item of savings in operating costs would apply only if both alternatives were actually built.

For this study, the capitalized earning rate is computed on an investment base total. This investment represents the recoverable value plus expected capital investment needed if the operation is to continue. Such items are as follows: working capital, realizable value of facilities that can be used elsewhere, and expected anticipated investment. The sum of these minus any shutdown costs is the realizable current capital investment.

The equivalent real dollars being earned is the profit after taxes from the operation; then the profits after taxes are considered to be the earnings on the realizable capital investment. If this earning rate is less than the company acceptable rate that the recoverable investment might earn elsewhere, the indications on an economic basis are that the operation should be discontinued. However, provision must be made to avoid customer displeasure and to alleviate the effects of shutdown on employees and the general public. In some procedures for discontinue decisions, the credit for depreciation being earned by the operation is included; this credit for depreciation is added to the profit to give an annual return. This annual return is divided by the investment to get some kind of earning rate index, but this really is the reciprocal of payout time and is not an earning rate.

THE INVESTMENT BASE

Clearly, there are many different values for capital upon which financial and economic decisions are based. Among these are initial investment, current investment, book value, net worth, total investment, process investment, stock market value, etc. Therefore, for a specific analysis, one must state specifically what base is being used and must consider how it may vary. Even the selection of one type of capital does not fix the base in dollars because of variations with time of either the base itself or the value of the dollar. It is the object of this section to discuss briefly a few factors that influence the base value.

In the general plan of economic analysis, the initial total investment is employed as a base for earning rates and payout times, even though the real base may change with time because of depreciation or inflation effects. This appears expedient because a constantly changing base introduces another complicating factor into an analysis that is already sufficiently complex. One compromise is to use one half the initial investment as an average base; here it is reasoned that the initial investment decreases to zero at the end of its life. However, this procedure ignores the fact that the depreciation account builds up and appears in the working capital part of the total investment. Thus, there is a compensation that tends to keep total investment constant. As a matter of fact as time passes in a successful company, with growth the retained earnings tend to increase the total investment involved.

In finance, the book value or stock market value of the stock appear to be generally accepted; some allowance for changes is made because of variations in the economic climate.

There is an increasing belief that in a dynamic economy a certain amount of inflation, amounting to an average of 2 to 3 per cent annually, must be recognized. As was pointed out in Chapter 4 the economy must expand and whether this is inflation, a raising of standards of living, or just more of everything is a question still to be answered. Some proposals have been made to add a separate factor to allow for the real dollar effect of this steady expansion of the economy. It would appear more expedient to combine all such factors into the one going value for money (since inflation effect is a compounding one like interest) and establish such rates of return for economic analysis at those levels that would automatically allow for this growth. This would appear more practical than to become involved in complex accounting procedures.

UNCERTAINTY CONSIDERATIONS

Since no one can predict the future with certainty, the best projections of future occurrences must be based upon past history. Where a pattern of happenings in the past has been established and this information is expected to be repeated in the future, the stochastic analysis with relative probability may be applied to good advantage in models for decisions. However, economic decisions are often made where there is no probability background to arrive at expected values. In these circumstances, management selects procedures on the basis of experience and judgment; in addition, management's intuitive understanding of economic concepts is utilized to a certain extent in weighting the various factors that may influence a decision.

The statistician can determine the confidence limits for an estimate of costs, earnings, payout time, etc., which may influence a decision. These limits are ascertained by obtaining a number of separate independent estimates of the various economic dollar components involved. For each of these estimates the variances can be established. Then, the variances of each component may be pooled to obtain a composite variance that can be employed to estimate a final answer. Thus, a final estimate with confidence limits may show that there is a 90 per cent probability that the annual profit will be between $400,000 and $450,000. This means that 9 out of 10 times, one can expect up to this range of profit.

In connection with economic enterprises, there is always some return from the depreciation allowance, even though no profit may be made. Thus, in a consideration of the probability of success as part of the economic analysis, there are two parts that make up the annual return, and each is weighted by its prospective component of probability. For example, suppose that a consensus of opinions has revealed that a proposal should earn an estimated profit 80 per cent of the time; that is, the probability that it will earn the profit is 0.80, and the probability that it will earn no profit is 0.20. On this basis, an expected long run annual return, R, is:

$$R = YP_Y + D \qquad \text{annual dollars} \qquad (16\text{-}6)$$

where Y = the annual net profit
P_Y = the long run expected probability that Y will be obtained (that is, the probability of success)
and D = annual depreciation

Then, the zero time present worth of the annual return in Eq. (16-6) may be compared with the initial investment to determine if the forecast for the proposal will prove it to be economical.

SUMMARY

As the relative pressures for funds are evaluated, management must supply capital to meet essential demands. Where the available capital within the company is insufficient, management can utilize outside sources (such as bond issues or stock sales). However, if capital cannot be raised to meet all requirements, some of the demands must be eliminated. Thus, a policy must separate out the acceptable demands for which capital can be supplied.

In setting a policy, management must include in its deliberations all the economic and financial factors discussed in this text plus other basic business considerations. These other considerations may involve changes in:

> *the money market* (the time value of money).
> *the general economic climate* (these changes may affect sales and returns).
> *the labor market* (these are caused by the economic climate or by expansion of operations into new geographical areas).
> *the growth pattern for the product* (these are caused either by the economic climate or by obsolescence of the product).
> *the competitive position of other manufacturers,* and
> *other industries affecting the potential market.*

All of these could influence either a decision or the future results of a decision.

PROBLEMS

16-1. A company has had research credits for the last two years of $85,000 and $115,000 respectively with next years predicted credits of $125,000. If research costs have been $100,000 per year, what should be the budget for next year on a 3 year cycle basis?

16-2. A research project is expected to produce an average annual earning of $80,000 over a five year life at a rate of 20 per cent. If the minimum acceptable rate is 12 per cent, what is the maximum five year research expenditure for this project for a profits tax of 52 per cent?

16-3. Compare the earnings per share for a company stock issue of 10,000 shares versus a 4 per cent bond issue to raise $300,000 of new capital. Present stock consists of 100,000 shares, the gross profits before any interest charges are expected to be $500,000, and the profits tax is 54 per cent.

16-4. Assume that in problem 16-3 the stock value remains constant and is essentially the total value of the investment; (a) What are the capitalized earning rates for proprietary earning based on total investment and equity capital? (b) If the gross profits are increased to $1,200,000 before interest, what are the earning rates?

16-5. What is the effect on annual return, on earning rate, and payout time if an added nonproductive investment of $60,000 is made to a plant investment of $450,000 with a gross profit of $100,000? Project life is 15 years and profits tax is 56 per cent. There are no additional operating costs associated with the nonproductive investment.

16-6. A company is planning a plant investment of a certain size for $1,200,000; the investment will produce net profits of $100,000 per year for 10 years. However, it is expected that during the sixth year and for every subsequent year another $100,000 of profits will be available if a second plant is built at an additional cost of $1,400,000. If instead a double-sized plant could be built initially for $2,000,000 with extra operating costs for the first 5 years of $20,000, would it be profitable? Assume straight-line depreciation and a 10 year study period. Compute the economic rates of return for the two alternatives by trial and error. Neglect salvage value and profits tax.

SELECTED BIBLIOGRAPHY

Only recent comprehensive texts are cited here. They all contain extensive documentation which may be consulted for extended coverage of the material in the present text.

Allen, R. C. D., "Mathematical Economics," St. Martins Press, New York, 1956.

Coppock, J. D., "Economics of the Business Firm," McGraw-Hill Book Co., Inc., New York, 1959.

Bowman, E. H. and R. B. Fetter, "Analysis for Production Management," Rev. ed., Richard D. Irwin, Inc., Homewood, Illinois, 1961.

Churchman, C. W., Ackoff, R. L., and Arnoff, E. L., "Introduction to Operations Research," John Wiley & Sons, Inc., New York, 1957.

Dean, J., "Managerial Economics," Prentice-Hall, Inc., Englewood Cliffs, N. J., 1957.

Daus, P. H., and Whyburn, W. M., "Introduction to Mathematical Analysis—with Applications to Problems of Economics," Addison-Wesley Publishing Co. Inc., Reading, Massachusetts, 1958.

Ferguson, R. D., and Sargent, L. F., "Linear Programming," McGraw-Hill Book Co., Inc., New York, 1958.

Grant, E. L., and Ireson, W. G., "Principles of Engineering Economy," The Ronald Press Co., New York, 1950.

Metzger, R. W., "Elementary Mathematical Programming," John Wiley & Sons, Inc., New York, 1958.

Morris, W. T., "Engineering Economy," Richard D. Irwin, Inc., Homewood, Illinois, 1960.

Samuelson, P. A., "Economics, An Introductory Analysis," 3rd ed., McGraw-Hill Book Co., Inc., New York, 1955.

Schweyer, H. E., "Process Engineering Economics," McGraw-Hill Book Co., Inc., New York, 1955.

Thuesen, H. G., "Engineering Economy," 2nd ed., Prentice-Hall, Inc., Englewood Cliffs, N. J., 1957.

APPENDICES

A. Glossary of Selected Terms

B. Estimated Life of Equipment

C. Tables for Economic Calculations
   - C-1. Compound Interest Factors, $s = (1 + i)^n$
   - C-2. Present Value of Annuity, $a_{n/i} = \dfrac{(1 + i)^n - 1}{i(1 + i)^n} = (1 - s^{-1})/i$
   (Also Called Present Worth Factor)
   - C-3. Continuous Compound Interest Factors, $s = e^{in}$
   - C-4. Continuous Present Value of Annuity, $a_{n/i} = (1 - e^{-in})/i = (1 - s^{-1})/i$, (Also Called Present Worth Factor)

D. Table of Random Numbers

GLOSSARY OF SELECTED TERMS

Accounts payable: The value of purchased services and materials which are being used.

Accounts receivable: The value of sold services and materials which have not been paid for.

Activities: A term used in model programming to identify variables of the same kind; these variables usually are different products.

Amortization: A plan to pay off a financial obligation according to some pre-arranged program.

Annual return: The sum of the annual earnings and depreciation. This is often called *net cash flow*.

Asset: An accounting term for capital owned by a company.

Average design rate: The cumulative production divided by the cumulative time during production.

Balance sheet: A tabulation of numbers which shows the owners of an enterprise how their capital is distributed.

Batch operation: A cyclic operation where no product is available until the complete batch is finished.

Book value: Original investment in an asset minus the accumulated depreciation.

Break-even chart: See *Economic production chart.*

Burden: Expense that must be allocated to various items of production; burden is essentially the same as an *indirect cost.*

By-product: A production item made as a consequence of the production of a main item. The by-product may have value in itself or as a raw material for reuse.

Capital: A term describing wealth which may be useful for business services or for the production of new items to satisfy human wants. See *Working capital.*

Capital ratio: The ratio of investment to sales dollars; it is the reciprocal of capital turnover.

Capital recovery: The procedure for providing an equivalent sum of original investment plus compound interest.

Capital turnover: The ratio of sales dollars to investment; it is the reciprocal of capital ratio.

Capitalized earning rate: Ratio of annual earnings to investment, neglecting time value of money.

Capitalized payout time: Ratio of investment to annual return, neglecting time value of money.

Capitalized value: A numerical value equivalent to variable expenditures computed according to a specified procedure.

Cash flow: The real dollars passing into and out of the treasury of a financial venture. *Net cash flow* is the *annual return*.

Challenger: A new installation that challenges a present installation (defender); replacement by the new installation is based on economic reasons.

Channels: The number of facilities available to service a queue.

Cobweb model: The interrelation between demand quantity and supply quantity.

Compound interest: The interest charges under the condition that interest is charged on any previous interest earned in any time period, as well as on the principal.

Compound interest factor: A mathematical expression denoted as s.

Constraints: A restriction applied to a mathematical model.

Continual repetitive cycle: A cyclic operation which continually repeats itself with no interim time.

Continuous compounding: A mathematical procedure for evaluating compound interest factors based on a continuous interest function rather than discrete interest periods.

Conversion: The ratio of units of product to units of raw feed stocks. This ratio may be greater than one if the dimensions of the product (number of dresses) is different from the raw stock (yards of cloth).

Conversion cost: The total cost of production minus the original-materials cost.

Cost of sales: The sum of variable and fixed costs representing the cost of getting the product to the consumer. Average unit cost of sales is the cumulative cost at any production divided by the output.

Criterion: Some parameter such as expense, profit, quantity, or the like, that depends on the operation and is subject to control directly or indirectly by those persons in charge.

Critical production rate: The production rate that yields the maximum profit in a given time period.

Cycle time: see *Total cycle time.*

Cutoff point: The rate or investment that management uses as a decision criterion.

Dead time: Same as *Interim time.*

Decision or decision making: A program of action undertaken as the result of (a) an established policy or (b) an analysis of the variables that can be altered to influence the final result.

Defender: An installation already in operation that must defend against replacement.

Demand rate: The rate of production required to meet demand for products.

Demand quantity: The relation between customer demand and price.

Depletion: The loss in value of an asset because of its physical removal.

Depreciation: The loss of value because of obsolescence or due to attrition. In accounting, depreciation is the allocation of this loss of value according to some plan. This may be a straight-line, a sinking-fund, a declining-balance, a sum-of-digits, or a use plan.

Deterministic: A term indicating no variation in the numbers being considered.

Dilution effect: Reduction in owners equity by additional sales of stock from the company treasury.

Direct cost: Any cost which can be directly related to the production; direct cost is the same as *variable cost.*

Discount interest: The interest charge taken at the present time for a debt to be paid in the future.

Discounted cash flow: See *Economic rate of return.*

Down time: The necessary time in a repetitive cyclic operation wherein the process must be out of production for turnaround operations.

Dumping: The practice of selling excess production at lower than the previously established price.

Earning: The difference between income and costs. In general, earning is synonymous with profit. Proprietary earning is net profit less financial cost.

Earning rate: The ratio of profit to investment. *Capitalized earning rate* does not consider the time value of money. *Economic rate of return* is the theoretical earning rate allowing for the time value of money and the tax on profits but neglecting the cost of borrowed money.

Economic balance: The relations among those factors that increase or decrease in a financial venture as a function of some production factor or parameter.

Economic earning rate: See *Economic rate of return.*

Economic good: Anything that is useful, transferable, and not abundant.

Economic lot size: The optimum amount of product made in one lot to give the minimum cost for a total production.

Economic model: A mathematical expression or tabular relation that expresses the interaction of the technical and economic variables applying to a specific problem.

Economic payout time: The number of periods, n, in the theoretical economy equation whereby the annual return at a given interest rate will equal the initial investment.

Economy production chart: A plot of revenue and cost factors against production output. The abscissa may also be investment.

Economic production lot size: See *Economic lot size.*

Economic rate of return: The theoretical earning rate allowing for the time value of money and the tax on profits but neglecting the cost of borrowed money.

Economic state: A broad term describing both the relationships among factors in the national economy and also its relative condition with respect to change.

Effective interest: The true value of interest rate computed by equations for compound interest rate for a 1 year period.

Elasticity: The relationships between quantity and price, both for demand and supply.

Equity: The owners' actual capital held by the company for its operations.

Exclusion chart: A real world model showing the relationship between price and demand.

Expected value: The average of a number of probable results over the long run.

Expected waiting time: This is the sum of the average waiting time plus the average time for service.

External funds: Capital obtained by selling stock or bonds.

Financial cost: The charges for use of borrowed capital.

Fixed assets: Real or material facilities that represent part of the capital in an economic venture.

Fixed costs: Those items of cost that are essentially constant regardless of output. Average fixed costs is the total of fixed costs divided by the total production.

Four-tenths factor: The general rule that prices for materials varies as the 0.4 power of the relative quantities.

Frequency distribution: A plot or tabulation of results showing the frequency at which they occur.

Frequency-distribution function. The mathematical relation describing relative frequencies as a function of the variable tested. A plot of such a function has an area under the curve equal to unity.

Future worth: An expected value of capital in the future according to some predetermined method of computation.

Gross national product: An economic indicator used to study the total business activity.

Holding costs: Charges for holding the product in inventory or in process.

Income statement: A tabulation of financial data describing the cash flows during a given time interval.

Incremental analysis: A procedure for economic analysis wherein only those factors that vary in the analysis are considered, since constant items affect only magnitude of the criterion and not the value of the optimum.

Incremental costs: The additional costs resulting from the production of one more item. It is the slope of the total cost line on an economic production chart. See *Marginal costs.*

Incremental revenues: The additional revenue resulting from the sales of one more item.

Indirect cost: That portion of fixed costs excluding management expense and selling expense as defined in this text.

In-process inventory: Hold-up of product in a semifinished state either because the process is stopped or because its nature requires material in various stages of completion.

Instantaneous production rate: The rate of production at any instant.

Interest: The cost for the use of capital. Sometimes referred to as the *Time Value of money.* See also *Simple interest, Compound interest, Effective interest, Nominal interest, Discount interest.*

Interim time: The time when an operation is out of production for reasons of choice or no demand.

Internal funds: Capital available from depreciation and profits.

Inventory: Quantity of material held in process or in storage.

Investment: Savings or capital allocated to a given operation. Total investment includes all initial capital involved. Process investment refers only to that portion directly relating to processing. Invested capital includes figures from a balance sheet corrected for deferred liabilities.

Investors method: See *Economic rate of return.*

Isoprice point: A point of common price for a given economic analysis where different procedures may be employed.

Isoprofit lines: A plot in linear programming where all procedures giving points on the line will yield the same profit.

Lead time: The interval of time between the placing an order, or the start of manufacture, and actual receipt of product. Same as *Make time.*

Leverage: The influence of debt on the earning rate of a company.

Liabilities: An accounting term for capital owed by a company.

Linear programming: A specific type of economic analysis where the mathematic models are linear equations.

Make time: Same as *Lead time.*

Marginal costs: The rate of change of costs with production or output. See *Incremental costs.*

Marginal profit: The rate of change of profit with production or output.

Marginal profit rate: The rate of change of profit with investment.

Marginal revenues: The rate of change of revenues with production or output.

Mark-up: The difference in sales price and total cost; may vary in interpretation.

Minimum acceptable rate of return: The minimum interest rate management policy selects as acceptable for a financial investment.

Model: See *Economic model.*

Noncontinual repetitive operation: A cyclic operation in which there may be dead time of varying intervals.

Monte Carlo: A procedure employing a random distribution of numbers to generate data that fit the same probability distribution of real data.

National income: An economic indicator used by economists to measure the total current dollars of business and government activity.

Net revenues: The excess of sales over costs excluding depreciation.

Net cash flow: See *Cash flow*.

Net sales: See *Revenues*.

Net worth: The capital of a financial venture that is accountable to the owners. It is the sum of the stockholders' investment plus the surplus.

Nominal interest: The number employed loosely to describe the annual interest rate. See *Interest*.

Operating efficiency: That proportion of the total time that an operation (including necessary down time) is actually producing or running.

Operations research: A general expression that includes a broad analysis of an operation to provide the best program of action based on economic considerations or any other criterion that may be selected.

Optimum policy: The procedure or numerical value of a criterion that produces the best result in balancing the least desirable (or more costly) factors against the more desirable factors.

Optimum rate: The production rate that yields the least costs.

Over-all average production rate: The cumulative production divided by the cumulative time including the down time for a cycle. Synonymous with *Over-all production rate*.

Payout time: The ratio of the investment to annual return to give capitalized payout time which omits the time value of money. The *Economic payout time* includes the time value of money.

Poisson distribution function: A special type of probability statement for discrete variables.

Present worth: The value at some datum time of expenditures, costs, profits, etc., according to some predetermined method of computation.

Present worth factor: A mathematical expression also known as the present value of an annuity of one.

Processes: A term used in model programming to identify variables of a similar nature, usually closely related but different procedures.

Production economy: The broad general area of real world economy that relates to the manufacture of goods, or the services utilized by or to satisfy human wants.

Production rate: The amount produced in a given time. See *Instantaneous production rate, Average design rate, Over-all average production rate*.

Profit: The difference between the returns from operating an enterprise and all of the costs for its operation. In this text *Gross profit* is the profit before deduction of profits taxes, and *Net profit* is the residue after deducting profits taxes from *Gross profit*.

Profitability: A term applied in a broad sense to the economic feasibility of a proposed venture or a going operation.

Profitability index: See *Economic rate of return*.

Proprietorship: Same as *Net worth.*

Protective stock: Same as *Safety stock.*

Queue: A number of sequential events that are waiting to be serviced. Examples are a waiting line, a machine assignment, etc.

Recovery: The ratio of pure feed units to pure product units.

Repetitive operation: One which repeats itself according to a time cycle. ·

Reserve stock: Inventory in excess of normal demands to take care of unexpected withdrawals from stock.

Return: The sum of the earning and depreciation. *Economic rate of return* is the theoretical earning rate.

Revenues: The net sales received for selling the product to the customer. Average revenue is the total revenues divided by the total sales.

Safety stock: Same as *Reserve stock.*

Salvage value: Value of all assets that may be recovered at end of the activity.

Savings: The excess of total income that is not consumed.

Semicontinuous operation: In a cyclic operation the procedure where finished product is made continuously when the manufacturing operation is in use.

Semicontinuous process: An operation cyclic in nature which has either interim time or dead time.

Simple interest: The interest charges under the condition that interest in any time period is only charged on the principal. Ordinary simple interest for less than one year is based on 360 days per year. Exact simple interest is based on 365 days per year.

Simplex method: A procedure used in linear programming.

Sinking fund: An accounting procedure computed according to a specified procedure to provide capital to replace an asset.

Six-tenths factor: The general rule that investment costs vary as the 0.6 power of the relative capacities.

State of a queue system: Refers to the number of calls in the service center plus the waiting calls.

Stochastic analysis: Studies based on probability with a distribution of results.

Stock-out: The condition where no product is available to supply demands.

Supply quantity: The relation between producer supply and price.

Surplus: The excess of earnings over expense which is not distributed to stockholders.

Tax credit: The amount available to a firm as part of its annual return because of deductible costs for tax purposes.

Time value of money: The expected interest rate that capital should or will earn.

Total cycle time: The elapsed time between identical points in consecutive cycles.

Turnaround: The procedure required to clean out, recharge, discharge, or maintain maintenance of an operating process.

Unrecovered balance: The amount outstanding in a venture, allowing for profits generated and including the time value of money.

Value of money: The expected interest rate that capital should or will earn.

Variable cost: Any expense that varies with output. Similar to *Direct cost.*

Wealth: Economic goods in a tangible or physical form such as equipment, real estate, and minerals which are useful and may be transferred.

Withdrawal rate: Same rate as *Demand rate.*

Working capital: The excess of current assets over current liabilities.

Yield: The interest actually earned on a bond or any other type of investment.

APPENDIX B

ESTIMATED LIFE OF EQUIPMENT

The following tabulation for estimating the life of equipment in years is an abridgment of information from "Depreciation—Guidelines and Rules" issued by the Internal Revenue Service of the U. S. Treasury Department as Publication No. 456 (7-62) in July 1962. The original publication should be consulted for exact accounting details.

	Life (years)
Group one. General business assets	
1. Office furniture, fixtures, machines, equipment	10
2. Transportation	
a. Aircraft	6
b. Automobile	3
c. Buses	9
d. General purpose trucks	4–6
e. Railroad cars (except for railroad companies)	15
f. Tractor units	4
g. Trailers	6
h. Water transportation equipment	18
3. Land and site improvements (not otherwise covered)	20
4. Buildings (apartments, banks, factories, hotels, stores, warehouses)	40–60
Group two. Nonmanufacturing activities	
(excluding transportation, communications, and public utilities)	
1. Agriculture	
a. Machinery and equipment	10
b. Animals	3–10
c. Trees and vines	variable
d. Buildings	25
2. Contract construction	
a. General	5
b. Marine	12
3. Fishing	variable
4. Logging and sawmilling	6–10
5. Mining (excluding petroleum refining and smelting and refining of minerals)	10
6. Recreation and amusement	10
7. Services to general public	10
8. Wholesale and retail trade	10

Group three. Manufacturing

1. Aerospace industry	8
2. Apparel and textile products	9
3. Cement (excluding concrete products)	20
4. Chemicals and allied products	11
5. Electrical equipment	
a. Electrical equipment in general	12
b. Electronic equipment	8
6. Fabricated metal products	12
7. Food products, except grains, sugar and vegetable oil products	12
8. Glass products	14
9. Grain and grain-mill products	17
10. Knitwear and knit products	9
11. Leather products	11
12. Lumber, wood products, and furniture	10
13. Machinery unless otherwise listed	12
14. Metalworking machinery	12
15. Motor vehicles and parts	12
16. Paper and allied products	
a. Pulp and paper	16
b. Paper conversion	12
17. Petroleum and natural gas	
a. Contract drilling and field service	6
b. Company exploration, drilling, and production	14
c. Petroleum refining	16
d. Marketing	16
18. Plastic products	11
19. Primary metals	
a. Ferrous metals	18
b. Nonferrous metals	14
20. Printing and publishing	11
21. Scientific instruments, optical and clock manufacturing	12
22. Railroad transportation equipment	12
23. Rubber products	14
24. Ship and boat building	12
25. Stone and clay products	15
26. Sugar products	18
27. Textile mill products	12–14
28. Tobacco products	15
29. Vegetable oil products	18
30. Other manufacturing in general	12

Group four. Transportation, communications, and public utilities

1. Air transport	6
2. Central steam production and distribution	28

3. Electric utilities
   a. Hydraulic · 50
   b. Nuclear · 20
   c. Steam · 28
   d. Transmission and distribution · 30
4. Gas utilities
   a. Distribution · 35
   b. Manufacture · 30
   c. Natural gas production · 14
   d. Trunk pipelines and storage · 22
5. Motor transport (freight) · 8
6. Motor transport (passengers) · 8
7. Pipeline transportation · 22
8. Radio and television broadcasting · · · · · · · · · · · · · · · · · · · 6
9. Railroads
   a. Machinery and equipment · 14
   b. Structures and similar improvements · · · · · · · · · · · · · · · 30
   c. Grading and other right of way improvements · · · · · · variable
   d. Wharves and docks · 20
   e. Power plant and equipment · · · · · · · · · · · · · · · · see item 3
10. Telephone and telegraph communications · · · · · · · · · · variable
11. Water transportation · 20
12. Water utilities · 50

APPENDIX C

TABLES FOR ECONOMIC CALCULATIONS

TABLE C-1. Compound Interest Factors,[a] $s = (1 + i)^n$

n	1%	2%	3%	4%	5%	6%	7%	8%	9%	10%	n
1	1.010	1.020	1.030	1.040	1.050	1.060	1.070	1.080	1.090	1.100	1
2	1.020	1.040	1.061	1.082	1.103	1.124	1.145	1.166	1.188	1.210	2
3	1.030	1.061	1.093	1.125	1.158	1.191	1.225	1.260	1.295	1.331	3
4	1.041	1.082	1.126	1.170	1.216	1.262	1.311	1.360	1.412	1.464	4
5	1.051	1.104	1.159	1.217	1.276	1.338	1.403	1.469	1.539	1.611	5
6	1.062	1.126	1.194	1.265	1.340	1.419	1.501	1.587	1.677	1.772	6
7	1.072	1.149	1.230	1.316	1.407	1.504	1.606	1.714	1.828	1.949	7
8	1.083	1.172	1.267	1.369	1.477	1.594	1.718	1.851	1.992	2.144	8
9	1.094	1.195	1.305	1.423	1.551	1.689	1.838	1.999	2.172	2.358	9
10	1.105	1.219	1.344	1.480	1.629	1.791	1.967	2.159	2.367	2.594	10
11	1.116	1.243	1.384	1.539	1.710	1.898	2.105	2.332	2.580	2.853	11
12	1.127	1.268	1.426	1.601	1.796	2.012	2.252	2.518	2.813	3.138	12
13	1.138	1.294	1.469	1.665	1.886	2.133	2.410	2.720	3.066	3.452	13
14	1.149	1.319	1.513	1.732	1.980	2.261	2.579	2.937	3.342	3.797	14
15	1.161	1.346	1.558	1.801	2.079	2.397	2.759	3.172	3.642	4.177	15
16	1.173	1.373	1.605	1.873	2.183	2.540	2.952	3.426	3.970	4.595	16
17	1.184	1.400	1.653	1.948	2.292	2.693	3.159	3.700	4.328	5.054	17
18	1.196	1.428	1.702	2.026	2.407	2.854	3.380	3.996	4.717	5.560	18
19	1.208	1.457	1.754	2.107	2.527	3.026	3.617	4.316	5.142	6.116	19
20	1.220	1.486	1.806	2.191	2.653	3.207	3.870	4.661	5.604	6.727	20
25	1.282	1.641	2.094	2.666	3.386	4.292	5.427	6.848	8.623	10.835	25
30	1.348	1.811	2.427	3.243	4.322	5.743	7.612	10.063	13.268	17.449	30
40	1.489	2.208	3.262	4.801	7.040	10.286	14.974	21.725	31.409	45.259	40
50	1.645	2.692	4.384	7.107	11.467	18.420	29.457	46.902	74.358	—	50
100	2.705	7.245	19.219	50.505	—	—	—	—	—	—	100

[a]Certain of these data were adapted from the tables in "Engineering Valuation and Depreciation," by A. Marston, R. Winfrey, and J. C. Hempstead, McGraw-Hill Book Company, Inc., 1953, and are used by permission.

TABLE C-1.   (Continued)

n	12%	14%	16%	18%	20%	25%	30%	35%	40%	50%	n
1	1.120	1.140	1.160	1.180	1.200	1.250	1.300	1.350	1.400	1.500	1
2	1.254	1.300	1.346	1.392	1.440	1.562	1.690	1.822	1.960	2.225	2
3	1.405	1.482	1.561	1.643	1.728	1.953	2.197	2.460	2.744	3.338	3
4	1.574	1.689	1.811	1.939	2.074	2.441	2.856	3.321	3.842	5.007	4
5	1.762	1.925	2.101	2.288	2.488	3.052	3.713	4.483	5.379	7.510	5
6	1.974	2.195	2.437	2.700	2.986	3.815	4.827	6.052	7.531	11.265	6
7	2.211	2.502	2.827	3.186	3.583	4.768	6.275	8.170	10.543	16.898	7
8	2.476	2.852	3.279	3.759	4.300	5.960	8.158	11.030	14.760	25.347	8
9	2.773	3.251	3.804	4.436	5.160	7.451	10.605	14.891	20.664	38.020	9
10	3.106	3.706	4.413	5.234	6.192	9.313	13.786	20.102	28.930	57.030	10
11	3.479	4.225	5.119	6.176	7.430	11.642	17.922	27.138	40.502	85.555	11
12	3.896	4.817	5.938	7.288	8.916	14.552	23.299	36.636	56.703	—	12
13	4.363	5.491	6.888	8.600	10.699	18.190	30.288	49.459	79.384	—	13
14	4.887	6.260	7.990	10.148	12.839	22.737	39.374	66.770	—	—	14
15	5.474	7.136	9.268	11.975	15.407	28.422	51.186	90.140	—	—	15
16	6.130	8.135	10.750	14.131	18.488	35.527	66.542	—	—	—	16
17	6.866	9.274	12.470	16.675	22.186	44.409	86.505	—	—	—	17
18	7.690	10.572	14.465	19.677	26.623	55.511	—	—	—	—	18
19	8.613	12.052	16.779	23.219	31.948	69.389	—	—	—	—	19
20	9.646	13.739	19.464	27.398	38.338	86.736	—	—	—	—	20
25	17.000	26.448	40.900	62.677	95.396	—	—	—	—	—	25
30	29.960	50.922	85.896	—	—	—	—	—	—	—	30
40	93.051	—	—	—	—	—	—	—	—	—	40
50	—	—	—	—	—	—	—	—	—	—	50
100	—	—	—	—	—	—	—	—	—	—	100

TABLE C-2. Present Value of Annuity,[a] $a_{n/i} = \dfrac{s-1}{is}$
(Also Called Present Worth Factor)

n	1%	2%	3%	4%	5%	6%	7%	8%	9%	10%	n
1	0.990	0.980	0.971	0.962	0.952	0.943	0.935	0.926	0.917	0.909	1
2	1.970	1.942	1.913	1.886	1.859	1.833	1.808	1.783	1.759	1.736	2
3	2.941	2.884	2.829	2.775	2.723	2.673	2.624	2.577	2.531	2.487	3
4	3.902	3.808	3.717	3.630	3.546	3.465	3.387	3.312	3.240	3.170	4
5	4.853	4.713	4.580	4.452	4.329	4.212	4.100	3.993	3.890	3.791	5
6	5.795	5.601	5.417	5.242	5.076	4.917	4.767	4.623	4.486	4.355	6
7	6.728	6.472	6.230	6.002	5.786	5.582	5.389	5.206	5.033	4.868	7
8	7.652	7.325	7.020	6.733	6.463	6.210	5.971	5.747	5.535	5.335	8
9	8.566	8.162	7.786	7.435	7.108	6.802	6.515	6.247	5.995	5.759	9
10	9.471	8.983	8.530	8.111	7.722	7.360	7.024	6.710	6.418	6.145	10
11	10.368	9.787	9.253	8.760	8.306	7.887	7.499	7.139	6.805	6.495	11
12	11.255	10.575	9.954	9.385	8.863	8.384	7.943	7.536	7.161	6.814	12
13	12.134	11.348	10.635	9.986	9.394	8.853	8.358	7.904	7.487	7.103	13
14	13.004	12.106	11.296	10.563	9.899	9.295	8.745	8.244	7.786	7.367	14
15	13.865	12.849	11.938	11.118	10.380	9.712	9.108	8.559	8.061	7.606	15
16	14.718	13.578	12.561	11.652	10.838	10.106	9.447	8.851	8.313	7.824	16
17	15.562	14.292	13.166	12.166	11.274	10.477	9.763	9.122	8.544	8.022	17
18	16.398	14.992	13.754	12.659	11.690	10.828	10.059	9.372	8.756	8.201	18
19	17.226	15.678	14.324	13.134	12.085	11.158	10.336	9.604	8.950	8.365	19
20	18.046	16.351	14.877	13.590	12.462	11.470	10.594	9.818	9.129	8.514	20
25	22.023	19.523	17.413	15.622	14.094	12.783	11.654	10.675	9.823	9.077	25
30	25.808	22.396	19.600	17.292	15.372	13.765	12.409	11.258	10.274	9.427	30
40	32.835	27.355	23.115	19.793	17.159	15.046	13.332	11.925	10.757	9.779	40
50	39.196	31.424	25.730	21.482	18.256	15.762	13.801	12.333	10.962	9.915	50

[a]Certain of these data were adapted from the tables in "Engineering Valuation and Depreciation," by A. Marston, R. Winfrey, and J. C. Hempstead, McGraw-Hill Book Company, Inc., 1953, and are used by permission.

TABLE C-2. (Continued)

n	12%	14%	16%	18%	20%	25%	30%	35%	40%	50%	n
1	0.893	0.877	0.862	0.847	0.833	0.800	0.769	0.742	0.714	0.667	1
2	1.690	1.648	1.606	1.564	1.528	1.440	1.361	1.288	1.224	1.111	2
3	2.402	2.323	2.246	2.174	2.106	1.952	1.816	1.696	1.589	1.407	3
4	3.037	2.913	2.798	3.092	2.589	2.362	2.166	1.997	1.849	1.605	4
5	3.605	3.432	3.274	3.128	2.991	2.689	2.435	2.220	2.035	1.737	5
6	4.111	3.889	3.686	3.498	3.326	2.951	2.643	2.385	2.168	1.825	6
7	4.564	4.289	4.039	3.812	3.605	3.161	2.802	2.507	2.263	1.883	7
8	4.968	4.638	4.344	4.078	3.837	3.329	2.925	2.597	2.336	1.922	8
9	5.328	4.946	4.607	4.303	4.031	3.463	3.019	2.665	2.379	1.948	9
10	5.650	5.216	4.834	4.494	4.192	3.570	3.091	2.715	2.414	1.968	10
11	5.938	5.452	5.029	4.657	4.327	3.656	3.148	2.752	2.438	1.977	11
12	6.194	5.660	5.198	4.793	4.439	3.725	3.191	2.779	2.456	1.985	12
13	6.424	5.842	5.343	4.910	4.533	3.780	3.224	2.799	2.468	1.990	13
14	6.628	6.002	5.468	5.008	4.611	3.824	3.249	2.814	2.478	1.993	14
15	6.811	6.142	5.576	5.093	4.675	3.859	3.268	2.825	2.484	1.996	15
16	6.974	6.265	5.669	5.162	4.730	3.887	3.283	2.834	2.489	1.997	16
17	7.120	6.372	5.749	5.222	4.775	3.910	3.295	2.840	2.492	1.998	17
18	7.250	6.467	5.793	5.273	4.812	3.928	3.304	2.844	2.494	1.999	18
19	7.366	6.550	5.878	5.316	4.844	3.942	3.310	2.848	2.496	1.999	19
20	7.469	6.623	5.929	5.352	4.870	3.954	3.316	2.850	2.497	1.999	20
25	7.843	6.873	6.097	5.467	4.948	3.985	3.329	2.857	2.499	2.000	25
30	8.055	7.002	6.177	5.517	4.979	3.996	3.332	2.857	2.500	2.000	30
40	8.244	7.105	6.233	5.548	4.996	4.000	3.333	2.857	2.500	2.000	40
50	8.304	7.133	6.275	5.554	5.000	4.000	3.333	2.857	2.500	2.000	50

TABLE C-3. Continuous Compound Interest Factors, $s = e^{in}$

n	1%	2%	3%	4%	5%	6%	7%	8%	9%	10%	n
1	1.010	1.020	1.030	1.041	1.051	1.062	1.072	1.083	1.094	1.105	1
2	1.020	1.041	1.062	1.083	1.105	1.128	1.150	1.174	1.209	1.221	2
3	1.030	1.062	1.094	1.128	1.162	1.197	1.234	1.271	1.310	1.350	3
4	1.041	1.083	1.128	1.174	1.221	1.271	1.323	1.377	1.433	1.492	4
5	1.051	1.010	1.162	1.221	1.284	1.350	1.419	1.492	1.568	1.649	5
6	1.062	1.128	1.198	1.271	1.350	1.433	1.522	1.616	1.716	1.822	6
7	1.072	1.150	1.234	1.323	1.419	1.522	1.632	1.751	1.878	2.014	7
8	1.083	1.174	1.271	1.377	1.492	1.616	1.751	1.896	2.054	2.226	8
9	1.094	1.198	1.310	1.433	1.568	1.716	1.878	2.054	2.248	2.460	9
10	1.105	1.221	1.350	1.492	1.649	1.822	2.014	2.226	2.460	2.718	10
11	1.116	1.246	1.391	1.553	1.733	1.935	2.160	2.411	2.691	3.004	11
12	1.128	1.271	1.433	1.616	1.822	2.054	2.316	2.612	2.945	3.320	12
13	1.139	1.297	1.477	1.682	1.916	2.182	2.484	2.829	3.222	3.669	13
14	1.150	1.323	1.522	1.751	2.014	2.316	2.664	3.065	3.525	4.055	14
15	1.162	1.350	1.568	1.822	2.117	2.460	2.858	3.320	3.857	4.482	15
16	1.174	1.377	1.616	1.897	2.226	2.612	3.065	3.597	4.221	4.953	16
17	1.197	1.405	1.665	1.974	2.340	2.773	3.287	3.896	4.618	5.474	17
18	1.209	1.433	1.716	2.054	2.460	2.945	3.355	4.221	5.053	6.050	18
19	1.221	1.462	1.768	2.138	2.586	3.127	3.781	4.572	5.529	6.686	19
20	1.221	1.492	1.822	2.226	2.718	3.320	4.055	4.953	6.050	7.389	20
25	1.284	1.649	2.117	2.718	3.490	4.482	5.755	7.389	9.488	12.18	25
30	1.350	1.822	2.460	3.320	4.482	6.050	8.166	14.88	14.88	20.09	30
40	1.492	2.226	3.320	4.953	7.389	11.02	16.44	24.53	36.60	54.60	40
50	1.649	2.718	4.482	7.389	12.18	20.09	33.12	54.60	90.02	148.4	50
100	2.718	7.389	20.09	54.60	148.4	403.4	—	—	—	—	100

TABLE C-3. (Continued)

n	12%	14%	16%	18%	20%	25%	30%	35%	40%	50%	n
1	1.128	1.150	1.174	1.197	1.221	1.284	1.350	1.419	1.492	1.649	1
2	1.271	1.323	1.377	1.433	1.492	1.649	1.822	2.014	2.226	2.718	2
3	1.433	1.522	1.616	1.716	1.822	2.117	2.460	2.858	3.320	4.482	3
4	1.616	1.750	1.896	2.054	2.226	2.718	3.320	4.055	4.053	7.389	4
5	1.822	2.014	2.226	2.460	2.718	3.490	4.482	5.755	7.389	12.18	5
6	2.054	2.316	2.612	2.945	3.320	4.482	6.050	8.166	11.02	20.09	6
7	2.316	2.664	3.065	3.525	4.055	5.755	8.166	11.59	16.44	33.12	7
8	2.612	3.065	3.597	4.221	4.953	7.390	11.02	16.44	24.53	54.60	8
9	2.945	3.525	4.221	5.053	6.050	9.488	14.88	23.34	36.60	90.02	9
10	3.320	4.055	4.953	6.050	7.389	12.18	20.09	33.12	54.60	148.4	10
11	3.743	4.665	5.812	7.243	9.025	15.64	27.11	46.99	81.45	—	11
12	4.221	5.366	6.821	8.671	11.02	20.09	36.60	66.69	121.5	—	12
13	4.759	6.172	8.004	10.38	13.46	25.79	49.40	93.90	—	—	13
14	5.366	7.099	9.393	12.43	16.45	33.12	66.69	134.3	—	—	14
15	6.050	8.166	11.02	14.88	20.09	42.52	90.02	—	—	—	15
16	6.821	9.393	12.94	17.81	24.53	54.60	121.51	—	—	—	16
17	7.691	10.80	15.18	21.45	29.96	69.90	—	—	—	—	17
18	8.671	12.43	17.81	25.55	36.60	90.02	—	—	—	—	18
19	9.777	14.30	20.90	30.80	44.70	116.0	—	—	—	—	19
20	11.02	16.44	24.53	36.60	54.60	—	—	—	—	—	20
25	20.09	33.12	54.60	90.02	148.41	—	—	—	—	—	25
30	36.60	—	—	—	—	—	—	—	—	—	30
40	121.51	—	—	—	—	—	—	—	—	—	40
50	—	—	—	—	—	—	—	—	—	—	50
100	—	—	—	—	—	—	—	—	—	—	100

TABLE C-4. Continuous Present Value of Annuity, $a_{n/i} = \dfrac{s-1}{is}$

(Also Called Present Worth Factor)

n	1%	2%	3%	4%	5%	6%	7%	8%	9%	10%	n
1	0.995	0.990	0.985	0.980	0.975	0.971	0.966	0.961	0.956	0.952	1
2	1.980	1.961	1.941	1.922	1.903	1.885	1.866	1.848	1.830	1.813	2
3	2.955	2.912	2.869	2.827	2.786	2.745	2.706	2.667	2.629	2.592	3
4	3.921	3.844	3.769	3.696	3.625	3.556	3.489	3.423	3.359	3.297	4
5	4.877	4.758	4.643	4.532	4.424	4.320	4.219	4.121	4.026	3.935	5
6	5.824	5.654	5.491	5.334	5.184	5.039	4.899	4.765	4.636	4.512	6
7	6.761	6.532	6.314	6.105	5.906	5.716	5.534	5.360	5.193	5.034	7
8	7.688	7.393	7.112	6.846	6.594	6.354	6.126	5.909	5.703	5.507	8
9	8.607	8.236	7.887	7.558	7.247	6.954	6.677	6.416	6.168	5.934	9
10	9.516	9.063	8.639	8.242	7.869	7.520	7.192	6.883	6.594	6.321	10
11	10.417	9.874	9.369	8.899	8.461	8.052	7.671	7.315	6.982	6.671	11
12	11.308	10.669	10.077	9.530	9.024	8.554	8.118	7.714	7.338	6.988	12
13	12.190	11.447	10.765	10.137	9.559	9.027	8.535	8.082	7.663	7.275	13
14	13.064	12.211	11.432	10.720	10.068	9.471	8.924	8.422	7.959	7.534	14
15	13.929	12.959	12.079	11.280	10.553	9.891	9.287	8.735	8.231	7.769	15
16	14.786	13.693	12.707	11.818	11.013	10.285	9.625	9.025	8.479	7.981	16
17	15.634	14.411	13.317	12.335	11.452	10.657	9.940	9.292	8.705	8.173	17
18	16.473	15.116	13.908	12.831	11.869	11.007	10.234	9.538	8.912	8.347	18
19	17.304	15.807	14.482	13.308	12.265	11.336	10.507	9.766	9.101	8.504	19
20	18.127	16.484	15.040	13.767	12.642	11.647	10.763	9.976	9.274	8.647	20
25	22.120	19.673	17.588	15.803	14.270	12.948	11.803	10.808	9.940	9.179	25
30	25.918	22.559	19.781	17.470	15.537	13.912	12.536	11.366	10.364	9.502	30
40	32.968	27.534	23.294	19.953	17.293	15.155	13.417	11.990	10.808	9.817	40
50	39.347	31.606	25.896	21.617	18.358	15.837	13.854	12.271	10.988	9.933	50

TABLE C-4. (Continued)

n	12%	14%	16%	18%	20%	25%	30%	35%	40%	50%	n
1	0.942	0.933	0.924	0.915	0.906	0.885	0.864	0.844	0.824	0.787	1
2	1.778	1.744	1.712	1.680	1.648	1.574	1.504	1.438	1.377	1.264	2
3	2.519	2.450	2.383	2.318	2.256	2.111	1.978	1.857	1.747	1.554	3
4	3.177	3.063	2.954	2.851	2.753	2.528	2.329	2.153	1.995	1.729	4
5	3.760	3.596	3.442	3.297	3.161	2.854	2.590	2.361	2.162	1.836	5
6	4.277	4.059	3.857	3.669	3.494	3.107	2.782	2.507	2.273	1.900	6
7	4.736	4.462	4.211	3.980	3.767	3.305	2.925	2.611	2.348	1.940	7
8	5.143	4.812	4.512	4.239	3.991	3.459	3.031	2.683	2.398	1.963	8
9	5.503	5.117	4.769	4.456	4.174	3.578	3.109	2.735	2.432	1.978	9
10	5.823	5.381	4.988	4.637	4.323	3.672	3.167	2.771	2.454	1.987	10
11	6.107	5.612	5.175	4.789	4.446	3.744	3.210	2.796	2.469	1.992	11
12	6.359	5.812	5.334	4.915	4.546	3.801	3.242	2.814	2.479	1.995	12
13	6.582	5.986	5.469	5.020	4.629	3.845	3.266	2.827	2.486	1.997	13
14	6.780	6.137	5.585	5.109	4.696	3.879	3.283	2.836	2.491	1.998	14
15	6.956	6.268	5.683	5.182	4.751	3.906	3.296	2.842	2.494	1.999	15
16	7.112	6.382	5.767	5.244	4.796	3.927	3.306	2.847	2.496	1.999	16
17	7.250	6.482	5.838	5.295	4.833	3.943	3.313	2.850	2.497	2.000	17
18	7.372	6.568	5.899	5.338	4.863	3.956	3.318	2.852	2.498	2.000	18
19	7.481	6.643	5.951	5.374	4.888	3.965	3.322	2.853	2.499	2.000	19
20	7.577	6.708	5.995	5.404	4.908	3.973	3.325	2.855	2.499	2.000	20
25	7.918	6.927	6.136	5.494	4.966	3.992	3.331	2.857	2.500	2.000	25
30	8.106	7.036	6.199	5.530	4.988	3.998	3.333	2.857	2.500	2.000	30
40	8.265	7.116	6.240	5.551	4.998	4.000	3.333	2.857	2.500	2.000	40
50	8.313	7.136	6.248	5.555	5.000	4.000	3.333	2.857	2.500	2.000	50

APPENDIX D

TABLE OF RANDOM NUMBERS[a]

(1)	(2)	(3)	(4)	(5)	(6)	(7)
16408	81899	04153	53381	79401	21438	83035
18629	81953	05520	91962	04739	13092	97662
73115	35101	47498	87637	99016	71060	88824
57491	16703	23167	49323	45021	33132	12544
30405	83946	23792	14422	15059	45799	22716
16631	35006	85900	98275	32388	52390	16815
96773	20206	42559	78985	05300	22164	24369
38935	64202	14349	82674	66523	44133	00697
31624	76384	17403	53363	44167	64486	64758
78919	19474	23632	27889	47914	02584	37680
03931	33309	57047	74211	63445	17361	62825
74426	33278	43972	10119	89917	15665	52872
09066	00903	20795	95452	92648	45454	09552
42238	12426	87025	14267	20979	04508	64535
16153	08002	26504	41744	81959	65642	74240
21457	40742	29820	96783	29400	21840	15035
21581	57802	02050	89728	17937	37621	47075
55612	78095	83197	33732	05810	24813	86902
44657	66999	99324	51281	84463	60563	79312
91340	84979	46949	81973	37949	61023	43997
91227	21199	31935	27022	84067	05468	35216
50001	38140	66321	19924	72163	09538	12151
65390	05224	72958	28609	81406	39147	25549
27504	39131	83944	41575	10573	08619	64482
37169	94851	39117	89632	00959	16487	65536
11508	70225	51111	38351	19444	66499	71945
37449	30362	06694	54690	04052	53115	62757
46515	70331	85922	38329	57015	15765	97161
30986	81223	42416	58353	21532	30502	32305
63798	64995	46583	09785	44160	78128	83991
82486	84846	99254	67632	43218	50076	21361
21885	32906	92431	09060	64297	51674	64126
60336	98782	07408	53458	13564	59089	26445
43937	46891	24010	25560	86355	33941	25786
97656	63175	89303	16275	07100	92063	21942
03299	01221	05418	38982	55758	92237	26759
79626	06486	03574	17668	07785	76020	79924
85636	68335	47539	03129	65651	11977	02510
18039	14367	61337	06177	12143	46609	32989
08362	15656	60627	36478	65648	16764	53412
79556	29068	04142	16268	15387	12856	66227
92608	82674	27072	32534	17075	27698	98204

[a]Taken from Table of 105,000 Random Decimal Digits issued as Statement 4914, File No. 261-A-1 by U. S. Interstate Commerce Commission, May 1949.

Table of Random Numbers (Continued)

(1)	(2)	(3)	(4)	(5)	(6)	(7)
23982	25835	40055	67006	12293	02753	14827
09915	96306	05908	97901	28395	14186	00821
59037	33300	26695	62247	69927	76123	50842
42488	78077	69882	61657	34136	79180	97526
46764	86273	63003	93017	31204	36692	40202
03237	45430	55417	63282	90816	17349	88298
86591	81482	52667	61582	14972	90053	89534
38534	01715	94964	87288	65680	43772	39560
13284	16834	74151	92027	24670	36665	00770
21224	00370	30420	03883	94648	89428	41583
99052	47887	81085	64933	66279	80432	65793
00199	50993	98603	38452	87890	94624	69721
60578	06483	28733	37867	07936	98710	98539
91240	18312	17441	01929	18163	69201	31211
97458	14229	12063	59611	32249	90466	33216
35249	38646	34475	72417	60514	69257	12489
38980	46600	11759	11900	46743	27860	77940
10750	52745	38749	87365	58959	53731	89295
36247	27850	73958	20673	37800	63835	71051
70994	66986	99744	72438	01174	42159	11392
99638	94702	11463	18148	81386	80431	90628
72055	15774	43857	99805	10419	76939	25993
24038	65541	85788	55835	38835	59399	13790
74976	14631	35908	28221	39470	91548	12854
35553	71628	70189	26436	63407	91178	90348
35676	12797	51434	82976	42010	26344	92920
74815	67523	72985	23183	02446	63594	98924
45246	88048	65173	50989	91060	89894	36036
76509	47069	86378	41797	11910	49672	88575
19689	90332	04315	21358	97248	11188	39062
42751	35318	97513	61537	54955	08159	00337
11946	22681	45045	13964	57517	59419	58045
96518	48688	20996	11090	48396	57177	83867
35726	58643	76869	84622	39098	36083	72505
39737	42750	48968	70536	84864	64952	38404
97025	66492	56177	04049	80312	48028	26408
62814	08075	09788	56350	76787	51591	54509
25578	22950	15227	83291	41737	79599	96191
68763	69576	88991	49662	46704	63362	56625
17900	00813	64361	60725	88974	61005	99709
71944	60227	63551	71109	05624	43836	58254
54684	93691	85132	64399	29182	44324	14491
25946	27623	11258	65204	52832	50880	22273
01353	39318	44961	44972	91766	90262	56073
99083	88191	27662	99113	57174	35571	99884
52021	45406	37945	75234	24327	86978	22644
78755	47744	43776	83098	03225	14281	83637
25282	69106	59180	16257	22810	43609	12224
11959	94202	02743	86847	79725	51811	12998
11644	13792	98190	01424	30078	28197	55583
06307	97912	68110	59812	95448	43244	31262
76285	75714	89585	99296	52640	46518	55486
55322	07598	39600	60866	63007	20007	66819

Table of Random Numbers (Continued)

(1)	(2)	(3)	(4)	(5)	(6)	(7)
78017	90928	90220	92503	83375	26986	74399
44768	43342	20696	26331	43140	69744	82928
25100	19336	14605	86603	51680	97678	24261
83612	46623	62876	85197	07824	91392	58317
41347	81666	82961	60413	71020	83658	0241
78994	36244	02673	25475	84953	61793	50243
04904	58485	70686	93930	34880	73059	06823
46582	73570	33004	51795	86477	46736	60460
29242	89792	88634	60285	07190	07795	27011
68104	81339	97090	20601	78940	20228	22803
17156	02182	82504	19880	93747	80910	78260
50711	94789	07171	02103	99057	98775	37997
39449	52409	75095	77720	39729	03205	09313
75629	82729	76916	72657	58992	32756	01154
01020	55151	36132	51971	32155	60735	64867
08337	89989	24260	08618	66798	25889	52860
76829	47229	19706	30094	69430	92399	98749
39708	30641	21267	56501	95182	72442	21445
89836	55817	56747	75195	06818	83043	47403
25903	61370	66081	54076	67442	52964	23823
71345	03422	01015	68025	19703	77313	04555
61454	92263	14647	08473	34124	10740	40839
80376	08909	30470	40200	46558	61742	11643
45144	54373	05505	90074	24783	86299	20900
12191	88527	58852	51175	11534	87218	04876
62936	59120	73957	35969	21598	47287	39394
31588	96798	43668	12611	01714	77266	55079
20787	96048	84726	17512	39450	43618	30629
45603	00745	84635	43079	52724	14262	05750
31606	64782	34027	56734	09365	20008	93559
10452	33074	76718	99556	16026	00013	78411
37016	64633	67301	50949	91298	74968	73631
66725	97865	25409	37498	00816	99262	14471
07380	74438	82120	17890	40963	55757	13492
71621	57688	58256	44702	74724	89419	08025
03466	13263	23917	20417	11315	52805	33072
12692	32931	97387	34822	53775	91674	76549
52192	30941	44998	17833	94563	23062	95725
56691	72529	66063	73570	86860	68125	40436
74952	43041	58869	15677	78598	43520	97521
18752	43693	32867	53017	22661	39610	03796
61691	04944	43111	28325	82319	65589	66048
49197	63948	38947	60207	70667	39843	60607
19436	87291	71684	74859	76501	93456	95714
39143	64893	14606	13543	09621	68301	69817
82244	67549	76491	09761	74494	91307	64222
55847	56155	42878	23708	97999	40131	52360
94095	95970	07826	25991	37584	56966	68623
11751	69469	25521	44097	07511	88976	30122
69902	08995	27821	11758	64989	61902	32121
21850	25352	25556	92161	23592	43294	10479
75850	46992	25165	55906	62339	88958	91717
29648	22086	42581	85677	20251	39641	65786

Table of Random Numbers (Continued)

(1)	(2)	(3)	(4)	(5)	(6)	(7)
82740	28443	42734	25518	82827	35825	90288
36842	42092	52075	83926	42875	71500	69216
38128	51178	75096	13609	16110	73533	42564
60950	00455	73254	96067	50717	13878	03216
90524	17320	29832	96118	75792	25326	22940
49897	18278	67160	39408	97056	43517	84426
18494	99209	81060	19488	65596	59787	47939
65373	72984	30171	37741	70203	94094	87261
40653	12843	04213	70925	95360	55774	76439
51638	22238	56344	44587	83231	50317	74541
69742	99303	62578	83575	30337	07488	51941
58012	74072	67488	74580	47992	69482	58624
18348	19855	42887	08279	43206	47077	42637
59614	09193	58064	29086	44385	45740	70752
75688	28630	39210	52897	62748	72658	98059
13941	77802	69101	70061	35460	34576	15412
96656	86420	96475	86458	54463	96419	55417
03363	82042	15942	14549	38324	87094	19069
70366	08390	69155	25496	13240	57407	91407
47870	36605	12927	16043	53257	93796	52721
79504	77606	22761	30518	28373	73898	30550
46967	74841	50923	15339	37755	98995	40162
14558	50769	35444	59030	87516	48193	02945
12440	25057	01132	38611	28135	68089	10954
32293	29938	68653	10497	98919	46587	77701
10640	21875	72462	77981	56550	55999	87310
47615	23169	39571	56972	20628	21788	51736
16948	11128	71624	72754	49084	96303	27830
21258	61092	66634	70335	92448	17354	83432
15072	48853	15178	30730	47481	48490	41436
99154	57412	09858	65671	70655	71479	63520
08759	61089	23706	32994	35426	36666	63988
67323	57839	61114	62192	47547	58023	64630
09255	13986	84834	20764	72206	89393	34548
36304	74712	00374	10107	85061	69228	81969
15884	67429	86612	47367	10242	44880	12060
18745	32031	35303	08134	33925	03004	59929
78934	40086	88292	65728	38300	42323	64068
17626	02944	20910	57662	80181	38579	24580
27117	61399	50967	41399	81636	16663	15634
93995	18678	90012	63645	85701	85269	62263
67392	89421	09623	80725	62620	84162	87368
04910	12261	37566	80016	21245	69377	50420
81453	20283	79929	59839	23875	13245	46808
19480	75790	48539	23703	15537	48885	02861
21456	13162	74608	81011	55512	07481	93551
89406	20912	46189	76376	25538	87212	20748
09866	07414	55977	16419	01101	69343	13305
86541	24681	23421	13521	28000	94917	07423
10414	96941	06205	72222	57167	83902	07460
49942	06683	41479	58982	56288	42853	92196
23995	68882	42291	23374	24299	27024	67460

INDEX